Kitplane
Construction

Kitplane Construction

Ron Wanttaja

Foreword by Dave Martin
Editor of *Kitplanes Magazine*

TAB Books
Division of McGraw-Hill, Inc.
New York San Francisco Washington, D.C. Auckland Bogotá
Caracas Lisbon London Madrid Mexico City Milan
Montreal New Delhi San Juan Singapore
Sydney Tokyo Toronto

pbk 5 6 7 8 9 10 11 12 13 FGR/FGR 9 9 8 7 6 5 4
hc   1 2 3 4 5 6 7 8 9 FGR/FGR 9 9 8 7 6 5 4 3 2 1

Library of Congress Cataloging-in-Publication Data

Wanttaja, Ron.
    Kitplane construction / by Ron Wanttaja.
        p.   cm.
    Includes index.
    ISBN 0-8306-3565-3 (p)
    1. Airplanes—Home-built. I. Title.
TL671.2.W26   1990
629.134′2—dc20   90-11290
                                    CIP

Acquisitions Editor: Jeff Worsinger
Book Editor: Norval G. Kennedy
Director of Production: Katherine G. Brown
Book Design: Jaclyn J. Boone
Cover photograph courtesy of Glasair.

Contents

Foreword

THE TIME WAS THE EARLY 1970s, shortly after the Experimental Aircraft Association (EAA) moved its annual summer fly-in to Oshkosh in the center of Wisconsin. A reporter covering the week-long convention for a large aviation magazine was not kind in his review. He noted the long lines of open-cockpit, steel-tube-and-fabric airplanes built by EAA members in their workshops, and he saw few *experiments* among these little airplanes licensed in the experimental category. What he perceived was tinkerers with a welding fetish who dabbled mainly in 1930s aircraft technology. Compared to the sleek output of new airplanes from Wichita (Cessna and Beech) and Lock Haven (Piper), the homebuilt airplanes at Oshkosh were laughable.

Times change.

By the late 1980s, the Big Three light aircraft factories—Cessna, Beech and Piper—had all but closed their single-engine production lines. Under a new owner, Piper reintroduced a line of small airplanes, but as this is written, the success of the venture is in doubt.

Virtual collapse of the conventional American lightplane industry relates to skyrocketing costs due in part to product liability suits. For example, every Cessna built in the 1930s that is still flying—and there are some—continues to carry the threat of a suit.

Making matters worse is the system used by the Federal Aviation Administration (FAA) to certify new aircraft. Certification costs millions of dollars for even the simplest aircraft, but once a plane has been approved, its type certificate can be modified to include improvements. As a result, few all-new factory-built light aircraft have been introduced since the 1960s. Instead of adopting new materials and state-of-the-art aerodynamics, the industry remained entrenched in 1940s technology. Annual airplane price hikes were based mainly on increasing engine size, occasional aerodynamic cleanup, and the yearly paint scheme change and interior trim upgrade.

By the early '70s, the EAA membership included aerodynamicists Jim Bede and Burt Rutan. Together, they helped move personal aviation in a totally new direction. Bede's claim to a place in history results from his introduction of kit-built aircraft: the successful BD-4 (a fast, boxy four-seater) followed by the tiny single-seat BD-5, which is fired the imagination of thousands. (The jet version of the BD-5 is flown in airshows and commercials as the *Coors Bullet*.) The BD-5 project failed, leaving some 3,500 kit buyers without essentials like engines and landing gear when Bede declared bankruptcy in 1979, but the seed of the idea of homemade airplanes built from kits had been planted.

Burt Rutan is best known as designer of Voyager, the homebuilt airplane that circled the world nonstop and unrefueled in 1986. Rutan was virtually unknown outside EAA circles until he introduced the revolutionary VariEze in 1975. A larger version of the two-seat, all-plastic, tailless *canard* homebuilt airplane followed: the Long-EZ. (The canard is the small wing ahead of the main wing that gives these aircraft a tail-first appearance in flight.)

Dick Rutan (Burt's brother) and Jeana Yeager—the pair that later flew Voyager around the world—set records in Long-EZs, and long lines of these popular aircraft soon popped out of garages and workshops and perched nose-down at the EAA's yearly Oshkosh convention in an area marked *EZ Street*. (The retracting nosegear is raised so that these planes kneel when they are on the ground, solving a center-of-gravity problem when no one is in the cockpit.) Rutan's designs were plans-built projects—never available as complete kits—but other designers filled the demand for similar configurations in kit form.

The shift in a high-tech, multimillion-dollar industry away from corporate, factory-built aircraft to even-more-sophisticated airplanes built by *consumers* in their workshops might be unprecedented. Yet for the last several years, the FAA has registered more new light single-engine aircraft in the experimental (amateur-built) category than with standard (factory-built) airworthiness certificates.

The high-tech, mostly composite kitbuilt aircraft that are featured on magazine covers and articles in the nonaviation press are one end of a spectrum of new homebuilt aircraft that includes some excellent open-frame designs that look like ultralights, plus a good assortment of airplanes built the old-fashioned way with welded-steel fuselages, wood or metal wings and covered with fabric.

Who would have dreamed as recently as 10 years ago that homebuilt aircraft would be viewed favorably by thousands of pilots, the FAA and the general public? The result of economic and other conditions leading to the present situation is reflected in the most recent annual directory issue of the homebuilt aircraft magazine I edit: More than 350 *current* projects are listed, and most are available in some type of kit.

Yet the careful potential builder/pilot can ask hundreds of pertinent questions. Are these aircraft safe? How do they compare to factory-built aircraft? Can I build an airplane? *Should* I build one?

These are the types of difficult questions that Ron Wanttaja's book will help you answer—before investing money in any kit project. His approach is a logical, step-by-step analysis of the pluses and minuses of homebuilding and homebuilt-aircraft ownership. He might convince you that building or even owning a homebuilt is not for you. But as you read this book, you will gain insight into every aspect of homebuilt-aircraft decision-making—from *whether you should* to *which project* all the way to deciding *who will make the first flight* of the new homebuilt and how to do it.

This is a book of nuts and bolts. As a homebuilder myself—one who has both abandoned and completed successfully several homebuilts—I can assure that Ron Wanttaja's detailed descriptions of building in the main construction media—composite, welded-steel-tube-and-fabric, wood, sheet metal, and aluminum tubing—provide as accurate a flavor for what is involved in building as you can get outside of the workshop.

No important step is left out. And if you *do* decide to build, Wanttaja's construction hints that might not be found in the kit manual should speed the building process a bit.

People often ask me for advice on whether they should build an aircraft. I respond by first asking several questions: "What have you built in the past? Furniture? A house? Electronics? Did you enjoy the process?" For potential builders whose only real motivation is to own the finished product, I advise a lot of caution. They might be much better off buying somebody else's aircraft. On the other hand, if Ron Wanttaja's description of aircraft-building makes you want to buy tools and join the fray, you might be among the group that has found that the pitfalls of homebuilding, which Ron Wanttaja describes in detail, are handsomely compensated by the joy of building and flying your own aircraft. Good luck in your decision!

Dave Martin
Editor, *Kitplanes Magazine*

Acknowledgments

I'D LIKE TO THANK the following people for their help:

Colleen Reece, for her faith and encouragement;

Cecil Hendricks, EAA technical counselor, for answering hundreds of damn-fool questions at all hours of the day and night;

Ed Ullrich, for the same, plus developing and printing nearly every photograph in this book;

The gang on USENET's Rec.Aviation, for their advice and raucous fellowship;

The company representatives who supplied data and photos and answered dozens of questions, especially Jim Metzger of Avid Aviation and Bill Sprague of Stoddard-Hamilton;

The men and women of EAA Chapters 26 and 441, especially those who opened their workshops to my camera;

And finally, I'd like to thank Lisa, my bride, for putting up with eight months of distraction and foul moods; may our next 10 years together be as interesting as our first decade.

Introduction

THIS BOOK IS A COMPLETE GUIDE to choosing and building a kit homebuilt aircraft, or kitplane. The implied simplicity of a kit appeals especially to those who have no experience in aircraft construction or repair. But the first-time builder generally has no idea of the skills or techniques necessary, nor the actual time and money required.

This problem manifests itself in several ways. For example, most tube-and-fabric airplanes (Avid Flyer, RANS, and Kitfox) are delivered with prewelded fuselages that appear to be ready for assembly. Not true; it might take the builder dozens of hours of preparation until he or she is ready to start attaching other components.

Or consider the assembly of control cables. A typical kitplane's plans might say: "Build a cable 24 inches long with shackles on each end." But what if the builder doesn't know how to make safe Nicopress fittings? There are cases where builders crimped Nicopress sleeves with pliers, which is a fatal practice.

Kitplane Construction isn't intended to replace the kit's instructions. Nor does it provide information on building a cockpit latch or landing gear assembly. This book presents the basic skills required to build composite, metal monocoque, tube-and-fabric, and wooden airplanes.

This book is helpful in three ways: first, the book provides sufficient background to help choose the kit that matches the prospective builder's preferences and existing skills; second, the information presents a solid background on construction techniques, which helps the builder understand kitplane plans, and reduces uncertainty and error; third, the book gives the builder enough knowledge to recognize problems and ask intelligent questions.

1

Beginnings

BREATHES THERE A PILOT who never made a model airplane?

I doubt it.

Who could resist the models shelves at Ben Franklin, Woolworth's, or Kreske's? The multicolored boxes featured pictures of bombers soaring, fighters scrapping, and evil huns going down in flames. Sure, there were more model cars than airplanes and the boats took up a lot of room, too. But the section with the airplane models always had that special look (Fig. 1-1).

Deciding was agony. Get an ME-109? A P-51 to protect your B-17? Or a Cessna with CAP decals? Finally, the allowance savings were doled out and the precious package tied to the bike for the trip home.

The magic moment came when the last decal was slipped into place. You'd hold the plane steady and stare into the cockpit. It was like you could feel the stick in your hand, and see the wings stretching to either side. The roar of the engine. The whoooooosh of the slipstream. You could almost taste the pure, delirious, joy of flight.

Then it was time for a strafing run across the desktop, or a low-level mission against the family dog.

When you were a kid, all you had to worry about was getting glue on your fingertips. Mistakes could be painted over. If all else failed, a botched kit could always fall victim to that great allowance-waster, Fourth of July fireworks.

It's one thing for a 10-year-old to blow his allowance on the newest product by Revell or Monogram. But now as an adult you're considering dropping a year's salary on a kit for a homebuilt airplane. Page through the homebuilders' magazines. Every advertisement extols the virtues of a kitplane, from a fabric-covered fun plane to a fiberglass speedster. The phrases "easy-to-build," "Requires only average skills," and "Average construction time: 400 hours" leap out at you.

The 10-year-old in you says "Go for it!" But decades of life's lessons have made you more cautious. How much do you really *know* about aircraft construction? Can you rivet? What does "prepreg" mean? Where are you going to get an

Fig. 1-1. The model shelves of the local hobby store launched many a homebuilder.

engine? How much is everything going to cost? Do you know enough to even ask the right questions?

It's tough to get straight answers in the kitplane field. Kit advertisements can stretch the truth just like any other ad. Most articles in homebuilder magazines are aimed at those projects already well along: problem solving rather than information for beginners. They don't answer the basic question of the first-time builder: "Can I successfully build a homebuilt airplane?"

I can't answer that and neither can anyone else.

But what I *can* do is provide enough information to allow you to make that decision. I'll help you select the kitplane that is appropriate to your needs and skills. I'll provide an honest and realistic appraisal of the amount of time, money, and tools required. I'll show you the requirements of aircraft-quality workmanship. And finally, I'll demonstrate basic building operations of the various types of construction.

This last point requires some explanation. Don't look in the contents and expect to find a chapter entitled "How to Build a Glasair" or "How to Build an RV-6." To exceed the detail given in the manufacturer's plans, this entire book would have to be dedicated to a single kitplane.

Instead, detail is given on the typical operations required by various construction types. By reading the composite construction chapter, for example, you'll see how to prepare the surfaces for bonding, how to make the layups, and how to prevent or correct typical errors. Similar chapters explain metal, tube, and wood construction.

Maybe you've already decided on composite construction, or a simple tube-and-fabric airplane. Why bother with chapters on metal or wood construction? Few kits use only one construction material. Metal airplanes, for example, usually have composite cowlings and fairings. Composite airplanes make extensive use of steel and aluminum fittings, panels, and brackets. The Sea Hawker even uses balsa wood in its wings. Wooden kitplanes, and many tube-and-fabric types, use all materials: wood, metal and composites.

By the end of this book, you'll know how to tell a good kit from a bad one. You'll have a fairly accurate estimate on the total cost of the whole project. Engine selection will be easier. You'll handle tools with new confidence. During construction, fewer surprises will sap your budget and enthusiasm.

Best of all, your chance of successfully completing the airplane will be far higher.

With the explosive growth of the kitplane market in the last 10 years, one might think the concept of kit aircraft is quite new. It isn't, really. The modern kitplane benefits from a motley assortment of hits and flops.

HISTORY

In the years prior to World War I, all aircraft were essentially "homebuilt." They might have been made by a factory, but the techniques used were the same as the shade-tree aircraft builder. Plans for "production" aircraft were readily available, and Demoiselles and Bleriots popped up in places the factories had never heard of. The only things needed were mechanical aptitude and piano wire from the neighborhood hardware store. Little separated the production aircraft from the homebuilt.

This situation didn't change during the Great War; the greatest advance was the mass production of aircraft engines, which benefited the homebuilder as well. Design innovations arose, but manufacturing took just shade-tree mechanics and a well-stocked general store.

You don't believe it? In the last 20 or 30 years, a number of people have built authentic replicas of Sopwiths, Nieuports, and SE-5As. The actual plans were obtained from either the descendants of the original factories or such sources as the Imperial War Museum. Using authentic engines, these builders complete true reproductions. Even subscale replicas, as seen in Fig. 1-2, are often true to the original in construction method.

The splintering away from the "homebuilt" roots was a by-product of the designer's striving for faster and bigger machines. By substituting cast or machined fittings over those made out of sheet stock or angle iron, the designers could greatly reduce weight and increase strength. Construction began to require presses, milling machines, and custom forgings and castings. Financially, the processes only made sense for large production runs. This left the homebuilder behind.

Fig. 1-2. Construction of Dave Gauthier's ³/₄-scale Fokker Triplane is similar to the original. The fuselage is welded steel tube; the spars and ribs are wood.

But this splintering process had advantages as well. For example, manufacturers needed something stronger and more reliable than piano wire, and thence came the development of strong, flexible aircraft cable for bracing and control wires. Some detractors might say that aircraft homebuilders have been riding the coattails of the "legitimate" aircraft industry. But homebuilders still lead the way, witness the composite revolution of the '70s and '80s. Aircraft such as the VariEze and Glasair have led the way for development of composite production aircraft like the Beech Starship (Fig. 1-3), and, indirectly, the B-2 stealth bomber.

Fig. 1-3. The Beech Starship 1 might not have been developed if Burt Rutan's VariEze hadn't sparked the acceptance of canard aircraft.

The first mass marketed kitplane was the brainchild of a man named Ed Heath. In the late '20s and early '30s, he sold a number of his "Heath Parasol" kits (Fig. 1-4), which initially used a converted motorcycle engine as a power-plant. The Parasol had a ready-made market because most other production air-craft were aimed at the wealthy sportsman rather than the everyday fun flyer. You could buy the completed aircraft for $975 or a kit for $199 (Fig. 1-5). After several ownership and name changes, Heath Aircraft eventually became the nucleus of the famous Heathkit company. They don't make airplanes anymore. Pity.

Fig. 1-4. The Heath Parasol was the first mass-marketed kitplane. This is a Super Parasol, slightly larger and more powerful than the original.

For its day, the Parasol wasn't bad. Seventy mph cruise, stall speed about 25, and about 45 miles per gallon. The Parasol kit resembled several of today's kits: built from plans, from material packages, or from a complete kit including pre-welded fuselage and tail feathers.

By the 1940s, homebuilding had been nearly regulated out of existence. Gov-ernment regulations and the depression quashed many production aircraft, and kitplanes had even less hope. The regulatory stranglehold was broken with the introduction of the amateur-built category in 1947. Homebuilding picked up. The New York glider manufacturer Schweizer even offered kit sailplanes.

In the late '50s, Ray Stits produced a kit of his SA-6B Flut-R-Bug, a two-seat tandem shoulder-winged monoplane. This kit included everything except the engine and propeller. The steel-tube fuselage came prewelded. Twenty-seven were sold at $1,100 each, as well as more than 1,200 sets of plans. The small number of kits sold doesn't reflect on the quality of the aircraft, or the lack of a market. Then as now, it's often cheaper to buy a used production aircraft than a homebuilt kit. Back then, a Flut-R-Bug kit cost the same as a good used Aeronca Champ, and required years of building time.

To be successful, the kitplane designer obviously had to offer pizzazz and superior performance at a low price. Jim Bede promised that, and more, in 1969.

Fuselage	$32.87	Instrument and cowl support	$ 5.12
Controls	9.15	Tail unit	9.50
Pilot seat	.90	Landing gear	41.32
Tail skid	1.27	Center wing supports	3.11
Cowling	17.55	Outer wing supports	10.52
Engine mounting	3.44	Wings, gas tank, ailerons	76.00

TOTAL COST IF PURCHASED SEPARATELY (varies slightly with market) $247.00

SPECIAL PRICE ON COMPLETE BILL OF MATERIAL, $199.00; BOXING, $5.00

FOR GROUP PRICES SEE BLUE PRICE LIST

All Prices Subject to Change Without Notice

General Specifications of the "Parasol"

Span	25 ft.	Stabilizer area	5.5 sq. ft.	Useful load	300 lbs.
Chord	4 ft. 6 in.	Rudder area	3.8 sq. ft.	Gas capacity	5 gals.
Angle of incidence	4 degrees	Length over all	17 ft.	Oil capacity	6 qts.
Wing area	110 sq. ft.	Height over all	6 ft.	High speed	85 m.p.h.
Aileron area	10 sq. ft.	Weight, empty	285 lbs.	Landing speed	28 m.p.h.
Elevator area	5.2 sq. ft.	Rate of climb (first minute)	600 ft.	Cruising radius	200 miles

Skiis

The use of skies on your airplane doubles the usefulness of the craft. These handy appliances which you can attach in ten minutes can be bought ready built for $25.00 or you can buy the material and blueprints for $10.75 and make them yourself. You are never "snowed in" with skiis.

Complete "Parasol" Prices

COMPLETELY ASSEMBLED, READY TO FLY AWAY AT CHICAGO.

Equipped with Heath B-4 motor and Walnut prop	$975.00
Equipped with wheel brakes	add 35.00
Equipped with motor starter	add 25.00
Equipped with metal propeller	add 35.00
Heath Parasol without motor	640.00
Crating complete plane	18.00
Boxing, motor only	3.00

Notice

ALL ORDERS MUST BE ACCOMPANIED BY AT LEAST ONE-THIRD IN CASH; BALANCE C. O. D. OR SIGHT DRAFT.

Export orders must be accompanied by Cash in Full, and a reasonable amount for crating and shipping charges must be included. Personal checks must be certified.

The duty on aircraft material to most foreign countries, such as Canada, is in the neighborhood of 30 per cent.

Anyone that places an order for $100.00 worth of merchandise from our price list will be supplied with a set of blueprints free or will get credit for the blueprints already purchased.

Any parts that are found defective may be returned to us upon receiving shipping instructions from us.

In case of damage in transit, notify your express or freight agent at once and have him verify damage, as our responsibility ceases with delivery of merchandise in good order to the common carrier.

Price, Complete Bill of Material, $199.00

HEATH AIRCRAFT CORPORATION

LINCOLN 6196-6197   **1721-29 Sedgwick Street**   **Chicago, Illinois**

Fig. 1-5. The Parasol was the first modern kitplane, at least from the packaging point of view. EAA

He announced the development of the BD-5 Micro (Fig. 1-6), with a design goal of 215 mph on a 35-hp two-stroke snowmobile engine. Barely 13 feet long, with a wing span of just 17 feet, this single-seat bullet seemed to match every pilot's dream. The promotional material listed only 300 hours to build the kit!

Thousands of homebuilders plopped $500 deposits towards the BD-5 kit. The project consisted of a number of subkits, and the contract called for the payment of the entire balance ($2,000 total) upon delivery of the first subkit.

Controversy soon erupted. No one really doubted Jim Bede's ability to design fast airplanes. After all, his BD-1 homebuilt had just entered production as the American Yankee, and the Yankee went like scat. But Bede's claims for

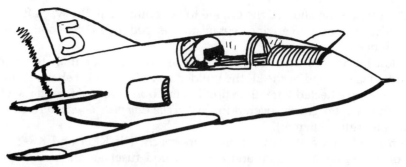

Fig. 1-6. The BD-5 took the aviation world by storm, promising 200 mph on 35 horsepower. Engine problems prevented its success. Jet-powered versions are common on the airshow circuit.

BD-5 performance seemed too good to be true. Skepticism grew to the point where *Flying* magazine offered a reward for the first BD-5 to exceed 200 mph in level flight.

Sales continued. Many initial subkits were delivered, obligating the buyers to remit the balance. But deliveries slowed. Bede held off delivering the powerplant subkits.

The BD-5 had engine problems. Pusher aircraft require special attention to engine cooling, because the prop blast isn't available. The Micro's first engine suffered from overheating and seizing. In addition, the BD-5 used a variable-speed reduction drive and shaft to transmit power from the centrally-mounted engine to the tail propeller. Problems with such a drive shaft is a recurring theme throughout the history of homebuilding. Several other engines were tried, but other problems cropped up.

Yet the hype continued. They demonstrated a jet version, the BD-5J. They announced a sailplane version, the BD-5S, with a wingspan 10 feet longer. Plans were made to sell the ready-to-fly BD-5B for $4,000. But they still couldn't field a reliable propeller-driven model.

Buyers became restless waiting for the rest of the kit. If a builder insisted, Bede would supply the remainder of the subkits, based upon whatever engine was currently in vogue. A few owner-built BD-5s were completed with other engines, only to fall victim to the same cooling and drive shaft problems the factory suffered. An alarming percentage of crashes ended with a fatality; a BD-5 pilot sits in the very front of the airplane with little crash protection.

The company "danced on the verge" for a few years, then went bankrupt in 1979. Many hopeful builders lost their deposits and some were stuck with partial kits. All were disillusioned. If it hadn't been for three men, the kit aircraft concept could have suffered a fatal blow.

Burt Rutan worked for Bede for a few years during the early 1970s, then broke off to build his own designs. After success with his unconventionally-configured yet conventionally-constructed VariViggen, he designed the VariEze: "very easy" to build and fly. The Eze was a pusher, like the BD-5, but its tail-first *canard* configuration allowed the engine to be mounted all the way aft. By avoiding a midships-mounted engine, Rutan sidestepped the BD-5's cooling and drive shaft troubles.

Rutan introduced another innovation as well. Rather than the traditional built-up wing of wood or metal, the builder carved the airfoil shape from closed-cell foam using a heated wire, then applied multiple layers of fiberglass. (The history of Rutan's designs is covered in TAB's *The Complete Guide to Rutan Aircraft*, by Don and Julia Downie.)

Ken Rand took a slightly different approach. His KR-1 and KR-2 (Fig. 1-7) were of conventional configuration, with wood fuselage and wing spars, but

Fig. 1-7. The KR-2 offers surprising performance from a VW engine.

used foam and fiberglass to achieve a sleek, fast, and tiny airframe (A KR-1 stands only $3^{1}/_{2}$ feet high, with 17-foot wingspan). Powered by a converted Volkswagen engine, a well-built KR cruises at a blistering 180 mph.

Even as the Bede empire tottered, the VariEze and the KR series stepped in to feed the growing market for a fast-building sport aircraft. Kits appeared, consisting of the material packages similar to those offered by Heath 40 years earlier. Other composite designs and kits appeared, such as the Polliwagen and the Quickie.

It took John Christensen to introduce the modern kitplane. He was a competition aerobatic pilot whose expertise in semiconductor manufacturing technology allowed him to retire at the age of 32. His Christen Eagle (Fig. 1-8) was developed for one simple reason: Curtis Pitts wouldn't sell him the rights to the Pitts Special. Christensen decided to market a kitplane that matched the Pitt's aerobatic prowess, yet included the creature comforts a customer with $25,000 to burn would expect. Note that this is $25,000 in 1970s dollars, when a loaded Chevy Camaro went for one-quarter of the Eagle's price; the price soon shot upwards. Eagle sales continue today and Christensen finally obtained the rights for the Pitts Special.

Fig. 1-8. The Christen Eagle is a Cadillac kit at a Cadillac price. Often, the well-to-do sportsmen could afford the cost, but not the time, and hired mechanics to assemble the kits for them. Like the BD-5J, the Christen is a fixture at airshows.

The Christen Eagle kit is Complete, with a capital C. Building time is estimated at 2,500 hours, which is much more than the VariEze, for example, but the kit contains *everything* required to build. The parts come bubble-wrapped to cardboard backing, and even the razor blades to remove the parts are included. John Christensen single-handedly set the tone for the modern kitplane industry.

The floodgates opened. Tom Hamilton, working from a gravel strip, nicknamed "The Pig Farm" from its former use, developed the Glasair. Instead of the moldless technique used by Rutan, the Glasair kit delivered molded composite fuselage, wing, and tail surface components very similar to those of plastic scale models. The Avid Flyer arrived for those pilots wanting a fast-building kit with Piper Cub-like performance. The ultralight movement produced a market for Air Recreational Vehicles (ARVs), aircraft with ultralight simplicity but closer in appearance and appointments to a regular airplane. Leaders, variations, and competitors arose, as did a support industry of suppliers and dealers.

More than 90 years since the Wright brothers began tinkering in their bicycle shop, homebuilding goes on. The tinkering might be with epoxy or aluminum rather than than bamboo and muslin, but the spirit is the same.

THE EXPERIMENTAL AIRCRAFT

Before going any further, let's first point out what's wrong with homebuilts.

Don't take it personally. Homebuilts are great; I fly one, and I'm building another. But my eyes are open wide to the dangers of the hobby. Don't step into kitbuilding with the attitude that the airplane will be the same as a factory model. By understanding the problems of homebuilt aircraft, you can take steps to avoid the dangers.

The Factory Aircraft

All United States aircraft must be *certificated* as airworthy. Let's look at the garden-variety factory-built general aviation aircraft. These are certificated under the *normal* or *utility* categories, factory aircraft for short.

The certification process for factory airplanes is neither easy nor cheap. The manufacturer must demonstrate that the aircraft meets certain stability standards, is controllable through all normal phases of flight, can withstand a certain amount of positive and negative G-loading, and the like. When the FAA is satisfied, it awards the aircraft a type certificate in the appropriate category.

But the manufacturer faces market pressures as well. For example, a fuselage with just a single door on the passenger's side is stronger, lighter, and easier to build. But most pilots prefer doors on both sides. The history of general aviation is chock-full of design features instigated by the marketing department.

For instance, Cessna added "Omni-Vision"—a rear window for the cabin—to the 150 in 1964. This was actually a good change, enhancing safety and reducing the tunnel-like feeling of the earlier models. It did hurt aerodynamics a bit, though. Cruise speed dropped 3 mph.

Two years later, they made another change. The large, refrigerator-shaped vertical stabilizer gave way to a sleek swept version. It reduced rudder effectiveness and increased the total length of the aircraft by 21 inches. Cruise dropped another 2 mph. Aerodynamically, it made no sense. But as for marketing, 1966 was a year of record-setting sales. Figure 1-9 compares the '65 and later versions of the 150.

Fig. 1-9. The Cessna in the foreground is an early-model 150 with non-swept vertical stabilizer. Compared to the later models in the background, the early model is less stylish but offers greater rudder effectiveness and higher cruise speed. Marketing dictated the change, and sales skyrocketed.

Aircraft design is a study in compromise. The designer must take the performance objectives, customer preference, and marketing inputs and generate a successful aircraft design. He or she cannot hope to please them all; rather, the designer's goal is an aircraft that each can accept with minimal grumbling.

The aircraft's type certificate is the buyer's assurance that the plane meets FAA standards. However, the type certificate also places requirements on the owner.

First and foremost, most maintenance must be performed by a licensed *airframe and powerplant mechanic* (A&P). Annual inspections require an A&P with *inspection authorization* (IA). Any parts installed must be approved under a *technical service order* (TSO).

Not all problems will be discovered during the test phase. Existence of a type certificate grants the government certain powers. When safety-related problems arise, the FAA can issue an *airworthiness directive* (AD). Compliance with the provisions of the AD is mandatory.

Prior to the award of the type certificate, an aircraft is tested in a variety of configurations. After certification, any changes that affect the flight characteristics or operation must be fully documented and proven safe, and aircraft operation with the changes is restricted until the FAA awards a *supplemental type certificate* (STC).

Common subjects for STCs include upgrading to a larger engine, installing speed modifications such as gap seals and special fairings, and tailwheel conversions of trigear aircraft. Changes with only minor effect (such as a radio installation) can be approved as a minor alteration, but otherwise, any change to the engine, airframe, or control system requires an STC.

To the uninitiated observer, this process can be drawn to ridiculous lengths. A number of years ago, an inventor developed an improved aircraft oil filter. The filter element was an ordinary roll of toilet tissue! The element didn't last very long, but one would expect it would be dirt-cheap to replace.

Alas, 'twas not to be. Any component installed in a type-certificated aircraft's engine must be TSO'd, and the average supermarket doesn't sell TSO'd toilet paper....

The Experimental Category

The *experimental* category offers an escape from the restrictions of type certification: no STCs, no ADs, and flight test requirements are far less complex.

The experimental category offers an escape from the other drawbacks of factory aircraft as well. Design compromises can be taken on the side of performance. The marketing side fades a bit. Individual aircraft can be optimized to the owner's requirements.

The experimental category isn't just the traditional homebuilt sport aircraft. There are several groups under the category, including:

- *Racing* aircraft intended for speed competition.
- *Research and development* aircraft undergoing flight test for either pure research or for certification purposes.
- *Exhibition* aircraft used for special purposes, such as movies or air shows. An example might be a highly-modified Stearman used in an aerobatic routine.
- *Amateur-built* aircraft built for recreational and educational purposes.

All groups come with restrictions. Aircraft certified as exhibition, for example, can only be flown at shows, to/from shows, or for necessary pilot proficiency flights. You can't legally take anyone for a joyride. Similarly, under the research and development classification, you must present the FAA with your proposed test schedule. Flights outside that schedule are prohibited.

The amateur-built category has the most relaxed rules. Aircraft in this category are restricted against flight over congested areas (except during airport operations), flight for hire, or night/IFR flight unless specifically approved. In addition, the builder must perform at least 51 percent of the work required to complete the aircraft. The builder is legally the manufacturer of the aircraft, and receives a *repairman certificate* that allows him or her to perform all inspections and maintenance on that aircraft.

One misconception concerns the amount of leeway granted in the experimental category. Let's say you bought a Cessna 152 with a run-out engine for about $4,000. Rex Taylor developed a conversion for a Honda auto engine that produces more than 125 horsepower at the propeller. The "Mini-Merlin" is far cheaper than a rebuilt Lycoming O-235 engine, and more powerful and fuel-efficient to boot. Why not replace that run-out engine with the Mini-Merlin, and certify the 152 in the experimental category?

Forget it, unless you can convince the FAA that you've built more than 51 percent of the Cessna. Lotsa' luck. At best, the research and development category is available, but the restrictions severely curtail any possibility of fun-flying. Don't even think about trying to get normal category approval. The certification process costs manufacturers millions of dollars.

Stability and Handling

On the surface, it appears the experimental amateur-built category has all the advantages. It's cheaper to build an airplane than buy a new one. By performing all maintenance, you save the cost of an A&P.

Homebuilts usually offer performance advantages as well. The typical kitplane is faster and climbs better than a comparable factory-built airplane. And with few exceptions, factory planes are stodgy and boring.

But there's a dark side to experimental aircraft, one that must be understood before construction is started.

Federal requirements that a type-certificated airplane must meet have been mentioned. Many of those requirements have to do with stability and handling.

Sporty handling is one reason homebuilts are popular. Most feature sensitive controls and a "fighter-like" feel. Some have control forces so low that they don't even need cockpit-adjustable elevator trim. But such sensitivity is gained at the cost of *stability*.

Remember, aircraft design is the art of compromise. Designers of production aircraft usually sacrifice performance in the interest of meeting the FAA's handling and stability requirements. Kitplane designers can decide not to compromise. The gentle stall characteristics of the Cessna series aren't an accident; the airfoils and wing shapes selected might not be the fastest, but they ensure good stall behavior. A 172 might not demonstrate the control response of a Pitts Special, but it won't start rolling if you glance down to unfold a chart.

There are various types of stability. Pitch stability, for instance, is defined as the willingness to return to the trim airspeed when disturbed. Assume an aircraft is trimmed out at 100 mph; pull back the stick momentarily and release. An aircraft with positive stability would soon reassume 100 mph. One with neutral stability will maintain whatever speed was reached before the stick was released. Aircraft with negative pitch stability keep slowing until the stall occurs.

Production aircraft *must* demonstrate positive pitch stability. But these regulations don't apply to homebuilts. *Aviation Consumer* magazine tested one of the most popular kitplanes:

"[It] clearly has no desire to maintain its trim speed—a fault that would immediately flunk it for FAA certification.

"The poor little autopilot trimmed its brains out trying to hold altitude, but never did manage anything better than an endless 500-fpm roller-coaster ride.

"The stall itself—at least the nibble-at-the-edge variety we tried—was docile enough . . . [The kit manufacturer's test pilot] said he preferred not to try full stick-to-the-stop-and-hold-it stalls, as required by FAA certification testing, without a parachute."

Scared? Don't be. Just don't expect Cessna-like handling or stability in a homebuilt. The aircraft aren't unflyable, in fact, the above aircraft is a real joy to fly, a mini-fighter. The article quoted above also says: ". . . we think (it) is a fine airplane overall. If by magic we could suddenly conjure up the spare time . . ."

A few last points on the stability and handling issues:

- The time to find out about handling quirks is before you buy the kit. Get a test flight. Read pilot reports. One magazine had these comments from a kitplane's designer: "I won't permit deep stalls . . . I don't think you can get it out of a spin . . . As far as unusual-attitude flight testing, we're not under an obligation to keep some crazy from killing himself." I don't know about you, but this doesn't give me a warm, fuzzy feeling. Note that it is a different airplane than the earlier set of quotes.
- A proper flying checkout is a must. Don't assume you can test-fly your new homebuilt with just a few Warrior hours under your belt. Find someone with the same model and get some copilot time. Some kit manufacturers offer test rides at the factory.
- If you plan on equipping your aircraft for IFR flight, select a kitplane with adequate stability.

Other Drawbacks of Homebuilts

Homebuilts differ in other characteristics as well.

The maximum allowable stall speed for type-certificated small aircraft is 66 mph. There is, of course, no such rule for homebuilts. One way to achieve a high cruise speed is to install a tiny wing, which, of course, increases stall speed. Some versions of the BD-5 stall at over 100 mph.

In addition, your favorite kitplane might not stall as gently as the trainer you learned to fly in. It might break more sharply, tend to drop a wing, and require more altitude to recover.

It also won't have a stall horn. Type-certificated planes must have some sort of warning that occurs at least 5 mph above stall. The requirement can be satisfied with either an aerodynamic buffet or a mechanical system like a horn. Conscientious kitplane designers try to ensure some airframe shuddering prior to stall, but it isn't required by regulations.

Homebuilts, being generally smaller than their factory counterparts, require even more care in the *weight and balance* department. For a worst-case example, examine the CG envelope of the Cessna 150 versus that of the BD-5. The 150 has an allowable *center of gravity* (CG) range of 6.5 inches, but the BD-5's is 2.5 inches, about four fingers wide.

The BD-5 is an unusual case, still, most homebuilts are "short-coupled," that is, they don't have much tail *moment-arm*. Because their handling habits are generally worse to start with, they can get absolutely appalling when loaded out of limits.

Sometimes, the very uniqueness of some homebuilt designs works against them. The Quickie, for example, was a Rutan design featuring the main landing gear mounted on the tips of the canard. A good design solution, perhaps, and definitely an eye-catcher. More than 1,500 one- and two-seat Quickies have flown. However, more than a thousand have been involved in accidents, mostly ground loops. The survivors are rapidly being converted to a more stable gear configuration.

The last major difference is engine selection. While the larger kitplanes use certificated engines, the smaller homebuilts and ARVs tend to use non-aircraft engines such as Volkswagens and Rotaxes. These are fine engines and with reasonable care they can approach the reliability of Continentals or Lycomings.

But most lack one thing the TSO'd engine has: a *dual redundant ignition system*. These uncertificated engines have one coil, one distributor, and only one plug per cylinder.

Catastrophic engine failures, such as broken connecting rods, are rare compared to breakdowns in accessories such as fuel pumps or magnetos. The most common reason of engine stoppage is fuel starvation, but it's usually caused by pilot error.

A *magneto failure* in a certificated engine is cause for alarm, but doesn't seriously affect performance. Each cylinder still has one working plug. Magnetos do fail, occasionally. I've had two mag failures in 18 months. Both times, I made it home on the remaining mag.

A component failure will kill the engine with single-ignition. One failed plug on a certificated engine probably won't be noticeable except during *runup*. A Rotax 532-powered aircraft with a bad plug becomes a glider. My VW Bug has limped home on three plugs, but I wouldn't want to try it in a VW-powered plane. Coil or distributor failure means a forced landing.

The same logic applies to all auto-engine conversions. Reliable though the engines might be, they invariably sport only single ignition. One mitigating circumstance is the use of *electronic ignition systems* in many of these conversions.

These automobile units are reliable and rugged, and work well on aircraft. Troublesome mechanical components such as points are eliminated, and the rest of the system is improved as well. One system uses a separate coil for each spark plug, thus reducing the impact of a failed coil.

However, electronic ignition must be provided with electrical power; the system does not generate its own electricity as a magneto does. Generators fail rather more often than magnetos do, and the pilot would then have to detect the failure and land before the battery gives out. And even electronic ignition is subject to certain single-point failures that can cause engine failure.

As a point of interest, there are a couple of dual-ignition versions of the Volkswagen. These either supply a new set of cylinder heads or mount the second plug in the existing head. Preferred practice is to fire one set of plugs with a standard aircraft magneto, and the other set with an electronic ignition setup. Also, the Rotax 582 has both dual-ignition versions and oil injection.

More detail on engines is given in chapter 3.

The Advantages

The advantages of owning and flying a homebuilt aircraft are noteworthy.

Less Expensive to Own. Routine maintenance for a Cessna 150, including annuals, costs at least $500 a year. By earning a repairman certificate, you pocket this money.

Less Expensive to Operate. Homebuilts generally get better performance on smaller engines. While autofuel STCs are available for a number of certificated aircraft, some fuel systems have difficulty with additives. A homebuilt can legally run on whatever the builder says it can, as long as the fuel system is designed to handle the autofuel additives.

Fig. 1-10. A Murphy Renegade offers open-cockpit thrills at a fraction of the cost of an antique aircraft.

Parts and accessories installed on homebuilts do not need a TSO. This can save a considerable amount of money. The pull-cable that activated the starter on my old 150 broke. Cessna wanted $85 for a replacement, yet a local autoparts store sold an identical item for $5. I had to buy an "official" Cessna unit.

In contrast, one magneto failure in my Fly Baby was due to a broken carbon brush in the distributor. A VW unit turned out to be identical, and worked fine.

Lower-Cost Performance. To get the performance of some of the kitplane speedsters on the market, one would have to buy a Bonanza or other complex aircraft. The fixed-gear Glasair FT, for instance, cruises at more than 200 mph.

Unique Configurations. If you want an open-cockpit biplane for less than $40,000, a homebuilt is your only option (Fig. 1-10). A full-size P-51 Mustang might cost half a million dollars; Mike Loehle's 3/4 scale replica sells for less than $10,000.

Ego Gratification. Arrive at the gas pumps in a 150, and nobody notices. Roll up in a Murphy Renegade, though . . .

Homebuilt Safety

After all this discussion on how homebuilts are trickier to fly, how do the numbers really stack up?

Very nicely. Eliminate accidents that occur during the first 40 hours and the homebuilt rate is roughly equal to production aircraft. We're worse in some categories, but better in others.

Structural failure is not a major accident cause. If built with reasonable attention to the plans, the modern kitplane is more rugged than the typical factory job.

One category where homebuilts have a worse record is "reckless or careless" operation. Our sporty little planes attract more than their share of those who buzz girlfriend's houses, or engage in improper low-level aerobatics. Countering this problem, our numbers are lower in the "Continued VFR flight into IFR conditions" category. Most of our airplanes are intended for fun flying on nice days; homebuilders simply don't tend to fly when the weather's rough.

When the test phase is completed, the typical homebuilt is as safe as a factory production airplane.

THE KITBUILDING PROCESS

I know, you're ready to pick out a kit and start building. But let's examine a few other subjects.

Types of Construction

To some people, the term *kitplane* is synonymous with *composite airplane*. Several types of construction common to kit aircraft exist. Let's look at composite first.

A composite material is *nonhomogeneous*, it doesn't consist of solely one component. Wood is a composite because not only does it contain air bubbles, which allow it to float, but the differences in grain provide the same effect. As far as aircraft construction is concerned, a composite material is artificially made by bonding materials together. It doesn't have to be high-tech. One early example was an aluminum and balsa sandwich. Plywood is a composite material because it consists of several thin layers of wood held by glue.

Composite construction works by combining two (or more) materials whose advantages complement each other and whose deficiencies cancel out. In the simplest form, strong, but flexible, fiberglass cloth is soaked in stiff, but brittle, epoxy (itself a mixture of resin and hardener) to form strong-and-stiff components like cowlings and wheelpants.

If such a composite isn't strong enough, layers of cloth-and-epoxy can be separated by a sheet of foam. Composite sandwiches like this form the basic structure of a number of kitplanes.

How does it work? Imagine an ordinary piece of paper. It's easy to tear and fold, but resists stretching. Imagine, now, a sheet of styrofoam. The foam resists bending and crushing, but crumbles when stretched.

Glue a sheet of paper on either side of the foam. Apply a bending load, and the tension strength of the paper resists. Inserting the foam has changed the bending moment (to which the paper has little resistance) to a stretching moment.

The advantage of composite construction is the easy workability of the component materials. The fiberglass cloth can be cut by a scissors, the foam can be cut or shaped by ordinary hand tools or a hot wire "cheese slicer," and the epoxies are easily mixed and applied.

The two basic types of aircraft composite construction are *molded* and *moldless*. The first uses a mold to define the shape of the structure; fiberglass layups are made directly on the mold, allowed to harden, then removed; the mold is then ready to make another piece. These are your fundamental composite kitplanes; where fuselage halves and wing panels are supplied with the kit. Because they're expensive and time-consuming to make, molded construction is best suited for mass-produced kitplanes. Examples include the Glasair III (Fig. 1-11), the Pulsar, and several other aircraft.

Moldless composite aircraft can be built in a number of ways. Most common is to build the basic structure out of wood, glue foam to the structure, then carve out the desired shape and apply the fiberglass and epoxy. Or, one can eliminate the wood and carve the shape from foam alone. The moldless method is used for the Long-EZ, KR-1 and -2, and the War Air Replicas aircraft.

Metal monocoque construction is used by aircraft manufacturers from Piper to Boeing. A thin sheet of flat metal is pretty flimsy. But roll the metal into a wide tube and it becomes stiff. Attach some bulkheads inside to prevent the tube from collapsing under heavy loads, and you've got a light, strong structure ideal for aircraft.

The metal of choice is aluminum that is *alloyed* with other metals to optimize its characteristics. Typical homebuilts using this material include the Van's Aircraft RV series (Fig. 1-12), the Zenair line, and the Thorp T-18 Tiger.

Glasair III

Cruise: 282 mph
Landing distance: 900 feet
Payload: 900 pounds
Seats: 2
Construction: Composite

Fig. 1-11. The Glasair III is a typical molded composite kit aircraft.

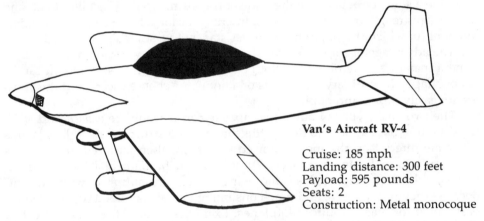

Van's Aircraft RV-4

Cruise: 185 mph
Landing distance: 300 feet
Payload: 595 pounds
Seats: 2
Construction: Metal monocoque

Fig. 1-12. Van's Aircraft is the largest-selling source of metal-monocoque kits. A wide variety is available: single seat, tandem, side-by-side, taildragger, or tricycle gear.

Tube-and-fabric construction is another of the traditional ways to make light aircraft. The main structural shape of the fuselage is defined by a metal *truss*. The wing can be built in one of several methods. The *spars*, for example, can be solid wood, built-up wooden *boxes*, or *metal extrusion*. The *ribs* can be stamped metal, cut plywood, built-up shapes, or even foam. Fabric is applied to the structure, hence the term *ragwings*, then sealed with dope to produce an enclosed, streamlined shape. Commercial tube-and-fabric lightplanes include the Piper Cub series and the Bellanca Citabria.

Tube-and-fabric kitplanes use prewelded steel tubing, like the Avid Flyer (Fig. 1-13), or aluminum tube with *gussets* or *extrusions pop-riveted* in place, like the Murphy Renegade.

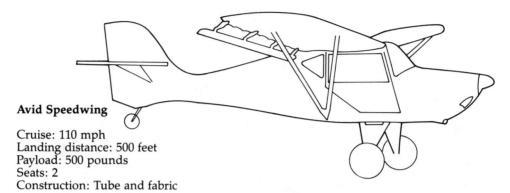

Avid Speedwing

Cruise: 110 mph
Landing distance: 500 feet
Payload: 500 pounds
Seats: 2
Construction: Tube and fabric

Fig. 1-13. The sporty little Avid Commuter is one of several related tube-and-fabric aircraft from Avid Aviation.

Wooden aircraft are built just like scale balsa models. The fuselage structure consists of *longerons* and *bulkheads* glued into the proper shape. Like tube-and-fabric aircraft, the wings can be built several ways and are either sheathed in plywood or covered with fabric. Wooden kitplanes include the Loehle 5151 Mustang replica and the Sequoia Falco (Fig. 1-14).

The preceding is only a general guide because a number of airplanes use a combination of methods. Wing construction is often a different method than the fuselage. The Prescott Pusher had an unusual combination of steel-tube, composite, and metal-monocoque construction (Fig. 1-15).

Generally, designers keep to one mode of construction throughout. Not only does it make the parts list simpler, but it reduces the construction time because the builder won't have to learn two unrelated skills.

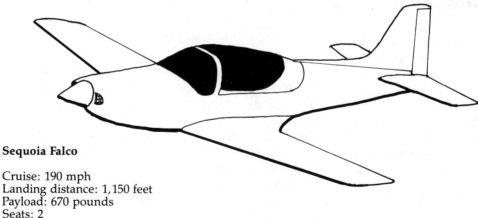

Sequoia Falco

Cruise: 190 mph
Landing distance: 1,150 feet
Payload: 670 pounds
Seats: 2
Construction: Wood

Fig. 1-14. Wooden kitplanes are rare, but the Sequoia Falco is no slouch when it comes to looks or speed.

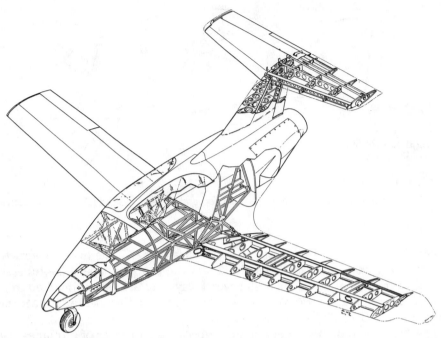

Fig. 1-15. The ill-fated Prescott Pusher kit was a hybrid of construction types. The basic airframe was steel tube inside a nonstructural composite shell. The wing was all-aluminum.

Advantages/Disadvantages

Each type of construction method has its own advantages and disadvantages, both during construction and afterwards.

Composite construction is the most controversial. There's no question that the most streamlined shapes are produced by composites, and that it's far easier to bond two fuselage halves than to *jig up* bulkheads and drive 10,000 rivets. Composites don't rot like wood; they don't corrode like metal. And no one doubts the strength of composite aircraft.

Curiously, its very strength works against it. Controversy rages regarding the crashworthiness of composite airframes. Composites have no "give." A metal aircraft slightly deforms on impact and absorbs some of the crash forces before they can affect the occupants. Composite structures maintain their shape against high forces, then shatter, allowing those forces to be transmitted to the passengers. Yet this doesn't always seem to be true. For instance, a composite aircraft prototype crashed into a housing development after an engine failure. Two houses, a van, and the plane were wrecked, but the aircraft occupants walked away.

However, another important point is repairability. Major damage to a structural component will usually require replacement of the entire component. One must then hope that the kitplane manufacturer is still in business, and still retains the molds for one's particular aircraft model.

Yet another drawback is temperature sensitivity. Some composite formulations lose strength when warmed excessively, such as might happen if the plane sits outdoors in the sun for long periods. The FAA requires that certificated composite aircraft, mostly sailplanes, be painted white to reduce this problem.

Speaking of sensitivity, some composite kitplane builders develop allergies to the materials used. The introduction of safer epoxies has reduced this occurrence, but always follow handling and safety instructions.

One thing that can't be escaped is the odor. The epoxies have a strong chemical smell and require excellent ventilation of the workspace. Also, these chemicals must be used within particular temperature ranges, and the workshop might have to be heated during the winter months.

Metal monocoque and tube-and-fabric construction methods have several advantages over composite. Crashworthiness is good, aluminum or steel allergies are almost unheard of, and aluminum or fabric-covered aircraft can be painted any desired color. However, metal is not as easily formed into the swoopy curves necessary for high-speed aircraft. Aluminum can be bent into complex shapes, but the necessary skills take time to learn.

On the plus side, aluminum doesn't care what the temperature is, so one doesn't have to heat the workshop, as long as the builder doesn't mind bundling up. However, if the exterior skin (aluminum or fabric) is installed in the cold, wrinkles can appear during the summer. Unlike composites, there are no special restrictions on exposure to the elements. But outside storage in coastal areas can accelerate the corrosion process.

Approved procedures exist for repair of damaged monocoque or tube aircraft. However, if a metal tube kit comes with a prewelded fuselage, you won't acquire the skills necessary to repair the fuselage should it get damaged. For persons both accident-prone and pain-sensitive, aluminum construction uses many sharp tools and creates sharp edges on metal.

Wood has a combination of the advantages and disadvantages of the other construction modes. Like composites, wood construction requires a climate-controlled workshop and the finished aircraft must be protected from the elements.

Fig. 1-16. The Fisher Super Koala is an example of a quick-building wooden aircraft.

Like metal construction, wooden aircraft have good crashworthiness, and approved damage repair methods assist reconstruction. Because of the nature of the material, wooden kitplanes include fewer precut parts, and generally require more work on the part of the builder. Additional skills might be required because the builder must learn to make *scarf joints* in plywood, *gusset and nail* other joints, and the myriad other tasks of the woodworker. However, this problem is lessened in modern kit aircraft like the Fisher Super Koala (Fig. 1-16).

Some advantages of wood are subjective in nature. Sawdust is a far more pleasant aroma compared to composite epoxies, and is easier to vacuum than metal chips. Wooden airplanes seem more solid and quieter than other types. While wood rot is still somewhat of a problem, modern preservatives drastically reduce the danger.

The 51 Percent Rule and You

Recall the various subgroups under the experimental category; amateur-built experimental aircraft cannot be flown for hire, and are subject to certain flight restrictions. The other big restriction, of course, is the 51 percent rule.

The rule exists to prevent the sale of nearly-ready-to-fly aircraft that do not meet the requirements of normal or utility category. We've already explained the process involved in certificating a factory aircraft, and how this process ensures stable and predictably-handling aircraft. Without the 51 percent rule, a manufacturer could bypass the regulations by merely requiring the installation of a few small parts.

The rule also prevents poorly-engineered modifications of existing planes. Anyone can take a pencil and sketch a low-winged Cub, or a clipped-wing 150. Without careful analysis of stresses and loads, such modifications are usually fatal. The 51 percent rule prevents such casual surgery to perfectly good airplanes.

That doesn't mean you can't use components of existing aircraft. Few FAA inspectors will complain if the entire firewall-forward section of a factory airplane appears on your homebuilt. Several older homebuilt designs, like the Breezy, even use complete wings from factory planes. However, these homebuilts are complex enough to make up for the existing parts. If you're building a Wag-Aero Sport Trainer (J-3 Cub replica) check with your FAA inspector if you plan on using a large number of parts from a wrecked Cub.

The FAA's interpretation of the 51 percent rule has evolved over the years. John Christensen made hurried changes to the Christen Eagle kit when an FAA representative decided the Eagle violated the rule. He received approval by including a rib-building kit and components rather than completed ribs.

Since then, the FAA's attitude has gradually relaxed. After all, molded composite kits don't contain ribs in the traditional sense, just cut-out pieces of foam. Certification of compliance with the 51 percent rule has been centralized as well, therefore individual FAA regions cannot ban a particular kit from their region.

While it's up to the kit manufacturer to ensure that their product meets the rule, you're the one who'll suffer if the FAA refuses to license your airplane as amateur-built. The aircraft will never be allowed to fly. All you could ever expect to recover is the cost of the kit, and kitplane manufacturers go out of business with regularity.

But there's another way the 51 percent rule can cause you problems, although not with the FAA. Let's take a look at the rule from another perspective. Let's say a kit manufacturer proudly states, "Our kit meets the 51 percent rule." So you order the kit, and when you open the box you find:

- Sketches of an aircraft design.
- The deeds for a bauxite mine in Oregon and an iron mine in Minnesota.
- Six spruce logs and a chainsaw.
- A dead cow and instructions for how to make glue from the carcass.

This meets the 51 percent rule—and how!

Truism Number One. "The 51-percent rule means that the builder must perform *between* 51 and 100 percent of the total work."

This makes it hard to compare kits because they all meet the 51 percent rule, but which kits require more work? Careful selection analysis is required. Hints on determining actual construction methods are in chapter 2.

The Kit Manufacturer's Dilemma

The cheapest (legal) way to acquire an airplane is to purchase a set of plans, then convert raw materials into aircraft components. The most expensive way is to buy a manufactured model like a Piper or Cessna.

By regulation, the kitmaker can do between 0 and 49 percent of the work. Obviously, the less work left for the builder, the better most buyers will like the kit. But the kit manufacturer has to charge more, which will reduce interest in the product. Figuring out the break-even point is a major source of ulcers in the kit industry.

Truism Number Two. "If it costs a kitplane manufacturer X dollars to include a step that reduces building time, the additional cost to the builder at least doubles X dollars."

In other words, a manufacturer decides to predrill critical holes, and that costs him $100 per airplane, the price of the kit increases by at least $200.

When subcontractors are involved, the cost to the kit buyer increases geometrically. Super Subcontractor Incorporated supplies wing spars to the Quikbuild Kit Company. Quikbuild decides to reduce the building time by having the holes predrilled through the spars and inserting bushings into the landing gear mounts. It costs Super Subcontractor $250 to perform the operation, and they bill Quikbuild for $500. The price of Quikbuild's kit then rises by $1,000 to cover the additional expense.

Sometimes it's worth it. Minor changes at the manufacturer's level can cause drastic reductions in building time. Hamilton-Stoddard Aircraft greatly reduced the building time between the Glasair I and Glasair II kits. The kit price also doubled, but much of that was due to inflation between 1981 and 1988. However, some of the rise must be laid at the door of the more complete kit.

The Builder

Beyond examination of homebuilts in general, and kits in particular, it's time to examine the poor shmuck that has to build the darn thing.

It's hard for a prospective builder to understand what kitbuilding is really like. The kit manufacturers make it look so easy: "Only 600 hours to construct, using only average workshop skills." If that's the case, why aren't more kitplanes flying? Why do you see ads for partially-completed kits?

Subsequent chapters explain the physical requirements of building; for now, let's concentrate on the mental preparations required.

Two basic points must be kept foremost in your mind during the construction process:

1. You are building an aircraft in which you plan to fly at heights and speeds incompatible with survival if major failures occur.
2. A kitplane project requires a major financial investment on your part. Houses cost more, and so do many cars. But inattention and errors during the acquisition and construction phases could cause a major financial loss with little chance of recoupment.

To reiterate something mentioned earlier: *I am NOT trying to talk you out of building a kitplane.* Rather, you must understand the physical, emotional, and financial costs involved before you are committed. Thousands of men and women have built experimental aircraft even before the advent of the modern kitplane. In all likelihood, you can too. I hope to prepare you for the effort.

The Personal Cost of Kitbuilding

Too often, when a kitbuilder is asked how much his pride and joy cost, the response is "Twenty thousand dollars and one marriage."

How free is your life? How much of it can you spend in the garage for the next couple of years?

Clearly, if you're the married sort, some type of compromise will be necessary. We'll assume your spouse is at least not hostile to your building an airplane, otherwise you have *big* problems.

Spousal resentment of a project stems from two points: money and neglect. It's been said that an airplane is a hole in the sky into which you throw money. A homebuilt project is even worse, and it's not yet even a hole in the sky. One's spouse can become justifiably testy if the money needed for a new couch or car tires somehow gets appropriated for a navigational *transponder*. The solution, of course, is careful budgeting, and the understanding that funds set aside for the kitplane might be raided for emergencies.

But that'll set back the completion date, you say?

One of the most absurd things you can do is set a deadline for completion. It seems standard practice to take the advertised construction time and divide by the number of weeks until the next Oshkosh fly-in to determine the hours you'll work per week.

But while you're toiling on the kitplane, the grass still grows; the kids still have to be taken to soccer practice; you haven't seen Auntie Grizzelda for 10 years; your family would like to see more of you than just a mound of sawdust at mealtimes.

A deadline can do two things to you. First, it can make you rush and make mistakes. Second, falling behind schedule is depressing. It becomes harder and harder to face the thing in the garage. Work grinds to a halt.

Set a realistic goal for the number of hours per week spent on the project. What's realistic? Count on no more than 11 hours on the weekend, and maybe three hours on each of three weekday evenings. That leaves enough time for family activities on the weekend, and keeps two weekday evenings open.

Of that 20 hours a week, count on only 15 or so of useful work. A lot of time is spent preparing, cleaning up, and so forth. Note that you don't have to be restricted to the schedule because any unexpected free time can be converted into workshop hours. If dinner will be delayed by half an hour, duck off to the workshop and buff those aluminum parts, or deburr those holes.

Don't underestimate your spouse, either. Many husbands and wives have spent hundreds of hours bucking their mate's rivets, or mixing epoxy for the layup crew. Building a kitplane can be a family activity if the rest of the family looks upon it that way. If your lifemate pitches in willingly, you've got a gem beyond compare.

Setting Goals

Obviously, with only 15 effective hours per week, it'll take a while to complete all but the most basic kitplane. Common wisdom says that the kitplane manufacturer's estimate of construction time is half the actual time; if the manufacturer says 500 hours, it'll take closer to 1,000. If you can get only 10 effective hours of work in per week (a pretty realistic figure), that's two years to complete the kit.

How do you maintain enthusiasm? Simply forget that you are building an airplane.

Instead, you're building aircraft subassemblies. Set goals based on the completion of individual items. It's like flying a cross-country. If you expect to climb to altitude, spot your destination, and fly directly to it, you won't get far. Instead, you fly to intermediate locations on the way to the target.

If the plans say to build an aileron, don't look beyond its completion. You've got all the time in the world, after all, making an aileron is easy. When it's done, admire it for a few moments, then carefully store it and look in the plans for the next project.

Glamour vs. Drudgery

There's a famous saying in the homebuilt world: "The last 10 percent takes 90 percent of the construction time." The airframe is the easiest part of 99 percent of homebuilt projects. For example, one tube-and-fabric design has a prewelded fuselage available. The plane takes 1,200 hours to build, but buying the pre-welded fuselage only cuts 100 hours from the total time.

What gives?

Truism Number Three. "Eighty percent of the work and 95 percent of the sheer drudgery is involved with subsystems, not structure."

Structural construction is the glamour-child of homebuilding. Few appreciate a finely crafted canopy latch, but everyone loves to run their hands over a just-completed fuselage. A wide-open garage door revealing a bunch of nondescript parts doesn't grab the neighborhood's attention like a sleek white fuselage on shiny aluminum landing gear.

Imagine you're building the newest composite kit. Bond the fuselage halves together by applying epoxy and cloth to either side of the join line. Bond a few bulkheads inside as well. After a few hour's work, the casual observer might think the fuselage is done.

But let's follow the installation of one subsystem: wheel brakes. First, let's assume the landing gear, wheels, and rudder pedals are in place. Follow the progress in Fig. 1-17.

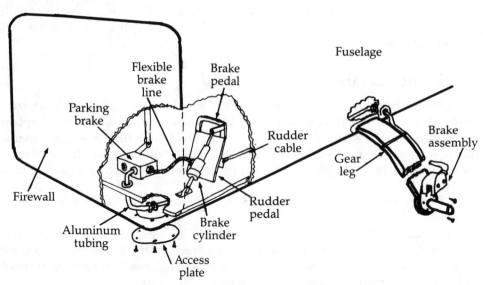

Fig. 1-17. No matter what construction method a kitplane uses, the brake systems are nearly identical.

The wheel incorporates a brake disk, but the brake assembly itself must be installed on the gear leg. Because the leg is just a blank slab, you have to drill the holes to mount the brake assembly. This demands precision because if the brake assembly is askew, wheel shimmy, poor braking, and abnormal pad wear can result. To be on the safe side, temporarily clamp the assembly in place, then jack the plane off the floor so the wheel can be spun to check pad clearance. You'll need a hydraulic hand pump or similar device to work the pads while the wheel is spinning.

Once the brake assembly is bolted in place, it's time to install the brake master cylinders. The kit manufacturer should have supplied the brackets and everything to mount the cylinders to the rudder pedals. Once the cylinder is mounted, you've got to run a brakeline through the parking brake valve to the wheel. On this aircraft, the line must run down through the cockpit floorboards, outboard to the fuselage side, aft through two bulkheads, out the bottom of the fuselage, then along the gear leg to the wheel cylinder. All in all, about four feet of brake line with at least four bends. Some kits might give the exact instructions for brake line routing; lets assume this kit says, "route the brake line from the pedal to the wheel."

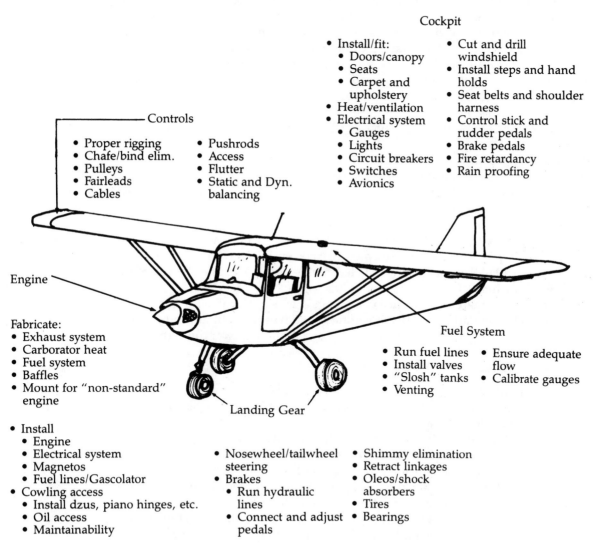

Cockpit

- Install/fit:
  - Doors/canopy
  - Seats
  - Carpet and upholstery
- Heat/ventilation
- Electrical system
  - Gauges
  - Lights
  - Circuit breakers
  - Switches
  - Avionics
- Cut and drill windshield
- Install steps and hand holds
- Seat belts and shoulder harness
- Control stick and rudder pedals
- Brake pedals
- Fire retardancy
- Rain proofing

Controls

- Proper rigging
- Chafe/bind elim.
- Pulleys
- Fairleads
- Cables
- Pushrods
- Access
- Flutter
- Static and Dyn. balancing

Engine

Fabricate:
- Exhaust system
- Carborator heat
- Fuel system
- Baffles
- Mount for "non-standard" engine

- Install
  - Engine
  - Electrical system
  - Magnetos
  - Fuel lines/Gascolator
- Cowling access
  - Install dzus, piano hinges, etc.
  - Oil access
  - Maintainability

Fuel System

- Run fuel lines
- Install valves
- "Slosh" tanks
- Venting
- Ensure adequate flow
- Calibrate gauges

Landing Gear

- Nosewheel/tailwheel steering
- Brakes
  - Run hydraulic lines
  - Connect and adjust pedals
- Shimmy elimination
- Retract linkages
- Oleos/shock absorbers
- Tires
- Bearings

Fig. 1-18. Building the structure of a kitplane is only the beginning of the job. No matter how fast the airframe goes together, the myriad little details of the systems will take most of the construction time.

You don't want to do it as one piece. If it ever broke in service, replacement would be difficult. Lay out a piece of rope to figure out the length of each segment, then use a tubing cutter on the stock supply of brake line. Each segment end must have a pipe fitting installed, requiring flaring with a special tool. Brake lines are soft aluminum, but they can be ruptured or pinched if carelessly bent, or if you push too hard trying to work them around tight quarters. You'll have to make a guess at the bend, try to install the line segment, then withdraw it and adjust the bend angle. You might have to redo one or two pieces that develop crimps or come out too short.

Once the rest of the interior is installed, the brake lines will be tough to get at. So, whether the plans call for it or not, you should protect the lines from any hazards, and make sure you can get to them when necessary. For instance, add chafe protection where the line passes close to any metal fittings, and install grommets anywhere the line passes through a bulkhead or the side of the airplane. Add access/inspection panels where necessary because five years from now, you might want to remove brakelines without disassembling the airplane.

When the line is connected, fill the brake system with hydraulic fluid, bleed the air, and test for leaks.

Done? Sigh, wipe the sweat off your brow, then start all over on the *other wheel's* brake!

There's not a kit in the world delivered with the brake system installed. Similar complex processes must be performed on other components and subsystems: controls, radios, instruments, doors, cockpit heat, fuel systems, and the like. And it's a whole 'nother problem forward of the engine mount. Figure 1-18 gives a brief summary of necessary subsystem work.

The worst byproduct of subsystem work is the apparent lack of progress. A buddy will stop by and glance at the project: "Haven't been working on it, huh?" Even if you point out the newly-installed brakes, he'll examine the completed system: "So what?" To a non-builder, it looks easy.

It's tough to keep your enthusiasm going. Again, make all your goals short-term ones. If one particular goal is getting you down, leave it and build something else. Or contact another builder and have them help you through.

The Builder as Inspector

There's one major ability necessary to complete a safe homebuilt: The willingness to look at a newly-completed part, spot an error, say, "This is garbage," then chuck it in the trash and start over again.

Homebuilders generally build two airplanes: One in the hangar, and one in the scrap bin. Everyone makes mistakes, especially when starting out. I cut and drilled three rudder spars for my project before I was satisfied.

Modern kitplanes are, for the most part, wonders of fool-proof design. The designers work overtime to reduce the chance of the builder making a mistake.

But if you blow it, scrap it. It's your life on the line. If the part is made from raw materials, start over again. If it's a prefabbed kit item, call the manufacturer and order another one. All legitimate kit manufacturers sell individual parts to builders for precisely this reason. If many folks seem to stumble over the same step, a revised design quite often results.

Does that mean only perfect workmanship is acceptable? No. Replacing a part for purely cosmetic reasons is up to you. Small errors might be fixable, or might not affect strength or operation. Contact the kit manufacturer if you have any doubt, and/or talk to an EAA technical counselor.

But remember, *you* are the final quality inspector. Read the plans carefully, execute the steps as called for, and ensure every part installed in your aircraft is the best you can do.

Modifications

Builders occasionally modify their aircraft during construction. Nothing scares kitplane designers more. As mentioned earlier, aircraft design is a study in compromise, and an unknowing change made by the builder might reduce safety margins to dangerous levels.

For example, take an ARV kit designed to be powered by a Rotax 503 engine. Some people don't like two-stroke engines. Because the Volkswagen four-stroke engine is approximately the same horsepower, why not use it instead of the Rotax?

In the first place, the VW weighs at least 40 pounds more than the Rotax. Most ARVs weigh less than 400 pounds, so the empty weight has increased by 10 percent. The builder might be tempted to boost the gross weight to maintain the same empty weight; this is perfectly legitimate because legally, the builder is the manufacturer and can claim whatever gross weight he wishes.

Yet the extra weight decreases the G-load factor. If the aircraft at the design gross has a limit load factor of +5.5 Gs, increasing the gross might drop the load factor to less than 5. The stall speed increases, and the forward shift in CG increases the elevator loads.

There are other problems as well, but you get the point. This does not mean that you cannot make *any* changes. But check with the kit manufacturer first, and don't be too let down if they refuse to approve it. They get inundated with suggestions, and their liability status is precarious enough without granting quick approval to some stranger's idea.

The Dark Side

Careful inspections during construction and no unauthorized modifications seem like reasonable precautions. Yet too often, the wreckage of a homebuilt aircraft reveals evidence of violation of those basic precepts:

- A wing separates in flight. Investigators find the builder *never installed the wing bolts*; the *holes* hadn't even been drilled. The aircraft flew 15 hours with only a layer of fiberglass holding the wing in place.
- A wing disintegrates on a high-speed pass. Upon investigation, it is found that the builder had skimped on fiberglass cloth.

- A small homebuilt spins in on final during its first flight. Examiners determine the CG to have been at least two and a half inches behind the aft limit, in part due to unauthorized modifications. In fact, the builder expended considerable effort to *hide* the CG problem—he "fudged" the weight and balance paperwork presented to the FAA.

Why? What made these builders commit such fundamental errors as omitting wing bolts, or to modify their airplanes past the limits and beyond? No one will ever know. Each answer died in the respective wreckage.

Your airplane won't be perfect; you will make mistakes. Expect them; correct them. Don't let it get to you, just redo the part or ask for advice.

The Skills Required

The biggest question you're asking yourself right now is whether you have the skills to build the aircraft. That's not actually as serious a consideration; problems are more likely to occur due to deliberate actions rather than accidentally due to lack of skill.

But what level of skill is required? Many kitplane ads use the term "average skills." What constitutes average? Twenty years ago, most high-school boys took shop classes, and learned the basics of machine work and welding. In our school, shop students melted aluminum scrap, cast rods in sand molds, and turned cufflinks from the castings.

Things have changed. Today's high-school student spends more hours behind a computer keyboard than a lathe.

What does "average skills" mean to a kitplane manufacturer? They spend much of their time immersed in homebuilt aircraft construction. Bandsaw operation is second nature. They know which size drill bits are used for rivet holes. (*Of course* the holes have to be deburred, everybody knows that!)

As far as I'm concerned, "average skills" means you know that the pointy end of the drill makes the holes. It means you know nothing about aluminum alloys, epoxies, AN hardware, or fiberglass cloth. That's what this book is about: presentation of the basic techniques and processes in a form that those with "average skills" can understand. With this help, and a little practice, you'll be ready to tackle just about any kit. One of the benefits of kitbuilding is that you'll end up with "above average" skills. Another benefit is that you will have the tools and experience to tackle those household projects you used to shy away from.

Can a person with "average skills" build the modern kitplane? Most certainly. The better kitplanes come with critical holes drilled, and complex parts are prefabricated. Successful completion requires careful study of the instruction and diagrams, and cautious execution of each construction step.

But your ultimate goal isn't to build an airplane. That's far easier than you think. Your goal is to build a safe airplane. Read the instructions carefully, and build the plane exactly according to the plans.

A QUICK NOTE ON CHANGES

A lot can happen between the time an author writes a book to the time you read it. As this book goes to press, Wheeler Technology has closed its doors. While the product is apparently of good quality, the loss of two prototype aircraft within a year put the company in dire financial straits. The announcement said the closure would be temporary. Time will tell. Be advised, then, that companies and products are subject to change without notice.

The following chapters also make extensive reference to the cost of kits, components, and tools. Specifying prices in a book is an awkward practice; prices might rise drastically in the time between writing and publication, or the book might remain in print for several years without revision. For example, gasoline prices shot up 40 cents a gallon in a short time.

Yet I dislike imprecise terms: "relatively inexpensive" or "fairly costly." It's easier to write and easier to research, but that is a disservice to all readers, instead of those who don't buy the book the minute it is released.

Therefore, prices will be quoted and rounded up for simplicity. To provide some comparison for later readers, I've selected a standard to which both home-builders and authors can relate:

As I write this book, a basic fast-food hamburger costs seventy-five cents (US). This is the standard burger; not the bigger ones with lettuce, tomato, etc.

If I say a part costs $100, and burger prices have risen to $1.00, you can figure the current price of the part as $100 \times (1.00/.75)$, or about $133. Buy a few catalogs and compare, just to be on the safe side.

Enough preparation.

Let's pick out your kit!

2

Selection

HOW DOES THE AIR FORCE PICK A NEW AIRCRAFT?

First, they make a careful analysis of the projected mission: What top speed is necessary? How fast should it cruise? How much payload? Maneuverability?

These requirements are then included in a Request For Proposals (RFP) released to the aerospace industry. Interested companies, like Northrop, Grumman, and Boeing, analyze the requirements and each submits a design proposal. These proposals show how the company's design will meet or exceed the RFP requirements, and how much the aircraft will cost.

The Air Force System Program Office (SPO) in charge of the project analyses the submissions. The SPO critically examines each proposal in light of the mission requirements, the total cost, the development time before the aircraft is operational, and whether the manufacturer can actually perform to the level promised.

In the government world, this process is slow and expensive.

Things are simultaneously easier and harder for a private citizen choosing a kitplane. While you don't have to satisfy a Pentagon functionary, family finances don't stretch as well as the military budget. And while kitplane manufacturers strive to please, it isn't cost effective for them to make a design change to please a single customer.

But the basic process is sound. There are certain performance levels the aircraft must meet; your budget is not unlimited; and you'd like to be flying within a certain amount of time. And you don't want to send your money to a bankrupt or larcenous company.

This chapter will help pick the plane that's right for you. Even if you've already picked out the kitplane you want to build, you'll learn how much it'll cost to build and a few other interesting details.

DETERMINING YOUR MISSION

It's easy enough to pick a kitplane, the "eenie, meenie, miney, moe" method, for example. Or the prospective builder can page through a favorite magazine until a design catches his or her eye. After all, an airplane is an airplane, right?

Too often, a kitplane is picked for the wrong reason. A builder thinks he needs a fast airplane, and picks the fastest one advertised. It is discovered too late the builder's private strip is too short for the hot little homebuilt. Or he finds the designer made the plane go fast by minimizing the cockpit size, and it's too small for a king-sized builder.

The result is a half-completed project junked, or a low-flight-time kitplane up for sale.

To avoid such problems, first determine your basic mission, then derive the performance needs and other requirements.

The Mission Statement

There are as many "missions" as there are pilots.

Some pilots just like to fly for the pure joy of it, sometimes called "cutting holes in the sky." These folks could get by with a small, single-seat, simple aircraft with good handling characteristics. The 5151 Mustang is an example. However, many want to take spouses and friends flying, so two-seaters like the ProTech PT-2 and the Fisher Classic (Fig. 2-1) are their choice.

Fig. 2-1. The Fisher Classic kit costs $4,000 more than the similar Fisher FP 404 single seater, but the extra seat is worth it in utility and market value. However, the FP-404 can be licensed as an ultralight, if necessary.

Such basic airplanes aren't enough for people who want room for radios and baggage, higher cruise speeds, and longer range. Some still need the capability to operate from short grass fields, but others intend to use their kitplanes for serious travel between large cities. The RV series, the Glasair, and the Questair Venture are aimed at these folks.

Other needs exist. While many kitplanes are capable of light aerobatics, a few builders need competition-level performance. The Christen Eagle still is in production. Or, some want to splash around a local lake in a homebuilt amphibian like the Sea Hawker.

But is the mission really right for you? Will you outgrow the simple little "puddle jumper?" Are you truly interested in aerobatics? Do some serious thinking about how your needs might grow. It might take five years to build your airplane; it would be a pity if you outgrew it before it flies.

Sometimes it's tough to determine a single mission. For example, you'd like a fun little knockaround aircraft, but still require fast cross-country travel. Keep in mind that good cross-country aircraft can be rented at practically any FBO. Buy the little Murphy Spirit just for fun, and count on renting a Piper Arrow or equivalent for heavy-duty traveling.

To start, come up with a single sentence expressing your basic mission: "Carry one passenger on long cross-countries between major airports." "Solo competition aerobatics." "Nostalgic-looking biplane for local hops." "Fast-building kit for moderate cross-countries with light aerobatic capability." "Operate from short private airstrip, with high-speed travel to nearby destinations." "Simple to fly for kid's lessons."

Determining your mission might seem trivial, yet it is an important step in figuring out which homebuilt fits your requirements for the least money.

A Few Notes on Irrationality

"Mission?" some are wondering. "I just want to own a Finkerbean Special." In other words, you're not making a logical, rational decision based upon careful analysis of needs and features.

There's nothing wrong with that!

Few of us *have* to own an airplane. The kitplane is being bought with your discretionary income and isn't expected to "pay its way." Some people might buy a sports car, others a boat. We just prefer airplanes.

However, there are things you expect to do with your Finkerbean Special. Maybe your fantasy is flitting to that little mountain strip near your favorite fishing hole. Or packing the wife and kids aboard and winging your way to Grandma's at 250 knots.

Before you spring the bucks, shouldn't you at least make sure the plane matches your needs?

One of the rites of manhood is the buying of the first car. Considerable effort goes towards finding *exactly* the right set of "wheels" to fit one's trans-pubescent self-image. When found, it is bought, whether Mom and Dad approve or not.

Reality soon intrudes. The engine leaks oil like a grounded tanker. The chassis is sprung, the shocks are shot, the electrical system shorts out. Have you ever seen a shiny older car at the side of the road, with a stricken teenager tenderly stroking the steering wheel, waiting for Dad and a tow rope? Too often, that first car is one's introduction to the need for rational decisions, instead of emotional ones.

It's bad enough with a $500 car. Do you want to go through the same process on a $25,000 kitplane? With so much money, and thousands of hours of construction time at stake, shouldn't you make sure the plane is what you need, and not just what you want?

But hey, it's your money and your decision. If your analysis finds Brand X better meets your needs than the Finkerbean Special, but you still prefer the Finkerbean, go for it!

ALTERNATIVES

Before spending all that money and time building a kitplane, why not look at some alternatives?

Is This Kit Really Necessary?

Looking for a kitplane for fast cross-country travel? How does this sound: Four Seats, 160 mph cruise, retractable gear, and fully IFR-equipped. Cost is $25,000, including midtime engine, constant-speed propeller, instruments, radios, paint, wheels, and upholstery.

Building time? Zero hours because the plane is a used Mooney, of mid-60s vintage (Fig. 2-2).

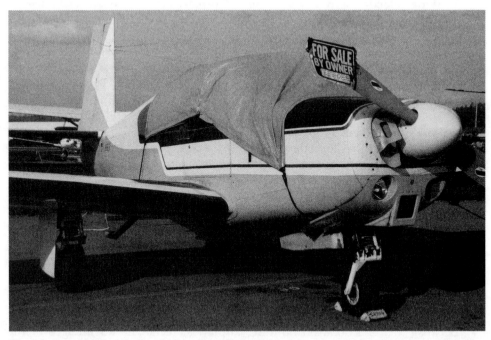

Fig. 2-2. If you'd rather fly than build, why not spend the same amount of money on a used factory aircraft rather than a kitplane? Performance might be slightly less, but you'll be flying immediately instead of next year. And the doors probably leak no more than that of an average kitplane.

Note that $20,000 will also buy a Lancair kit, but without engine, propeller, avionics, etc. You'll end up paying at least another $15,000. The Lancair will cruise faster. But $10,000 faster?

Homebuilts became popular for two reasons. First, they offered features unavailable in production aircraft, such as aerobatic capability or open cockpit flying. Second, their performance was out of proportion to their cost. Back when a new Cessna 150 was priced at $10,000, a Thorp T-18 could be built for half the cost and cruised 50 percent faster. The Thorp was designed for a Lycoming O-290G taken from a military-surplus ground power unit.

Things are a bit different now. Kit aircraft aren't cheap, and inexpensive surplus engines are long gone. Kitplanes still offer greater performance for the money, but the performance increase over production aircraft isn't as great as it used to be.

If you're looking *solely* for an aircraft with particular features, you're better off buying a good used factory aircraft. Here are the relative advantages of used production aircraft over kitplanes:

1. **Better overall flying characteristics.** Even older production planes are less demanding to fly than certain kitplanes. As mentioned in chapter 1, some kitplanes get great performance at the expense of handling.
2. **Instant availability.** Drop the money on the barrelhead, and you can fly away in your new airplane, rather than toiling in the garage for the next five years.
3. **Wide selection and a wide range of prices.** Find the exact airplane you need. The selection of kitplanes is still rather limited, and prices are set by the manufacturer.
4. **Known resale.** If you need to sell the airplane, market value is easily determined. Kitplane resale varies; sometimes the owner can't get more than the cost of the kit itself.
5. **Loan availability.** You can get a loan to buy one, using the plane itself as collateral. Unless you have other property to become collateral, it's nearly impossible to get a loan to buy a kitplane.
6. **Off-the-shelf maintenance.** If it needs repair, you can buy ready-to-install parts. While parts for older aircraft might be harder to obtain, it's still easier than trying to remember how a certain kitplane part is made. In case of major damage, a wing or tail section can be bought from salvage and bolted into place.
7. **No emotional resistance to outside tiedowns.** After spending several years building a kitplane, most builders want inside hangarage for their pride and joy. This is expensive at most airports. But few pilots have compunctions about leaving a used production airplane outside. This can significantly reduce fixed costs of ownership, for example, a closed hangar at a local airport rents for $150 a month; tiedowns are $100 less.

Because this is a book on building kitplanes, recall the advantages of the homebuilt aircraft:

1. A new aircraft for the price of a used production model.
2. Save maintenance costs by legally performing your own inspection and repairs. Also, parts prices for production aircraft are generally exorbitant.
3. The plane shall have exactly the equipment you want, in exactly the right paint scheme.
4. Planes that come with individually-available subkits have the equivalent of interest-free financing because you can buy a subkit when you have the money for it.
5. If finances get tight after completion, the plane can be disassembled and stored at home to save tiedown and insurance costs.
6. Performance is generally better than equivalently-priced used aircraft.
7. Styling and appearance are far better than stodgy production models.
8. Significant personal satisfaction

The advantages and disadvantages of buying a used factory aircraft are shown in Table 2-1, and a similar table for kitplanes is included as Table 2-2.

**Table 2-1. Advantages and disadvantages of
buying a used factory aircraft instead of a kitplane.**

ADVANTAGES	DISADVANTAGES
Easier to fly/No surprises	Expensive maintenance
Flyable upon purchase	Older equipment
Loans easily available	Little selection on options
Wide selection of types	Ordinary appearance
Wide range of prices	High-cost certified parts
Known resale value	
Parts availability	
No special flight restrictions	

Table 2-2. Advantages and disadvantages of kitplanes.

ADVANTAGES	DISADVANTAGES
New aircraft for used price	Danger of poor construction
Better performance	Take long to build
Lower maintenance costs	Hard to finance
Cheaper parts	Liability upon sale
Gradual purchase of subkits	Unknown resale value
Low cost inactivation	Flight restrictions
Attention-getting	
Personal satisfaction	

There is one major breakpoint to determine whether you should buy or build: If your sole intent is obtain an airplane to fly, you're better off buying a used production plane. You'll have it immediately, and if your tastes change, it can be sold and another type bought.

Building a kitplane is a challenge—many consider the actual flying of the aircraft rather anticlimactic. Unless you are prepared for the grind and mess of construction, and unless you can look at the building process as an end unto itself, you're better off shopping in *Trade-A-Plane*.

Operating Issues

Beyond constuction expenses, operational expenses must be considered. The monetary aspects are covered in chapter 3, but what about the personal costs?

Once the builder of a kitplane is awarded a repairman certificate, he or she is fully authorized to perform all maintenance and inspection of the aircraft built.

The cost savings are obvious, at least $300 for the annual inspection (even more for retractable-geared aircraft), and probably at least that much more in labor costs for routine maintenance.

The drawback is sweat because annual inspections are hard work! The annual of a simple aircraft like a Cessna 150 takes an A&P at least four hours; a simple homebuilt should take at least that long. Complex planes take longer. For instance, retractable geared aircraft must be set on jacks for gear tests.

The repairman certificate makes it legal. But being legal isn't the same as being safe: Are you competent to maintain and inspect the aircraft? Sure, you know this specific airplane inside and out, far better than any A&P you might hire.

But that A&P knows aircraft; he or she knows how they wear, and can recognize the subtle signs of corrosion or impending failure. I completed an annual one year, and had an A&P friend look it over. He found a crack in the control stick. Not a major one, and not one that caused anything beyond a small amount of play. But with years of experience inspecting hundreds of airplanes, he was suspicious enough to chase down the problem.

Also, most homebuilders don't really know engines. We buy them and install them, but don't really understand maintaining them. A pilot can be a shade-tree mechanic with the family car and risk nothing more than a little shoe leather. Aircraft engine maintenance requires a little finer attention to detail.

For safety's sake, EAA recommends that every other annual be performed by an A&P. Take advantage of the experience contained in your local EAA chapter. EAA technical counselors are A&Ps or experienced homebuilders. They can give advice on the annuals, or perhaps direct you to a A&P who'll help for little or no cost.

The holder of a repairman certificate can probably save at least five hundred dollars a year compared to hiring an A&P for all maintenance and inspections. For the most part, safety records support the adequacy of the repairman certificate. However, many homebuilt accidents are of the "for want of a nail" variety, forgotten safety wires or lock nuts.

The FAA has issued a notice of proposed rule making in the late 1980s regarding a primary aircraft category. If adopted, this category will allow greater owner involvement in maintaining production aircraft. If enacted, the operational cost advantages of the experimental category won't be as pronounced.

Buying Used or Partially-Completed Kitplanes

If you can't face the construction process, but the appearance or performance of a particular kitplane is still attractive, there are a couple of options open. Consider buying a used kitplane. Prices can be quite reasonable. In fact, certain flying homebuilts can be bought for little more than the total cost of construction.

There are several reasons. First, some folks like the building process more than flying. Some have built five or more aircraft, and quite often are craftsmen with excellent workmanship. Second, the occasional builder doesn't do his homework before starting construction, and the resultant airplane doesn't meet his needs. Finally, a builder could own his creation long enough to get tired of it and wish to move on to something else.

A pilot's proverb: "You're better off buying a used snake than a used homebuilt . . . ."

Obviously, significant problems could exist with the aircraft. The structure is already closed up, making detailed inspection difficult. It's quite possible for a number of hours to be flown without a hidden flaw making itself known. Recall the case mentioned in chapter 1, a homebuilt flew 14 hours without wing bolts.

A prepurchase inspection is vital. Find someone with experience on the type of aircraft for sale, preferably someone who has built the same model. They should pass judgment on the workmanship and adherence to the construction plans. While some deviations are minor, changes to the control system, rigging, or basic structure should set the alarm bells ringing.

An inspection by an A&P would also be a good idea, because they're experienced in rapid assessment of airframe and engine condition. However, for Rotax-powered aircraft like the Avid Flyer (Fig. 2-3), check at the local ultralight center for someone who knows two-stroke engines.

The aircraft should be test-flown by yourself and the experienced builder you brought. You can judge if the plane is right for you; the experienced builder will determine if the plane flies like it should. Examine the logbooks. An aircraft that is flown regularly probably flies well; a plane that sits a lot might have problems.

Single-seat airplanes like the RANS S-9 (Fig. 2-4) present a problem. Will the owner let you test fly it? There are a number of cases of prospective buyers crashing single-seaters. If your experienced builder-friend arrives in his own version of the same model aircraft, the owner shouldn't have too many worries. Or, if the aircraft is otherwise acceptable, you might be able to buy the aircraft contingent on an acceptable test flight.

Maintenance is another issue. You *cannot* receive a repairman certificate because you didn't build the aircraft. The original builder can perform the maintenance, if he still retains the certificate and is willing. Neither is guaranteed. If

Fig. 2-3. A used kitplane can be a good deal, as long as you can verify proper construction and maintenance. However, finding a mechanic qualified to check out the Rotax engine might be a problem.

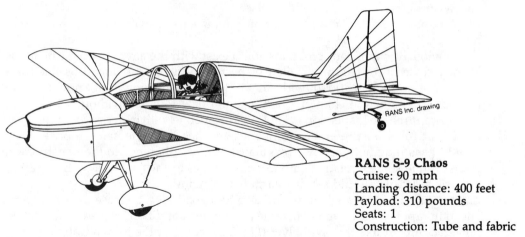

RANS S-9 Chaos
Cruise: 90 mph
Landing distance: 400 feet
Payload: 310 pounds
Seats: 1
Construction: Tube and fabric

Fig. 2-4. If you owned a hot little single-seater like the RANS S-9 Chaos, would you let a prospective buyer test fly it?

buying the plane from the original builder, a fresh annual inspection should be part of the deal. Have an outside mechanic inspect the aircraft as well, but licensed mechanics charge more for annuals than prepurchase inspections.

Don't fault the builder if he or she is unwilling to continue maintaining the aircraft after you buy it. The builder probably isn't a professional mechanic; the legal and other pressures are strong. If you buy from a friend, though, they might be willing to continue.

Any A&P mechanic is allowed to maintain and inspect homebuilt aircraft. One difference between homebuilts and production aircraft is that the A&P doesn't have to hold an *inspection authorization* (IA) to perform an annual on a homebuilt. That'll save about $50 a year. If the plane has an automobile or Rotax engine, it might be hard to find an A&P willing to be legally responsible for its maintenance and continued operation. Find the A&P before you buy the aircraft.

One interesting point is a gentleman's agreement between the FAA and the EAA. The owner of a homebuilt aircraft is allowed to maintain his airplane, as long as an A&P signs off the work *within the next 12 months*. If you can find a mechanic willing to work on this basis, you've got the best of both worlds: authorizations equivalent to a repairman certificate, with an experienced eye to keep the aircraft safe.

There's a thriving market in flyable homebuilts. Uncompleted kits are another kettle of worms. Buying and selling unfinished homebuilts is nothing new. It used to be said that 30 percent of homebuilt aircraft projects were eventually flown: 10 percent by the first builder, 10 percent by the second owner, and 10 percent by the third or subsequent owners. One Fly Baby ran through six owners until it was completed, 20 years after construction started.

A flying homebuilt has passed its most critical test and one can use its flying characteristics to judge how well it's built. An uncompleted project? Who knows what might be wrong with it. The sixth owner of that uncompleted Fly Baby found a critical problem with the wing bracing system. Who knows if the previous owners would have detected the problem before it was too late?

The second owner of a Sea Hawker amphibian kit thought the first builder had performed a crucial step in the construction of the wing spars. He hadn't. Fortunately, the second builder decided to perform a load test on the completed wings. The wings were designed for nine Gs. These wings failed at three.

Still, there are plenty of success stories in the purchase of uncompleted projects. But it's even more important for a knowledgeable builder to examine the aircraft. The less work that has been done on the project, the easier it is to check. Then again, the less progress, the less reason to buy the project rather than buying a new kit.

It's often a better idea to buy a new kit rather than even an unstarted previously-owned kit. There are a number of Glasair I kits that haven't yet been finished. Subsequent kits have been changed so the aircraft is easier and faster to build, and has better visibility and flying characteristics. A previously-owned Glasair I kit might be bought for less than $10,000, but it might be the wrong way if you're looking for a cheaper short cut.

There is one class of homebuilt for sale that is neither fish nor fowl. Ads occasionally appear saying: ". . . completed, signed off, ready for first flight . . .", or ". . . ready for paint and first flight."

In other words, the aircraft is finished, ready for its first flight and the builder is selling it!

Hmmmm . . . makes you wonder. There are legitimate reasons for selling the aircraft prior to first flight. The builder might have lost his medical. A spouse might have put a foot down and forbade the first flight. A divorce or family emergency might require liquidation of assets.

Or, of course, the builder might have no confidence in his own workmanship. It's up to you.

Remember, you must perform at least 51 percent of the work on the aircraft to receive a repairman certificate. A partially completed project might not allow you to qualify to do your own maintenance. Get a ruling from an FAA inspector prior to buying an unfinished project.

One other possibility is a professional builder. Some people buy kits, assemble them almost all the way, then advertise for a buyer. The purchaser gets his choice of color schemes and a ready-to-fly aircraft.

Strictly speaking, this isn't legal. The category is *amateur-built*, for "educational and recreational purposes," to quote the FAA. Technically, no person is allowed to build more than one model of a particular aircraft, otherwise, he's entered production.

But a number of companies operate openly, with no apparent interference from the government. Privately, FAA inspectors admit the professional shops produce far safer aircraft than the average builder.

It's an option to consider, if you have more money than time. As long as you can get the aircraft certified, things should be all right. But if the FAA decides to clamp down and start canceling airworthiness certificates, you could be hurting.

No matter how far along a partially-built kitplane is, or how low the price, keep your mission statement in mind. A good deal on the wrong kind of airplane is no bargain.

PERFORMANCE AND FEATURES

Still with me? That's the last time I'm going to try to talk you out of building a kitplane. Let's break the simple mission statement into hard requirements.

Performance Requirements

First, let's look at how much runway you have. If your mission calls for operation from "standard" airports only, there's no real takeoff/landing requirement. By standard, I mean 3,000–4,000-foot runway, a fairly clear approach, and the like.

But you plan to operate from smaller strips, therefore the takeoff/landing distance requirement is set by the shortest runway you anticipate flying from. To account for variations in technique and density altitude, subtract 10–20 percent from this distance to set the required runway length. Check the runway dimensions of likely destinations, too.

While stall speed usually has some relationship to handling, it's generally directly proportional to the kitplane's short-field capability. Therefore, there's little need to set a requirement for maximum stall speed. Be advised, though, if you're looking for a fast airplane, the stall speed is going to be high.

Once the aircraft has taken off, rate of climb becomes paramount. However, it's hard to set a required value, because *climb gradient* is more important than rate, but is rarely stated. If a 500-foot hill is located one mile from the end of the runway, a 60-mph plane with a 500-fpm climb rate will clear the hill, while a 180-mph plane with a 1,200-fpm rate won't. Set a rate of climb minimum, if it means that much to you.

Four interrelated parameters are cruise speed, range, fuel tankage, and fuel consumption. Refueling operations severely impact total time en route, therefore set the endurance requirement (usable fuel divided by fuel consumption at normal cruise power) first. Some kits have only two or three hours of onboard fuel, which is fine for local pleasure flights, but serious cross-country travel requires more. Again, set the minimum value based on your mission. Be realistic, how long do you honestly want to fly without a rest break? A typical value might be three hours; add another 30 or 45 minutes for reserves and round it up to four hours.

Now state the desired cruise speed. Note that *endurance* divided by *cruise speed* equals *range*, so no separate requirement for range is necessary. Local sport flying, of course, doesn't need a cruise speed requirement.

Compare your required performance against the advertised performance of available kits. Eliminate those completely out of the ballpark.

Believing the Numbers

Casting a jaundiced eye toward the performance figures supplied by the kit-plane manufacturers might be wise. How well can the figures be believed?

Few manufacturers actually lie, if the word lie is defined: "Published performance values known to be untrue." Let's take a look at one performance factor, cruise speed.

Seems pretty easy: take off, set cruise power, and read the gauge.

An airspeed indicator works by measuring the pressure differential between ram air and a source of neutral pressure. Most pilots consider the pitot tube the primary source of airspeed data, but the static vent is just as important. The makers of factory airplanes go through considerable effort to find the most accurate position for the static vent.

The easiest way to install a static system is to not install one at all, in other words, just vent the airspeed indicator to the cockpit. This approach appeals to most homebuilders and many kit manufacturers. However, the external airstream often creates a slight vacuum inside the cockpit. The lowered static pressure will cause the airspeed indicator to read higher than actual. The result is a "faster" kitplane, at least on paper.

And why should the manufacturer disbelieve it?

Next, take a look at that cruise power setting. How accurate is the tachometer? If it indicates 100 rpm low, the engine is running faster than indicated, and the cruise speed inches higher.

Finally, the manufacturer's own airplane is used to determine the performance figures. It was probably built by professionals and mounts a tuned and blueprinted engine. Gaps are nonexistent, surfaces are mirror-smooth, the weight is shaved. It's quite possible the radios and all nonrequired instruments have been removed.

The average owner-built kitplane can't match it. Once the first blush wears off excitement of the first flight, most kitbuilders have two regrets: heaviness and slowness.

How can the prospective buyer tell if the performance figures are realistic? The only sure way to find out is to fly formation with another aircraft and compare instrument readings. One aircraft I flew beside indicated 10 mph faster than mine did. But you're not likely to have the chance until it's too late.

The lesson is to take published performance figures with a grain of salt. We are stuck with the published figures; rarely does any other source provide such data. But don't select one plane over another merely because the manufacturer claims slightly higher performance.

Features

Performance isn't the only issue, though. Your mission might require that the aircraft have certain features. Not to mention your own biases.

What's your preferred configuration? High wing? Low wing? Biplane? Side-by-side seating? Tandem?

Generally, a low-winger gives better flight visibility for maneuvering in crowded airspace. A high wing aircraft is better for sightseeing (Fig. 2-5). Biplanes combine the disadvantages of both; you can't see what you're turning into or away from, unless the wings are aft.

Fig. 2-5. A high-wing kitplane like the Fisher FP-606 is a good choice if sightseeing is your mission.

Need IFR capability? Make sure candidate kitplanes have enough panel space for the necessary instruments. Is it a stable instrument platform? Read the reviews of each kit and take a thorough test flight.

Where will you keep the aircraft? If you expect to keep it at home and trailer it to the airport, folding wings, or at least quickly-removable wings, will be necessary. However, ask around because you'll find very few people actually keep their plane at home. For one thing, it adds a lot of hassle to the flying. The owner can't just drive out to the airport and fly; he's got to go home, hook up the airplane, tow it, unhook it, unfold the wings, check them carefully, and only then he can fly. After landing, the entire process is repeated in reverse. Do you want to go through this every time the skies call?

A more common use of folding wings is to store more than one aircraft in a hangar. This brings considerable savings, although a little bit of time is lost folding and unfolding.

Even if you don't use it often, though, it's a useful feature. Some areas have unflyable seasons; the aircraft can easily be brought home for several months. In these cases, removable wings are almost as handy as folders. When possible, pick a kitplane with the landing gear attached to the fuselage or a stub wing section. That way, the aircraft can still be rolled about with the wings removed.

How many seats? A simple little local cruiser can get by with just a single seat. But then, you'd like to give your friends a ride in your pride and joy. Besides, single-seat homebuilts are hard to sell when you wish to move on to something else. An exception, of course, is a spiffy little warbird replica (Fig. 2-6).

Fig. 2-6. Drive the Huns from the skies of Wisconsin in your own little warbird.

A multiseat kit is more complex than a single seater. You'll install dual controls, extra seats and support structure, and a pitch trim system. Because the plane is heavier, the structure is more complex. Larger, more costly engines are needed. It all adds up to more building time and greater expense. But you still build only one airframe and install one engine, no matter how many seats.

All in all, it's better to buy a two-seater than a single-seater, just for the increased resale value if nothing else. Again, it's easy enough to rent a production aircraft when necessary.

The seat configuration plays a significant role in performance. A tandem-seating aircraft (like a Cub or Champ) gives the pilot good visibility to either side and centerline seating. Because the aircraft can be narrower (only one person wide), tandem-seaters are theoretically faster. However, communications between occupants requires strong lungs or an intercom.

Side-by-side aircraft are more civilized. Both occupants can see the instruments and easily bellow into their corider's ear. Because aircraft engines are as wide as two people anyway, the performance cost is often negligible. The Van's RV-4 and RV-6 are good examples. With the same wing and engine, the side-by-side RV-6 cruises only 3 mph slower than the tandem-seat RV-4.

One last configuration consideration is the landing gear. Many kits are offered with the choice of *conventional* (*taildragger*) or *tricycle gear* (Figs. 2-7 and 2-8). Conventional gear is more rugged, and better suited to rough and short fields. It's easier to build because tailwheels can be simply attached to a leaf spring, bolted to the tail post, and connected to the rudder system via cables and springs. Nosewheel support structure is far more complex. Again, conventional geared aircraft are theoretically faster than trigeared types because tailwheels cause less drag than nosewheels.

Fig. 2-7. The Glasair design is available as a taildragger, or . . .

Fig. 2-8. . . . a tricycle gear configuration. Several other kits offer this option.

Most pilots are more comfortable with tricycle gear. There's no question that trigears are easier to operate and safer. A taildragger checkout is sometimes difficult to arrange, too. The theoretical performance disadvantage is minor to nonexistent. Looking at the RV-6 (taildragger) vs. the RV-6A (trigear), the nosewheel costs only 2 mph in cruise and 100 fpm in climb rate.

The cumulative effect is interesting. In going from the RV-4 tandem-seating taildragger to the RV-6A side-by-side tricycle gear, 5 mph in cruise, 75 feet of short-field performance, and 250 fpm of climb is lost. All on the same wing and engine. The RV-6A is 90 pounds heavier but has a 100-pound higher gross and 110 miles more range.

Performance and Features Summary

As of this point, you should know the approximate performance figures for your dream kitplane, as well as the preferred configuration and features. Use these parameters to narrow down the list of potential candidates.

Meeting a requirement isn't necessarily a pass-fail situation. Chances are, no single kit meets all your criteria, unless you have your heart set on one, and have biased the numbers in its favor. Shame, shame. In any case, the last section of this chapter includes a method to help decide close calls.

There are other factors involved in the building decision, though. It's time for the tawdry issue of money in this conversation.

COST OF CONSTRUCTION

That's "cost of construction," not "cost of kit." There's a considerable difference. To understand the construction costs involved, let's first look at what's included in the kit.

Types of Kits

Kits can be divided into three types: *materials kits, subkits,* and *complete kits*.

The materials kit is the cheapest. It contains the raw materials necessary for a plans-built aircraft. No work is performed by the kit supplier; in fact, the supplier often has no connection with the aircraft designer. The kit provides aluminum sheets, plywood, metal extrusions, and other materials that require considerable work to become aircraft parts (Fig. 2-9). Materials kits don't exist for molded composite airplanes.

If the materials kit doesn't do any work for the builder, what's the advantage? Basically, it's a painless way of obtaining the necessary hardware. The aircraft plans probably include a total list of materials, but rather than filling out an order form with countless lines, you can just order a single part number. There's always a price break for buying material in quantity, and a materials kit usually qualifies.

One has to watch out, though. Check the price list of materials, then compare with local prices. I know of at least one supplier whose aluminum prices are well above those I pay local distributors and I pocket the shipping charges.

Obviously, pilots looking for a way to build an aircraft quickly should pass up materials kits. While cowlings and other fiberglass parts are often available off-the-shelf, it's certainly not what most people consider a "kit."

The next step upwards are those designs that can be built from plans, but

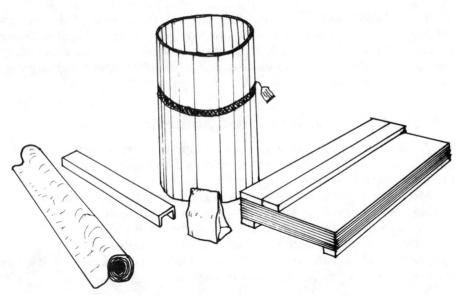

Fig. 2-9. A materials kit could consist of some rolled-up aluminum sheets, plywood, rolls of fiberglass, extrusions, and sacks of small hardware. The only advantage is the ability to order a single kit, instead of listing pages of parts numbers and quantities.

Fig. 2-10. Subkits often include much of the same stock items as the materials kit, but completed parts such as ribs and cowlings are often available. If the budget permits, the builder can greatly reduce his building time by buying additional subkits.

offer subkits to hasten construction. This is the most flexible method because the builder can buy whatever subkits that the budget allows, and build the rest of the airplane from scratch (Fig. 2-10). Or a builder unsure of his skills can buy subkits for critical or complex components. For example, a predrilled wing spar reduces the chance of error and helps ensure full strength of the final product.

Another advantage of the plans-and-subkits method is that the builder can still finish the aircraft if the kit manufacturer goes out of business. If the manufacturer of a molded-composite kitplane goes under and hasn't shipped a major component like the fuselage or wing, you're out of luck.

Some designs offer subkits only for those items that need welding or other special work. Other companies sell the entire aircraft in subkit form. Some molded composite designs can be bought as subkits, but all the subkits are required and the builder can't finish by plans alone. There's generally a small savings in buying all the subkits at once, in the form of a lower crating charge.

Of course, when the entire aircraft can be ordered as a single item, it becomes the last category; a complete kit (Fig. 2-11). Few are truly complete; even the best kits don't include paint or batteries.

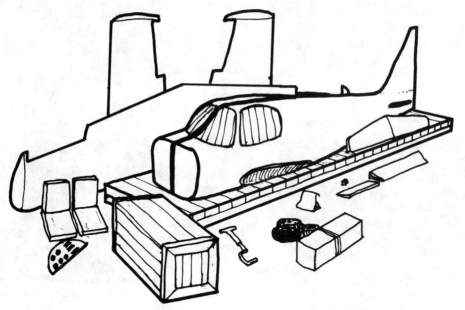

Fig. 2-11. The complete kit is everyone's dream, but they are the most expensive of the lot. Complete kits for metal airplanes often consist of merely all the subkits shipped at once.

Reducing the Kit Cost

The cost of a kit depends upon two factors: The cost of materials included in the kit, and the amount of work already done. Materials kits are cheap because the kit "assembler," which can hardly be called a "manufacturer," performs no work other than preparing parts for shipment.

Composite kits are the most expensive, but it's not because fiberglass is more expensive than aluminum or wood. Producing the molded components is a labor-intensive affair that is reflected in the kit price.

One nice feature of the subkit method is the ability to optimize construction time based on budget. Figure 2-12 shows the wing construction of the Murphy Renegade tube-and-fabric biplane. The plans show how to build the ribs. However, if the builder's budget allows, he or she can save a considerable amount of time by buying a rib kit.

This has always been a big advantage of homebuilding because the builder can save money by doing more work himself. Complete kits are great, but expensive. If money is tight, pick a design that can be built solely from plans and buy only the most necessary subkits.

Figuring out the cost of the kit is easy; knowing how much you'll spend before the aircraft flies is another thing.

Fig. 2-12. Factory-built ribs save a lot of construction time. The factory can probably build them lighter, as well. But budget builders can scratch-build for far less.

What the Kit Doesn't Include

The more goodies the kit provides, the more the kit costs. Perfectly obvious, isn't it? The problem is, how do you tell what the kit provides?

The most complete kits include descriptions like, "Everything but the paint and battery." That's not bad because it'll still cost a pretty penny more to build the plane, but at least you aren't stuck with any big-ticket items.

The biggest of the big-ticket items, of course, is the engine. Currently, most ARVs using the low-cost two-stroke engines include the powerplant. For planes using conventional aircraft engines, the additional cost is frightening. The engine is included in several of the more expensive kits, including the Questair Venture.

Engine prices can range from a few hundred dollars for a used two-stroke to well over $20,000 for the larger certificated engines. The kit for the Van's Aircraft RV-6 sells for about $9,000, and includes "just about everything needed except the engine, prop, instruments, and tires."

A new Lycoming O-320 engine will cost about $14,000. But a remanufactured, rebuilt, or used engine will drastically reduce the bite. If you insist on a new engine, several kit makers have arranged special discounts with Lycoming and Continental. Chapter 3 discusses engine options.

Let's look at what additional items might cost. Propeller? A Rotax prop for $250; thousands for a constant-speed model. Instruments? At least $500 for basic VFR, if you're buying new. Avionics? No kit includes radios or a transponder. Simple VFR birds can get by with just a hand-held transceiver, but even these sell for more than $300. By the time you add a nav receiver, LORAN, and a transponder, avionics cost can pass $2,500.

Most kits don't include these components and a great many others besides: Brakes? Paint? Upholstery? Epoxy and brushes? Crating and shipping? Find out what more you'll need before you order. The information package offered by the manufacturer should list what isn't included. If it doesn't, ask. If the manufacturer's answer is broad, such as "the fuel system" or "the interior," building will take considerably more money and time.

Ground Money

Everything in the previous section is "air money" for items that will fly. However, don't forget "ground money," which is the money paid for items necessary to build the airplane.

How well equipped is your workshop? The tools you already own probably aren't enough. For instance, if building an RV, you'll be doing lots of riveting. Do you own a compressor? That'll probably run between $200 and $500. A rivet gun will cost another $200, at least; add bucking bars, air hoses, rivet cutters, etc. Again, sometimes these items can be bought used for considerable savings (Fig. 2-13). Drill presses, band saws, table saws, grinders: it all adds up. Many kits require a solid and level workbench at least 10 feet long, which'll probably cost at least another $100.

Most kits require the same tools. Some of the smaller kits "can be built with simple hand tools," but you'd be better off getting a drill press, at least. The construction material greatly affects the tools required; wood seems to take the most, primarily in the "saw" category, of course. Metal is next, and composite kits seem to get by requiring the fewest. If you're building a composite kit in the cold country, adding heat to a garage might be a one-time and continual expense.

One advantage of the subkit system is the ability to bypass operations that require specialized tools. Bending a Wittman-type spring steel landing gear is beyond the capability of the typical home shop, so the availability of a prebent

Fig. 2-13. Buying power tools can seriously increase the cost of construction. Air compressors are vital for most kits, and run $200-$400. Haunt yard sales and the want ads to pick up older models like this one for $100 or so.

gear is a definite plus. By purchase of the proper subkit, you might be able to save money on tools as well as decrease overall construction time.

One last item: Don't forget to stay current during the building process. Too often, an aircraft's first flight is the builder's first flight of the past two years. Currency is a must; allocate a certain amount of money each month toward flight time. Or, drop flying altogether and plan on going through a thorough refresher just prior to the first flight. Count on needing at least 10 hours in the preceding month, if you go this route. Consider refresher training in an airplane similar to the one you are building: conventional gear or high performance.

Adding It Up

It would be nice if kits included everything from engine to screwdrivers, right down to a warm, comfortable, completely-equipped workshop. But they don't. Where most prospective kitplane buyers have only a general idea of the performance and features desired, all seem to know the limits of their budget. It's no fun to stretch your budget to buy the kit, then find out you'll need another $1,000 just to start building.

Table 2-3 is a worksheet to determine the approximate total cost of a typical kitplane project. Chapter 3 guides you through the selection process for the engine, instruments, and avionics, and chapter 4 covers tools. Determine the total cost for your configuration from the information in these chapters.

Table 2-3. Worksheet for determining cost of construction.

FACTOR	RANGE	YOUR KIT
Kit Cost		
Shipping	$0 (Pick it up yourself) to $1500	
Engine	$15,000+ (New A/C engine) $10,000+ (Reman) $ 8000+ (Quality rebuilt) $ 6000+ (Average rebuilt) $ 4000+ (Used) $ 2000+ (Auto engine conv.) $ 1500+ (Rotax)	
Propeller	$250 (ARV) $300-$1000 (Standard) $2000+ (Variable pitch)	
Wheels/Tires	$250 (ARV) to $1000+	
Paint	$200+ (DIY Enamel) $400+ (DIY Polyurethane) $2500+ (Pro job)	
Instruments	Basic VFR: $300-$1000 Add for IFR: $400-$2000+	
Avionics	$300+ (Handheld) $1000+ (Nav/Com) $1800+ (Nav/Comm with Glide Slope) $1500+ (LORAN) $1500+ (Transponder)	
Outside work (machine shop, welding, etc.)	$0-$???? (Probably less than $200)	
Ground Costs (Workbench, tools, etc)	$0-$1500 (Depending on kit, tools already owned, etc.) See Chapter 4 for breakdown of required and recommended tools	
	Total Cost to Build your kit:	$

CONSTRUCTION TIME

Another major factor that affects kit selection is the total time required to build the airplane.

Kit Type

Obviously, materials kits take the longest time to build, and complete kits the least time. The degree of completeness has a serious impact. Unless you order the non-included part in advance, you might not notice its absence until the part is needed. If there isn't another task ready, work stops until the shipment arrives. Is the engine included? Because of the high cost of aircraft engines, builders often wait until a "good deal" is found. That "good deal" might delay completion for months.

Does the kit include the engine mount? Getting a mount made might take weeks. Many kits don't include anything forward of the firewall. Kits that include the engine take far less time to build because all the nitty-gritty of power-plant installation (mount, fuel strainer, throttle linkages, etc.) are usually included as well.

Other Construction Time Factors

Beyond the kit type, there are four main factors that affect construction time: type of construction, complexity of the aircraft, detailed work included, and the amount of thinking the builder is required to do.

For equivalent aircraft, composite kits take the least time, followed by metal kits, and finally wooden kitplanes. This is part of the cost trade; composite kits generally cost more because a more completed aircraft is supplied. A Lancair costs more than an RV-6, yet its primary structure assembles far faster.

However, the complexity of the aircraft has a significant impact. The RV-6 has a simple, rugged, fixed gear. The Lancair has a mechanically-operated retractable tricycle system. The time a Lancair builder gains through composite construction might well be lost when it comes time for gear installation.

After all, one cannot expect to build a Glasair faster than a simple tube-and-fabric ARV. Let's look at just one small element, cockpit entry. The ARV might have only an open cockpit. But the fast glass bird's sliding window/door is necessary to allow the Glasair's high cruise. It'll take a number of hours to fabricate individual components, prepare the opening in the fuselage, assemble the structure, install, and adjust.

It's not a flaw of the Glasair design; it's just a reflection on the performance ranges. If your wish is for blistering cruise speeds, you'll pay for it with both a higher purchase cost and longer construction time. The Glasair has numerous systems; installing flaps, trim, retractable gear, and controllable pitch prop will take additional time.

Of course, the kitplane manufacturer can reduce the effect of additional complexity by including many prefabricated components. If an assembled gear linkage can be supplied, the time impact of retractable gear is reduced. A Pulsar can fly rings around a Baby Ace, but because of its modern kit design the Pulsar is finished far faster.

Some kits allow you to trade purchase price for the amount of work to be done. For example, Fisher Flying Products offers "quick build" options for most of their kits. For an additional $1,000, they supply prefabricated ribs, spars, wing tips, fuselage sides, bulkheads, formers, fin, rudder, horizontal stabilizer, and elevator. Neico Aviation claims a 700-hour reduction in building time for the fast-build version of the Lancair (Fig. 2-14).

Fig. 2-14. The Lancair is available in both regular and quick-build kits. The only difference is price.

If you've got the money, get the "quick-build" version because it will save a lot of time. Otherwise, trade building time for dollars.

The fourth factor is the amount of thinking the builder has to do. Imagine two kits that supply basic VFR instruments. Kit X includes a blank instrument panel. Builder X can lay out the panel however he chooses. Kit Y specifies the mounting location for each instrument.

Does kit X sound better? Yes, in some ways. After all, we're building a custom aircraft, and prefer to configure things exactly the way we want. But the time builder X takes to figure out a panel layout is time builder Y uses to actually install instruments.

Kit Y eliminates measurement time, too. Kit X includes instructions like, "cut a wire long enough to go from the instrument to the electrical bus." Kit Y says, "cut a wire 23 inches long and connect the instrument to the bus." For that matter, the maker of Kit Y could sell a pre-cut wire harness subkit and save even more building time.

One major function requiring builder decisions is engine installation. If the kit manufacturer specifies a particular engine, the kit design can be optimized and detailed installation instructions can be included. Otherwise, or if a nonstandard engine is chosen, the builder is on his own.

Builders buy kits for two reasons: builders are not qualified to design an airplane and a desire to complete the airplane as fast as possible. If you're essentially paying someone to design an airplane that can be quickly built, you want detailed instructions. Every time you have to wonder "Gee, how should I do this?" is time wasted.

This isn't to say that thinking is bad. Nor am I telling you to place your brain in park whenever you enter the shop. No kit matches the ideal; none is perfect. You must beware of inconsistencies and outright errors by questioning any step that doesn't seem right.

Fewer decisions means flying sooner.

Replicas

Replica aircraft take longer to build. The builder ends up doing a lot of things that do not affect flyability, but are required to make the plane look like another. For example, the Loehle 5151 uses an air-cooled Rotax 503 engine, but the builder must build a fake radiator air scoop underneath the fuselage.

The War Aircraft Replicas (WAR) use the same basic construction method as the KR series. A basic fuselage box is built, then foam is glued to the exterior and carved to shape. A KR-2 has a simple taper, but a WAR replica duplicates the complex curves of the original plane. The builder spends a lot of time with templates, carefully carving a scale shape.

World War I replicas have their own headaches, for instance, wheels. The actual airplanes had large-diameter wheels. These aren't available off-the-shelf anymore, and never did include brakes. Non-aircraft alternatives for the wheels are the only solution. But the builder can end up spending a lot of time designing and building brake assemblies (Fig. 2-15).

Fig. 2-15. Replicas can be fun, but require more work than normal designs. A complete brake system must be designed to fit the industrial tricycle wheels fitted to this CIRCA Nieuport 17.

Replica aircraft are classy and fun. But if you decide to build one, be resigned to a longer construction time and a few more hassles than the run-of-the-mill kit-plane.

The Manufacturer's Estimate

On the surface, this seems the easiest part of the process. After all, each kit manufacturer lists an estimated construction time. But if you thought verifying performance figures was complicated . . .

Cynical experience calls for doubling the published construction time. If you've never built an aircraft before, this is optimistic.

How does the manufacturer estimate construction time, anyway? In some cases, it's just a ballpark guess. Judging how long something will take to build is a science; few manufacturers have industrial engineers on the payroll. After all, building time is as much a marketing ploy as cruise speed. As mentioned in chapter 1, the published construction time for the BD-5 was 300 hours. Those BD-5s completed by private builders generally required more than 2,000 hours.

Fig. 2-16. An RV-4 fuel tank in its jig. More proof that woodworking skills are needed even on a metal airplane.

Again, the concept of "average skills" comes into play. To pilots, the ability to land an aircraft without breaking anything is considered an average skill. Yet to the common citizen, such a skill is quite extraordinary. It's that way with kitplane building as well. Someone with 10 years of riveting experience can't understand how a newcomer might take a while to get the knack.

The manufacturer's time estimate doesn't include correcting mistakes. New builders will make errors; some errors might not be costly in terms of dollars, but all increase construction time. Especially if the builder has to write the factory for a component to replace the ruined one.

Another item the factory estimate might not include is preparation, jig building, and cleanup time. You can't just glue two pieces of wood or composite structure together, or just drill and rivet aluminum, the surfaces must first be properly prepared and positioned. The jigs that hold the pieces together during joining must be accurately constructed before a component can be built (Fig. 2-16). Some kits incorporate the jigs within the shipping crate, which is a good reason for reading the "How-to-Open" instructions.

Finally, your shop will become extremely messy. Cleanups will range from an occasional squirt with an air hose to full-fledged GI parties.

The Final Analysis

Again, the kits that generally take the least time are composite. They are followed by the metal aircraft (either monocoque or tube) and finally by wood. Complex features of the aircraft (folding wings and retracting gear) add to construction time.

To get a feel for comparing the construction times between kits, look for the following:

- ☐ Completeness of kit.
- ☐ Fiberglass parts such as cowlings and wingtips.
- ☐ Predrilled parts.
- ☐ Covering envelopes (ragwings only).
- ☐ Use/inclusion of standard latches and hardware.
- ☐ Prebuilt wire harnesses.
- ☐ Preformed parts.
- ☐ Firewall-forward packages (engine, mount, firewall, and accessories).
- ☐ Upholstery and interior trim.
- ☐ Machined and welded parts.
- ☐ Special tools and building accessories, such as rubber gloves and mixing cups.
- ☐ Prebuilt jigs or jigless construction.
- ☐ Precut instrument holes.

The best way to find out how long a kit takes to build is to talk to someone who actually built one. Magazine articles are one source, but these aircraft are usually trophy-winners that the builder spends an inordinate amount of time on. Talk to the run-of-the-mill builders, instead.

Failing that, though, your best bet is to double the manufacturer's estimate. Of course, I'm being unfair to those who strive to advertise accurate times. My apologies, however, with no common basis to compare kits, it's best to assume the worst-case situation. If you finish in less time, great. Prevent heartburn by allowing a generous period for construction, rather than basing the schedule on suspect estimates.

THE PLANS

"Don't buy a pig in a poke."

I don't understand it, either. But if it means, "Don't buy a kitplane unless you have a chance to look at the complete set of instructions," I fully agree.

The instructions for any given kitplane can be obtained two ways. First, they can be borrowed from an acquaintance who is building the same aircraft. Second, they usually can be bought separately from the manufacturer. Often the instructions' purchase price will be deducted from the kit's price, should you decide to order.

Remember, though, it's a no-no to make a photocopy without permission of the copyright holder. It's especially wrong to build an plane from copied plans. I've never understood how some people can spend $10,000 building an airplane, and yet stiff the designer the lousy hundred bucks or so for the cost of the plans.

With the plans in hand, let's check 'em out.

Clarity

Do you understand them? It takes a special person to write kitplane plans. A brilliant designer doesn't necessarily have the knack for writing kitplane instructions. Some famous inventors had trouble with spelling and grammar.

Engineers often have particular trouble writing kitplane instructions. It's not that they can't use English, it's just tough for an engineer to write for a nontechnical audience. Technical language is the engineer's universe and specific terms within the universe have clear-cut, unambiguous meanings. Unfortunately the terms are completely incomprehensible to the layman. [The author is an engineer with a major U.S. aerospace firm.]

The instructions for each operation should include a complete parts list. Components assembled from previous operations should be identified (such as "Assembly A112") as should the part number, quantity, and description for each individual part.

Diagrams or pictures should provide unequivocal illustration of the construction of each component (Fig. 2-17). I prefer drawings to photos because cluttered backgrounds sometimes obscure the point being made.

Builder's Skills Required

Make no mistake about it, to build an airplane, you will *have* to learn new skills. Be it riveting, ripping capstrips on a table saw, or bonding fiberglass parts, there are going to be things you've never done before.

FLAP LEVER—Materials List and Builders Guide

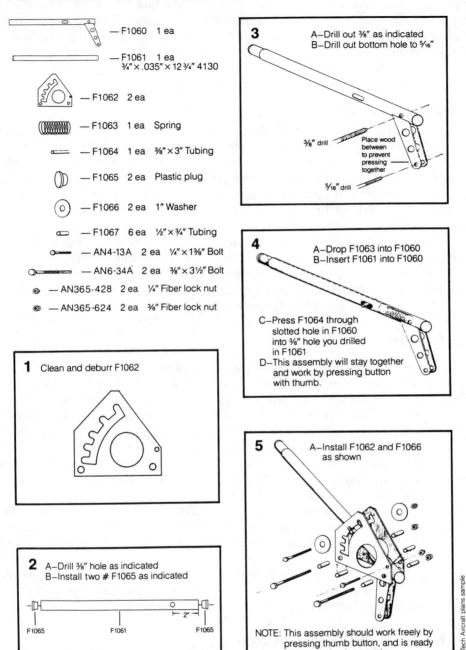

— F1060   1 ea

— F1061   1 ea
¾" × .035" × 12 ¾" 4130

— F1062   2 ea

— F1063   1 ea   Spring

— F1064   1 ea   ⅜" × 3" Tubing

— F1065   2 ea   Plastic plug

— F1066   2 ea   1" Washer

— F1067   6 ea   ½" × ¾" Tubing

— AN4-13A   2 ea   ¼" × 1⅜" Bolt

— AN6-34A   2 ea   ⅜" × 3½" Bolt

— AN365-428   2 ea   ¼" Fiber lock nut

— AN365-624   2 ea   ⅜" Fiber lock nut

1 Clean and deburr F1062

2 A—Drill ⅜" hole as indicated
B—Install two # F1065 as indicated

F1065   F1061   F1065
⊢ 2" ⊣

3 A—Drill out ⅜" as indicated
B—Drill out bottom hole to ⁵⁄₁₆"

⅜" drill

Place wood
between
to prevent
pressing
together

⁵⁄₁₆" drill

4 A—Drop F1063 into F1060
B—Insert F1061 into F1060

C—Press F1064 through
slotted hole in F1060
into ⅜" hole you drilled
in F1061
D—This assembly will stay together
and work by pressing button
with thumb.

5 A—Install F1062 and F1066
as shown

NOTE: This assembly should work freely by
pressing thumb button, and is ready
for installation in console.

Pro-Tech Aircraft plans sample

Fig. 2-17. A good set of plans includes a full parts list and step-by-step diagrams of each procedure.

However, one should establish exactly those skills that shall be necessary to build the aircraft. For example, look for:

Machining ("Turn the rod down to .875 inch." "Line-bore a half-inch center hole.") implies access to a lathe, and appropriate skills. While some of these processes might be done on other, simpler tools, it would be best to chase down a machinist.

Welding ("Tack-weld the rod in place." "Anneal the metal under low flame.") should not be considered an average skill. A typical area requiring welding is the landing gear.

Brazing ("Braze the washer in place.") is arguable because it's not that difficult, and a propane torch might suffice for small jobs. However, it does involve waving an open flame around the shop, which does require a modicum of attention.

Fiberglass mold making ("Glue foam in place, and carve the shape shown in the figure." "Protect the engine with plastic, then cover it with chicken wire and build up the cowling shape with plaster of paris."): these days, no one should be surprised if a kit requires some fiberglass work. However, much more time is required to make the mold than lay the glass. Making a fiberglass cowling from scratch might take 20 to 50 hours. Pick a kit that has major fiberglass parts (cowling, wingtips, fairings) supplied.

Again, subkit options allow bypassing operations that require a skill you don't have. If there's only one or two parts that require welding, buying a welded-parts subkit is certainly faster than learning aircraft-quality welding.

It's not unusual for a homebuilt aircraft to require any of these processes. But kitplanes shouldn't require welding; nor machining, nor making fiberglass wingtips from scratch. They aren't grounds for rejection of a design. But find out how much they'll affect your construction time and total cost.

THE MANUFACTURER

You're about to drop between $6,000 and $25,000 or more on an aircraft kit. What do you know about the company that takes your money? Depending upon the attitude and solvency of the manufacturer, your building time could go quick and smooth, or . . .

Kitplane manufacturers come and go. A few have become successful and prosperous. Others rose to the surface, then disappeared again with nary a ripple. A few surged aloft, then dropped back with a resounding splash.

A prospective buyer can prevent problems with the manufacturer.

Don't Be the First

Getting caught in the undertow of a dying manufacturer is depressing; builder support vanishes with the source of major components. Of all the BD-5 kits sold during the '70s, only a few have flown.

It's not just a long-ago historical phenomenon. A few years back, the Prescott Pusher (Fig. 2-18) was the darling of the kitplane world. A four seater with sharp

Fig. 2-18. The few Prescott Pusher kit-built versions that have been completed are widely reported to be poor performers, with 3,000-foot takeoff runs.

performance and handsome styling (looked a lot like a BD-5, for that matter), it hit the cover of every flying magazine. Many folks bought the first subkit, and others also sprung for the entire kit.

Then, about the time the first owner-built versions started to fly, the bubble burst. In a terse announcement, the owners announced the dissolution of the company. Those who had ordered subkits had 30 days to order the entire kit (for $35,000, less engine) before the line would shut down forever.

Why had these people been caught? It wasn't really their fault. There had been no real indication of a coming bust. They were hurt by being too fast; too eager to buy a well-marketed product of a hot new company. They wanted to be the first in the air with the latest aerial dreamboat.

How do you keep the same thing from happening to you?

Simple: Do not buy a kit produced by a brand-new company.

The temptation is overwhelming. They just announced the production of the Wiltfang 400, the slickest little job you ever saw. You want to buy a kitplane to be different; not to build yet another Kitfox or Glasair. You want to turn heads, not run with the herd.

But buying a kitplane early in the production run can turn your stomach. Four problems arise:

1. As mentioned, the manufacturer can go out of business. At best, you'll be left without any support. At worst, you might get stuck with a partial kit and have no way to finish it.

2. Nobody gets things right at first. Aircraft factories don't; hence there are airworthiness directives. Writers don't, either. No matter how many times this text is edited, no one would be surprised to find a minor typo.

   Kitplane manufacturers make mistakes, too. Sometimes the instructions are wrong. Sometimes two parts don't go together right. Builders might interpret instructions differently. How are these problems found?

   Simple. They depend on the first few builders pointing them out. Recall the impact of thinking on building time and imagine how much slower progress will be if every problem you encounter might be a design problem, and not due to your misunderstanding the instructions. Can you imagine the dead ends? The wait for redesigned parts? The frustration?

   The airplane I'm building has been through three redesigns of the rudder system, two of them instigated by in-flight failures of owner-built craft. No injuries resulted, but some new planes were dinged. A recent newsletter had a notification to beef-up some internal bracing. What about the early builders who have passed that stage, or have already-completed aircraft? They're penalized for being first.
3. The builder's support network isn't in place. One of the most invaluable aids to building is looking at a completed example. Or having someone with experience drop in to give you advice. If you're the first, no one can help you but the factory, and it might be thousands of miles away.
4. Things can only get better if the kit company remains solvent. The Glasair II is greatly improved over the initial models. Stability is better, performance is better, more options are available, and construction time is way down. Of course, the Glasair I sold for half the price of the current model . . .

Again, do not buy a kit from a newly-organized company. Wait a year or two while the kitplane is debugged by someone else.

Of course, there's a flaw in my argument: If no one buys the initial model, the company won't stay in business.

My recommendation is aimed at those who have never built an aircraft before; especially those with limited mechanical backgrounds. Some builders are A&P mechanics, others have restored older aircraft or are inveterate tinkerers. These folks can identify problems faster and can come up with sound solutions.

There has never been a shortage of those willing to buy an early production of anything *good*, be it kitplane, car, or electronic gadget. They know they might lose factory support.

But for pilots without the experience, and pilots who want to minimize the hassles of construction, wait. Things can only get better. If the airplane is good, it will succeed. If it isn't, you don't want to be involved.

Evaluate the Manufacturer

You should evaluate the manufacturer, as well as the kitplane itself. Here are some of the criteria:

1. What is the pricing strategy and delivery time for replacement parts? You will undoubtedly make some mistakes during construction. Make sure the network exists to buy replacements without waiting forever.
2. Do they supply detailed performance charts, like production aircraft? Some kitplanes give no further data than that supplied in promotional brochures. A detailed owner's manual (Fig. 2-19) is a good indication of thorough flight testing.

Fig. 2-19. In addition to detailed construction manuals, the better kits come with owners manuals rivaling those of production aircraft.

3. Does the manufacturer offer a newsletter? It's a positive sign if it does, but the newsletter (Fig. 2-20) should be more than a public relations sheet. It should answer common construction questions, provide insight into modifications and recommended changes, and assist in contacting other builders in your area.
4. Does anyone else outside the kit manufacturer offer a newsletter? That's an even better sign because a builder support network is in place. It also indicates the plane delivers its performance promises. However, it can also indicate a lack of confidence in the factory, or a manufacturer who doesn't care enough to publish his own newsletter. Study the tone of the newsletter to judge customer satisfaction.
5. Is there a help line to answer your questions? Don't expect a toll-free call, but builders should be given a number they can call for advice. Most problems occur during evenings and weekends, of course, and the best help lines will be open during these periods. Don't be surprised if they

Fig. 2-20. A good newsletter is one sign of a good kit.

aren't, because kit manufacturers are human too, and like weekends off. Sometimes the help line is only available during limited hours, like 9 a.m. to 4 p.m. on Saturdays.

6. Some kits have systems that allow the builder to do some of the construction in the factory before the kit is shipped. This is great because the factory lets you use the laborsaving tools and the correct jigs. Help is instantly available from the factory reps. Because these programs concentrate on the most critical sections, it helps ensure a strong and straight aircraft. However, it isn't free, and often must be arranged at the factory's convenience. Count on spending about two weeks of hotel and food bills at the factory location.

7. Ask for the names and phone numbers of those building the aircraft in your area. Find out their opinion of the factory's support.

OTHER EVALUATION CRITERIA

Several other areas should be considered.

Construction Material

Do you have any biases toward or against working with particular materials? If you've got a shop full of woodworking tools, and enjoy building furniture and cabinets, a wooden airplane kit might actually take less time to build. Otherwise, if you decide on a metal airplane, you'll be on the bottom of the learning curve. Similarly, if you already suffer from chemical allergies, a composite kit might make matters worse.

Don't have any preferences? Subsequent chapters discuss the techniques of working with each material. Study them; see if you start leaning in one direction or the other.

Mission selection might also affect type of material. Fast airplanes are invariably composite, metal monocoque, or wood; slow knock-around planes are usually tube-and-fabric.

Shop Requirements

The Glasair II has a one-piece wing 23.3 feet long. If your garage is only 20 feet wide, you're in trouble. Make sure you have space to build the kit before buying it. A two-car garage is sufficient for most; however, it's best to compare the size of the aircraft versus the space available. Yes, the Glasair wing could be built diagonally across the garage, but you won't have enough room left to build the fuselage.

Another important factor is handling the project. We've discussed the advantages of folding or removable wings, when it becomes time to carry the plane to the airport. Imagine the problems that arise when the plane does not disassemble. How are you going to haul a complete airplane, 20 feet long with 30-foot wingspan, from your workshop to the airport?

The easiest solution to this problem is to plan on doing final assembly at the airport. This requires the hangar rental, approval of the airport manager, and the like. It is not an ideal or cheap way to solve the problem.

I recommend picking a kitplane that allows the wings to be unbolted and the project legally carried on a standard trailer. Further information on shop size and arrangements is contained in chapter 4.

Maintainability

With your brand-new repairman's certificate, you'll be able to do all the maintenance yourself. But can you reach the components? Does the cowling quickly remove for access to the engine? Or do you have to remove the propeller to completely expose the engine? If you have to replace rudder cables, how easy is it to get access to the pedals? Are there access panels, or will you have to dive head-first under the instrument panel?

It's a serious consideration. Any moving component that you can't inspect is a snake, ready to bite. And should it be necessary to replace a component, you want wide access panels that allow removal and replacement. If there aren't too many visible on a particular model, ask the manufacturer to identify allowable locations.

Cabin Size and Comfort

Hopefully, you'll find a kitplane that fits comfortably in your workshop. Now, do *you* fit comfortably inside the airplane? Have you tried one out for size? Can you move the controls through their entire range? Can you see out? Throughout homebuilding's history, one factor has remained nearly constant: The original designer sizes the cockpit to fit himself. Too many designers have been little guys.

Peter M. Bowers designed his Fly Baby to fit his 6' 2" frame. Compensating for shorter pilots is easy; cushions can be added to the seat and the rudder pedals mounted farther back. Other homebuilt aircraft aren't so accommodating. The designer of my project was 5' 4"; I had to rearrange the cockpit floor for adequate legroom.

It's not just height, either. There are several homebuilts I can't fly because I have wide shoulders and can't even get my elbows in the cockpit. I'm not King Kong, either, a tad less than 6 feet tall!

Published numbers such as "accommodates 6'6" pilots" should be taken as mere guidelines. Try out the cockpit before you buy.

Check the visibility while you're inside. When you look to the left on some high-wing designs, and all you see is wing root. It's no fun flying an airplane hunched down to see sideways.

How Does It Fly?

While you're trying out the cockpit, arrange a test flight. Kit manufacturers give rides regularly; there should be no problem as long as the plane isn't a single-seater.

Often, the factory is thousands of miles away. Obtain their demonstrator air-show attendance plan. If you can't match their schedule, pay the airfare and fly to their location. Would you pay $20,000 for a car without so much as a test drive? Call first because some manufacturers have scheduled test ride dates.

A cheaper and easier option, of course, is to cadge a ride from somebody local. It's a little delicate; they've likely been deluged with requests and have no obligation to give anyone a ride. My Fly Baby shared a hangar with a grand champion homebuilt, and I didn't get invited for a ride in it until a full year had passed.

Eventually, you'll find yourself at the controls. You don't have to play "Mr. Test Pilot"; just fly the airplane to see if it's agreeable. Would your feet be comfortable on a long trip? If the plane uses a sidestick or other unusual control arrangement (Fig. 2-21), do you think you can adjust?

How much extra instruction will be needed before your first flight? Be honest; if you've flown nothing but 150s, there's no way you can step right into a White Lightning. Besides, experience in unfamiliar aircraft is excellent training for your coming test-piloting chores.

Fig. 2-21. The Zenair Zodiac's stick is mounted between the seats for access by both occupants. It's best to try out such an arrangement before committing to buy the kit.

THE BIG DECISION

By now, you probably know which airplane best fits your requirements. But what if you can't make up your mind? What if you can't decide between two similar airplanes? One is faster, but the other promises a shorter construction time. How do you pick which one is best for you?

One way is by "Weighting and Grading." Make a table of requirement categories, like cruising speed, range, wing position, gear position, etc. List *every* category important to you; add an "appearance" category if styling is important.

Alongside each category, list your desired values or configurations. Then assign a weighting factor for each category on a scale of 1 to 10. A weighting factor of 10 means that meeting your requirements in that category is vital, while a factor of one means the category isn't very important.

Now, for each kit examined, assign a grade to each category depending upon how well it meets the requirement. Assign a five if the requirement is met, going higher or lower depending upon how well the kit does. For instance, if your requirement is for a 200 mph cruise and the kit cruises at 210, you might assign a grade of six. A 120 mph cruise might get a grade of one.

The grades don't have to be proportional to the amount the kit misses or exceeds the requirement, but they must be consistent. If two planes cruise at the same speed, they should get the same score. If one is "uglier" than the other, it should be reflected in the "appearance" category, not under "cruise speed."

For each category, multiply the weighting by the grade. Then add all the category scores to determine the total score for the kitplane.

Table 2-4 shows a sample scoresheet using this process.

Was that any help? Or had you already made up your mind? No matter. Don't send in that order sheet yet. Let's look into some of the decisions that must be made, first.

Table 2-4. Sample kitplane evaluation sheet.

BASIC MISSION STATEMENT: Long range aircraft that can operate from 2000' grass strip

MUST HAVES: Certified engine, wingspan less than 25', at least two seats

KIT EVALUATED: Ulrich 2000

CRITERIA	PREFERED	WEIGHTING (1 TO 10)	THIS KIT	SCORE	WEIGHT X SCORE
CONFIGURATION:					
Wing Position	Low	3	Low	5	15
Gear	Trigear	8	Taildr.	2	16
Seating	SBS	10	Tandem	2	20
Seats	2	5	2	5	25
Styling		8	Nice!	10	80
PERFORMANCE					
Cruise Speed	150	5	120	3	15
Endurance	4 hrs	7	5	8	56
Take off Dist.	1500	9	800	10	90
Landing Dist.	1500	9	500	10	90
CONSTRUCTION					
Building Time	1000	4	1000	5	20
Type of Material	Metal	2	Tube + Fab.	4	8
Total Cost	$25,000	8	15,000	8	64
Completeness of plans		6	Poor	3	18
OTHER FACTORS					
Type of engine	Lyc./Cont	10	Cont.	5	50
Cabin comfort		8	Good	6	48
Flying qualities		9	Great	10	90
Aerobatic		1	yes	5	5

TOTAL SCORE: 710

Table 2-4. Sample kitplane evaluation sheet. (Continued.)

BASIC MISSION STATEMENT: <u>Long range aircraft that can</u>
<u>operate from 2000' grass strip</u>

MUST HAVES: <u>Certified engine, wingspan less than 25',</u>
<u>at least two seats</u>

KIT EVALUATED: <u> </u>

CRITERIA	PREFERED	WEIGHTING (1 TO 10)	THIS KIT	SCORE	WEIGHT X SCORE
CONFIGURATION:					
Wing Position	Low	3			
Gear	Trigear	8			
Seating	SBS	1 0			
Seats	2	5			
Styling		8			
PERFORMANCE					
Cruise Speed	150	5			
Endurance	4 hrs	7			
Take off Dist.	1500	9			
Landing Dist.	1500	9			
CONSTRUCTION					
Building Time	1000	4			
Type of Material	Metal	2			
Total Cost	$25,000	8			
Completeness of plans		6			
OTHER FACTORS					
Type of engine	Lyc./Cont	1 0			
Cabin comfort		8			
Flying qualities		9			
Aerobatic		1			

TOTAL SCORE: <u> </u>

3

Decisions

EVEN THOUGH YOU'VE PICKED OUT A KITPLANE, the decision process isn't over. How are you going to pay for your dream ship? What avionics and instruments should be installed? If the kit doesn't come with an engine, how do you find the right one at a reasonable price?

The material in this chapter, for the most part, shouldn't affect your choice of kit. But there are a few things to settle before making the chips fly, or even before signing the purchase order.

CAN YOU AFFORD TO OWN AN AIRPLANE?

People build kitplanes for a variety of reasons. Economy is one; homebuilts are cheaper to operate than new or used production aircraft. But owning a kitplane isn't free by any means.

It's quite possible to end up over your head even if you paid cash for the kit. The purchase money might have come from some special nest egg, or careful scrimping over a number of years, an inheritance, or whatever. A friend sold his house and bought a Questair Venture kit. He isn't married, of course.

So the nest egg is gone, but the plane must still be supported: insurance, hangar rent. Unless you've owned an airplane before, you're likely to be surprised at how much they cost to own. Even homebuilts.

Before you go much further, figure out if you can afford to operate the homebuilt you've selected. Let's add up the ownership costs for a typical kitplane.

There are two types of expenses, *direct* and *fixed*. Direct costs are those that accrue directly from flying the aircraft: fuel and oil bills, for the most part. Fixed costs are those you must pay even if the airplane doesn't fly. Tiedown or hangar rental and insurance are the major elements. If you finance the aircraft, the monthly payments must also be added to the fixed costs.

Fixed costs include the maintenance expenses; those required to keep the aircraft in a flyable condition. Typical maintenance expenses are the annual inspection and routine repairs. The base cost for a Cessna 150 annual runs around $350 for the labor, plus parts. A homebuilder with a repairman certificate saves the labor charges, and saves on those replacement parts that he or she can build. The kitplane owner is likely to pay more for an "aircraft" part than a certified mechanic because the mechanic probably gets a discount.

Accountants identify one other expense of owning an aircraft: lost interest. If the builder spent $20,000 to build the airplane, that money could instead have been invested and earned interest. This lost opportunity is lost income; hence, it's the same as an expense.

As far as I'm concerned, this "expense" is balanced by the simple joys of owning an aircraft. The freedom a personal plane grants is impossible to put a value on, but I'm willing to stipulate it's worth at least as much as the "lost interest." Let's look at those expenses that come out of actual pockets instead of out of a bank balance book.

Fixed Costs

Fixed costs are the most irritating aspect of aircraft ownership, as they continue whether the plane flies or not. It might snow 30 days out of the month, and on the 31st the driveway has to be shoveled, but you still have to pay the hangar, insurance, and other fixed costs. For the average owner flying a hundred hours or so a year, fixed costs *are* the major expense of ownership.

Storage, in the form of hangar or tiedown rents, hogs the lion's share. Tiedowns are cheaper, but hangars provide better protection. No matter what kind of kitplane, one element is common: the builder is loathe to tie it down outside. After spending thousands of hours and dollars, you'll want to hangar it as well.

Priced a hangar lately? The closed hangars at my airport rent for $150 a month, which is a steal compared to most major metropolitan airports. Hangars eight miles away at a controlled field run $250 on up, in a year, that's $3,000. If you fly 100 hours a year, the hangar rent alone amounts to $30 an hour.

Reduce the bite by hangaring with another aircraft and sharing the rent. Homebuilts are small; hangars are generally designed for the Cessna 210 or larger class of airplanes. If your kitplane will slip under another plane's wing, hangar cost is halved (Fig. 3-1).

Outside tiedowns are cheaper. Rent varies, depending upon the desirability of the location. Like hangars, tiedowns cost more at close-in major airfields with good security. A tiedown might run $100 a month at these fields, but a grass strip 20 miles away might have spots for $35. Whether this works for you depends upon the severity of the local weather and the type of kitplane: composite, wood, etc.

Don't be too eager to see a folding-wing aircraft stored in your garage as the solution of the hangar woes. Not only does it become a hassle (discussed in chapter 2), but most airplanes can't take a steady diet of being dragged over public roads. The highway vibration is far worse than takeoffs and landings, and a continuous 50 mph is far different from an occasional acceleration to takeoff speed. It causes premature wear and is hard on the instruments. Consider all the alternatives before deciding on keeping an airplane at home.

Fig. 3-1. Homebuilts are often small enough to share hangars, hence cutting the rent in half. Here, a Thorp T-18 snuggles nicely under the wing of a Cessna 140.

However, folding wings do make it easier to share a hangar (Fig. 3-2). Several kitplanes require only a few minutes to unfold the wings and prepare for flight.

But the cost of renting a hangar might be moot if none are available. The waiting list in my area is two years long. At one local airfield, closed hangars almost never become available. If you intend to keep your kitplane in a hangar, get your name in immediately. If your name comes up and your plane isn't ready, sublease the hangar to someone else.

Fig. 3-2. While keeping a folding-wing homebuilt at home is impractical in most cases, they make hangar sharing easier.

Insurance

Unless you've made some serious flying mistakes over the years, insurance is easier to get. Premiums vary with coverage, pilot experience, and the type of aircraft.

Straight *liability* (covering claims against you for the injuries or damage caused by your airplane) will cost $300 per year for a single seater and about $500 or so for the typical two-seat kitplane. Coverage can often be optimized to reduce the premium, for instance, if you just intend to fly family members, passenger liability coverage can be excluded. But in this litigous age, don't operate without some sort of liability coverage. It's not that expensive.

Hull coverage is different. "Hull" is the airborne equivalent to automobile collision insurance, where the insurance company pays for accident repairs. One major insurance company quoted yearly premium rates of 10% of the kitplane's value. An owner of a $15,000 Avid Flyer would pay $1,500 a year for hull insurance. Again, if flying 100 hours a year, that's $15 an hour just for the hull insurance.

Lower premiums can sometimes be found through an insurance broker instead of directly from the company. A local brokerage quoted an annual hull premium of four percent.

As the saying goes, "your mileage may vary." Higher deductibles reduce the premium, as will thousands of flight hours in your logbook. Features like retractable gear drive the rates higher. You can also ask for "not in flight" hull coverage. It doesn't pay off if you crash, but does cover your losses if the hangar catches fire or the plane is outside in a sudden hailstorm.

Many homebuilders don't carry hull coverage. They prefer to *self-insure*; if they break the airplane, they'll fix it out-of-pocket. Deciding to self-insure is like betting with the insurance company. Fly 10 years without "totaling" and you win. It's not for the faint of heart or those with expensive kitplanes. Also, if the airplane acts as loan collateral, the bank will require full insurance.

Self-insurance is more feasible with homebuilts than production aircraft because labor is a major factor in the cost of repair. When the owner has a repairman's certificate, the labor for any work, whether damage repair, routine maintenance, or annual inspection, is free. The accountant might argue that the time spent doing the annual could be better spent working for minimum wage at the local burger emporium, but we've already decided to ignore him.

But even if the labor's free, you'll have to make or buy replacement parts. Many parts have to be replaced at regular intervals. For instance, many ELTs must have the battery replaced every two years, and other parts must be replaced on a fixed schedule.

Otherwise, parts cost predictions are difficult because cost varies between types. A hydraulic failure in a retractable-gear aircraft might cost many hundreds of dollars, while torn fabric on a simple ragwing could be fixed for nearly nothing. While these costs are directly related to the operation of the aircraft, I lump them with fixed costs instead. A good term for them is "consequential" expenses, as they are only indirectly related to actually flying the aircraft.

Consequential expenses vary in an odd relationship with the aircraft's total hours. The first hundred hours or so will be costly, as problems will be discovered and repaired. And getting used to a new aircraft might result in a few minor dings.

After the bugs are ironed out, aircraft are relatively trouble-free for awhile. Then, as they age, repairs start creeping up again.

How much these costs run during the test period is anyone's guess. During the "steady as she goes" midlife of the aircraft, I average about $100 a year for engine and airframe parts for a simple homebuilt. The amount rises with the aircraft's complexity: radios, retractable gear, constant-speed prop, and the like. I'd guess consequential expenses for an average homebuilt might run $200 per year.

This isn't adequate for major expenses, but we'll establish a cash reserve under the direct costs.

If you've managed to get a loan to buy a kitplane, don't forget to include the monthly payments in the fixed costs. Other fixed costs are rather minor. Most states require annual aircraft registration. My home state charges homebuilts a flat $25, which is a pretty good deal, when production aircraft's rates are set by a sliding scale that starts at $50. Some state governments find ways to clap "luxury" taxes on aircraft, too.

A homebuilt has a major advantage over a production aircraft: Fixed costs can be completely eliminated during financial crises. If necessary, the aircraft can be disassembled and brought home for storage. Surrender the hangar and cancel the insurance. The kitplane can't be flown, but that saves direct costs as well. Mothball the plane until you can afford to fly it again.

But if it'll have to be sold, don't bring the airplane home. You'll get a lot more for a flying aircraft than one in pieces.

Table 3-1 gives the typical ranges of fixed costs, and best- and worst-case scenarios. Also provided is the equivalent per-hour rate for several annual utilization rates.

Direct Costs

Direct costs are easy to compute. Take the engine's per-hour fuel burn and multiply by the cost of fuel. Add a buck or so for oil, and there's the direct costs.

There is one major variable in the direct cost equation: cost of gas. The solution is to simply install an engine capable of running on 80 octane avgas. Of course, 80 octane is rare these days, but autogas isn't. Most aircraft that can operate on 80 octane will happily burn 87 octane regular car gas, or 92 octane unleaded.

The autogas solution isn't perfect. For one thing, 80 octane aircraft engines aren't made anymore. If you want a new engine in your dream ship, count on feeding the engine 100LL. If you're building a cross-country machine, forget the 80 octane engine, as only 100LL is available at most airports.

Autogas aircraft encounter many of the same problems as aircraft kept at home. It's a bit of a hassle. Feeding VW engines isn't so bad, with their 3.5 gph fuel consumption. But imagine keeping a 50-gallon tank filled; five-gallon jerry cans ain't gonna' do it, buddy.

If you do use autogas, use *good* autogas. Buy major brands from dealers with rapid turnover. The gas refiners optimize the mix for the season, and cut-rate dealers get special deals buying leftover winter gas in the spring. Avoid all gas

Table 3-1. Typical maximum and minimum fixed costs.

Fixed Costs

	Yearly		Monthly		Monthly	
	Min.	Max	Min.	Max	Case 1	Case 2
Storage:						
Hangar	$1,000	$3,500	$84	$290		200
Tiedown	500	1,000	42	84	55	
Insurance						
Liability	300	800	25	68	40	50
Hull:						
$10,000 Value	400	1,000	34	84		
$20,000 Value	800	2,000	68	168		
$40,000 Value	1,600	4,000	134	334		300
Loan Payments	0	6000 +	0	500+	0	150
Consequential Costs	100	?	10	?	10	20
Total Monthly Fixed Cost:					$105	$720

Case 1: Avid Flyer/Kitfox class, outside tiedown, no loan
Case 2: Lancair/Glasair class, closed hangar, $10,000 loan

	Flight Hours per Year				
	25	50	100	150	200
Case 1	$50	$25	$12	$7	$6
Case 2	345	172	86	57	43

Per-Hour Equivalent of Fixed Costs

suspected of adulteration. Rotax users should be especially wary, as two-stroke engines are very sensitive to gasoline quality.

To check for alcohol, use a small gradiated cylinder like a test tube or syringe. Fill it 20 percent full with water, then fill it the rest of the way with gas. Shake for thirty seconds, then let the water separate from the gas. The water will "swell" as it absorbs alcohol; if the tube is now more filled with water than before, don't use the gas in your airplane.

As far as oil is concerned, engines typically burn a quart every five or 10 hours. The oil should be changed every 25 hours for engines without oil filters, or at 50-hour intervals if an oil filter (not oil screen) is installed. Aviation oil costs approximately two bucks a quart, hence the oil costs come to about a buck an hour.

Routine fix up expenses are included under consequential costs, discussed earlier. However, occasionally things go bad. Very bad. A friend had to replace two engine cylinders in his Tri-Pacer. The total cost came to almost $2,500.

The best way to handle these problems is by establishing a *maintenance reserve*, where you set aside a fixed amount of money every flight hour. The basic purpose of this fund is to pay for the eventual overhaul of the engine. It also comes in handy for other expensive repairs.

Determining the rate to "charge" yourself for the maintenance reserve is relatively easy. It's based upon the TBO (*time between overhauls*) of the engine and the cost of overhaul. Any engine is going to need overhauling sooner or later. In the case of the Rotax, it's sooner but cheaper; 400 hours or so between overhauls, but a full overhaul costs only a few hundred dollars. An aircraft engine overhaul comes later (1,600 to 2,400 hours) and is much more expensive.

To determine the hourly rate for the maintenance reserve, take the overhaul cost and divide by the hours until overhaul (at the time of engine installation). It's as simple as that. Install a new Lycoming O-320 (TBO 2,000 hours), count on an overhaul cost of $5,000 (rather cheap, really) and build your maintenance reserve at the rate of $2.50 a flight hour. If you had installed a used Lycoming with 1,000 flight hours, the rate jumps to $5 an hour because the overhaul charge must be based upon only 1,000 flight hours.

The Rotax operators actually get the best deal. Their per-hour reserve rate comes to about a dollar an hour.

The maintenance reserve should be deposited into a separate account, to keep it isolated from other needs. Of course, it is available for worthier causes, especially if the spouse simply must have new living room furniture.

Or, you might decide to ignore the maintenance reserve. Most owners don't keep their aircraft that long; you'll probably sell the plane long before overhaul time approaches. Being more grasshopper than ant, my practice is dig up the money from somewhere when the time comes.

Keep in mind, though, that TBO is a rather nebulous term. Engines are further discussed later in this chapter.

Table 3-2. Per-hour and annual cost of ownership for various monthly fixed costs.

Monthly Fixed Cost	Hours/Year							
	25		50		100		200	
	Auto	100LL	Auto	100LL	Auto	100LL	Auto	100LL
$50	$35	$41	$23	$29	$17	$23	$14	$20
100	59	65	35	41	23	29	17	23
150	83	89	47	53	29	35	20	26
200	107	113	59	65	35	41	23	29
250	131	137	71	77	41	47	26	32
300	155	161	83	89	47	53	29	35

Per-Hour Cost

(Assumes 8 gal/hour, autofuel @ $1.40, 100LL at $2.10)

Monthly Fixed Cost	Hours/Year							
	25		50		100		200	
	Auto	100LL	Auto	100LL	Auto	100LL	Auto	100LL
$50	$880	1020	1160	1440	1720	2280	2840	3960
100	1480	1620	1760	2040	2320	2880	3440	4560
150	2080	2220	2360	2640	2920	3480	4040	5160
200	2680	2820	2960	3240	3520	4080	4640	5760
250	3280	3420	3560	3840	4120	4600	5240	6360
300	3880	4020	4160	4440	4720	5280	5840	6960

Total Yearly Cost

Adding It Up

Table 3-2 shows typical costs based on various annual utilization and the autogas/avgas decision. Fixed costs are the major expenses until utilization reaches about 100 hours per year.

One hundred hours per year works out to about two hours a week; about eight hours a month. That's quite a bit of flying, especially for the "knock-around" fun planes like Kitfoxes or Huskies. Your actual utilization will probably lie between 25 and 75 hours a year unless there are specific, regular flights you plan on making. The first year will be greater, of course.

At these low utilization rates, renting from the local FBO is sometimes cheaper. Exceptions are fast ships like Falcos and Ventures. Equivalent rental planes like Arrows or 182RGs are quite a bit more expensive; the break-even point between renting and flying a homebuilt is fairly low in these cases.

However, at such a low yearly utilization, the per-hour rate is about the same for simple or complex homebuilts. Owning a Lancair is cheaper than renting an Arrow, but a Super Koala will cost more than renting a 152. Our resident accountant sees no reason for owning an airplane at all.

But how important is convenience to you? If you wake up Saturday to blue skies, can you schedule a plane for the afternoon? For as long as you want it? Probably not.

How safe are the planes you're renting? Does the FBO fix every problem immediately? Or are they bandaged together until the next 100-hour? At least if you own it, you'll know the exact maintenance status. And rental aircraft are subject to considerable abuse.

Yes, in some cases, owning your own plane costs more; but we own airplanes because we *like* airplanes. We own them so we can fly them when we feel like, not when an FBO's schedule allows.

Sure, ownership is sometimes a hassle. But renting a plane is like riding the bus: You're locked to someone else's schedule, and you never know what the last rider did on the seat.

Don't base your decision solely on the per-hour cost. Work out your budget, and if you can afford ownership, go for it. The convenience and safety are worth it.

FINANCING

Go ahead, try it. March into a nearby bank and say, "I want to take out a loan to build my own airplane." Bankers work hard, and they enjoy an occasional laugh.

The direct approach might not be the best. Tales of prospective kitplane builders arranging financing are few and far between. I know of no one who has received a loan specifically to buy a kitplane.

Other avenues exist. Most loans require collateral; home equity, property, or other valuables. If you have something that can act as collateral, a loan shouldn't be too difficult. The bank merely wants to be sure of recovering its losses should the loan go into default.

Talking the bank into accepting the kit itself as collateral would be difficult because faulty workmanship would make it worthless. However, precedents for construction loans do exist (in house building, for example) so it might be worth a shot. The bank will want some sort of protection for the kit. Insurance for uncompleted kits is available, at a surprisingly low premium. The policies cover only non-construction threats: fire, theft, vandalism, etc.

Signature and *unsecured line-of-credit* loans are offered by many financial institutions. They're granted on the basis of a good credit history and the ability to afford payments. The loan officer bets whether you are a trustworthy character who pays his or her debts. Maximum loan values are generally 10–20 percent of your annual income. (More, if you're a third world country.)

Signature loans through credit unions seem to be the best. One credit union limits signature loans to $20,000; as of this writing the interest rate is 11.5 percent. The maximum term is eight years. Borrowing the maximum would yield a monthly payment of around $325. If you must finance, the signature loan holds the best promise.

The cheapest way of financing a kitplane is to buy subkits as your budget allows. This frees you of interest payments, although crating and shipping charges are higher than if the entire kit is purchased at once. As one might expect, the monthly payments play hob with the per-hour cost, as well.

ENGINE SELECTION

Unless you have a lot of hot-rodding experience, it's best to buy the engine specified by the kit manufacturer. Sure, you'd like to try one of those snazzy auto-engine conversions instead of a grumpy old Continental. But if you're out to get the plane flying with minimum difficulty and the least amount of time, stick with the recommended engine.

There are exceptions. Many kits are vague in engine installation details; a different engine might not be any harder to install. Or, perhaps someone locally has successfully used a particular engine, and you can copy his installation.

As mentioned last chapter, homebuilders often regard the engine as another "black box"; an item to be installed and not necessarily understood. Mounting an engine other than the recommended one causes more problems than just a change in engine mount. Some possible problem areas:

Center of Gravity (CG). Any change in weight or distribution will change the CG. As you learned in pilot training, out of limit conditions can drastically affect flyability.

Fuel Usage. If the same size tanks are used, installing a bigger engine might actually shorten the aircraft's range. You could install larger tanks, but the affect on CG must be considered and the extra fuel weight comes out of the useful load.

Cooling. Bigger engines produce more heat—it's simple physics. The cooling inlets and outlets on the cowling might be too small.

Access. A larger engine will take up more space under the cowling, making it harder to perform maintenance.

The following sections discuss the engine options open to the homebuilder: certified aircraft engines, auto engine conversions, and the Rotax series.

CERTIFIED AIRCRAFT ENGINES

Two major manufacturers, Lycoming and Continental, manufacture certified aircraft engines. A third, Franklin, has been out of business for years, so their engines are rarely seen on homebuilts. However, one from a Stinson or similar era plane occasionally appears, or a builder imports a Polish licensed-built Franklin. Also, a two-cylinder Franklin used on the Bellanca Champ (7ACA) during the early '70s sometimes appears on the nose of a small homebuilt. While the Franklin has its adherents, the parts situation is such to recommend against their use.

Other than the ARV category, all kitplane manufacturers specify certified aircraft engines for their products. It's a trifle odd, from one point of view. These light aircraft engines have evolved from engines introduced in the '30s and '40s. And they put these old-fashioned engines in our 21st century kitplanes?

Certified aircraft engines have one great advantage: proven reliability in an airborne installation. Auto-engine converters claim equal or greater reliability. They produce reams of calculations, and stacks of test-bed results.

But these engines haven't been flying for 40 years. The FAA and the NSTB track failures of certified aircraft engines; the resulting ADs and service bulletins continually improve the breed.

Aircraft engines are recommended because they are trustworthy, not because they are the most efficient, or the most powerful. Their faults and quirks are well known and if a problem arises, any competent A&P can track it down. When you experience a problem with an auto-engine conversion, it's up to you. The best you can expect is advice by long distance.

Aircraft engines aren't cheap. Prices can range from merely breathtaking to exoatmospheric. Depending upon your requirements, you might be able to buy an adequate engine for a third of the cost of a new one.

Engine Types

There are two major "flavors" of certified aircraft engines: Lycoming (Fig. 3-3) and Continental (Fig. 3-4). Both sides have their adherents; and both manufacturers have produced "lemon" engines.

One fast way of telling a Lycoming from a Continental is Lycoming's preference for locating the starter ring on the front of the engine, just behind the prop. Look into the front of the cowling; if you can see the crankcase, it's either a Continental or an older Lycoming without a starter.

If the cowling is off, the manufacturer's name is stamped into the valve covers; although the older Lycomings show an "L" in a hexagon. The final obvious difference is in the oil tank—small Continentals use a separate kidney-shaped tank, and larger Continentals have an attached oil pan that's smooth and rounded. Lycoming's oil tank is an angular pan bolted directly to the bottom of the engine. In fact, the residual heat from the oil tank to the carburetor is one reason Lycoming engines are less prone to develop carb ice than Continentals.

Fig. 3-3. Lycoming produces the O-320 and the O-360, which are the two most popular certified engines for kitplanes.

Fig. 3-4. Continental engines are used in planes like the Venture.

The manufacturer is about the only thing you can determine with a casual glance. For further information, check the engine data plate.

The type, configuration, and selected accessories are indicated by the engine's designation, such as O-360-C2A, TOL-200-A, etc. The designation's prefix shows the engine configuration, the numbers in the middle are the engine's displacement in cubic inches, and the suffix is used by the manufacturer to indicate small changes in configuration.

Common letters used in the prefix are:

O - Opposed cylinders
L - Inline cylinders
I - Fuel injected
T - Turbocharged
S - Supercharged
L - Left-hand rotation
G - Geared

Although different aircraft might use the same basic engine, slight differences appear. Changes in magnetos, the addition of a controllable pitch prop, and the like, result in a unique suffix to identify the variant. For example, the Cessna 152 mounts a Lycoming O-235-L2C, which is an opposed 235-cubic inch engine with Slick magnetos. The nearly-identical Cessna 150 mounted a Continental O-200A.

Selected Lycoming engines also feature the *dynofocal engine mount* (Fig. 3-5). The mounting holes for other engines are horizontal and parallel; the dynofocal

Fig. 3-5. A type 1 Dynofocal engine mount for an RV-4.

mount reduces vibration by "aiming" all mounting holes at the base of the propeller. There are two kinds, *Type 1* and *Type 2*. Type 2 is needed when prop extensions require a slight variation in the pointing of the mounting holes. Make sure you have the correct mount.

It's important to know the differences between engines of the same series, especially when looking for a good deal in used engines. The question is, how exactly must you match the plan's engine requirements? If the plans specify a Lycoming O-320-E1J, can you substitute a Lycoming O-320-H2AD, instead?

In this case, the -H2AD uses a dual magneto drive (two mags are essentially installed in one case). This has some reliability impacts, for if the single magneto drive fails, it kills both mags. However, there's another, bigger problem. The -E1J engine includes a governor for a controllable pitch prop; the -H2AD doesn't.

Hence, the substitution would probably be all right, as long as you weren't using a controllable pitch prop and were willing to trust the dual magneto.

Before starting your engine search, talk to the kitplane manufacturer and get a list of alternative engines. Some might require slight cowling modifications, others might get by with minor rearranging under the cowling.

The point made in the last chapter still applies. The more brain work you have to put into installing a different engine, the longer the construction time. But sometimes the extra work is worth it if you can find the engine at a good price.

Check the ADs issued for candidate engines. The Lycoming O-320-H2AD, for example, was the subject of many airworthiness directives. One hopes, by now, the problems are fixed. All Lycoming O-320s aren't tarred by the same brush—the suffix makes all the difference. The O-320-E2D (used in Cessna Cardinals and selected 172s) is very acceptable.

It's easier to track ADs based upon the aircraft the engine was originally installed in. If the seller doesn't know, guides are available from both Lycoming and Continental.

Every A&P with *Inspection Authorization* (IA) is required to maintain AD lists. If you have an "in" with an IA, that would be a good place to start. For that matter, any competent A&P should have some idea of the ADs affecting certain aircraft and engines. Your EAA chapter probably has at least one A&P.

Don't underestimate word of mouth. I've heard the Continental TSIO-360 suffers from some turbo problems and rarely makes TBO without a top overhaul, and I've been told a few negative stories about the Lycoming IO-360-E1A6D.

If you don't have the right contacts, a book called *The Aviation Consumer Used Aircraft Guide* is a good start. Selected local libraries carry it, but it's available through mail order from several places. The *Guide* gives a no-nonsense evaluation of most used production (and several homebuilt) aircraft, and usually includes an AD summary.

The December 1989 issue of *Light Plane Maintenance* contained an excellent article, "Four Cylinder Lycomings: A Homebuilder's Perspective" by Al Scott of Sequoia Aircraft. The issue is available from Light Plane Maintenance, P.O. Box 359135, Palm Coast, FL 32035.

Now that we know what we are looking for, let's look at the engine-buying process. There are three classifications: new/remanufactured, overhauled, and used.

Buying a New or Remanufactured Engine

The most expensive option is to buy a brand-new engine. They're offered direct from the factory and through dealers. The same engines are found at various prices, depending on how much the dealer is willing to discount.

Often, the kitplane manufacturer arranges an OEM (*original equipment manufacturer*) deal with the engine companies, which allows you to purchase new engines at a substantial discount. Check with the kit manufacturer.

Once one gets by the considerable obstacle of price, a new engine has several advantages. Lycoming's warranty pays parts and labor during the first year, and pro-rates parts costs during the second.

Remanufactured (or *zero-timed*) engines are probably the best deal. The original manufacturer takes a used engine, disassembles it, and replaces major components like cylinder heads, cylinder barrels, valves, and pistons. The remaining parts are checked, and those that don't meet new-part tolerances are also replaced.

Although the engine might have had thousands of hours upon arrival at the factory, it is now considered the same as a new engine, hence the term zero-timed, and receives the full new-engine warranty. It gets a new logbook, with no mention of the previous flight hours.

Zero-timed engines are as good as new, but cost around 20 percent less. However, you must supply a "core" engine. Unless you find an old junker cheap, this might add several thousand dollars to the total price.

Like new engines, remanufactured ones are offered through both the factory and dealers.

Buying an Overhauled Engine

Engines can be overhauled to two standards: *new* limits or *service* limits.

Overhauling to new standards is identical to the remanufacturing process, but because only the original manufacturer can zero-time an engine, any third-party work is only an overhaul. Even if the rebuilder exactly duplicates the factory operations.

An overhaul to service limits means the engine is completely disassembled and all parts checked. Those parts that don't meet the service limits established by the manufacturer are junked and replaced. These limits are looser than the new limits, but should still provide adequate margin to allow the engine to run to TBO.

In both cases, the engine is returned to service with the original logbook, carrying the same number of hours it had before the overhaul. This is why you'll see engines advertised as "2,500 TT (*total time*), 250 SMOH (*since major overhaul*)".

The majority of engine difficulties occur around the cylinders, not within the block. Problems with the cylinder head, valves, or even the cylinders themselves can result in a *top overhaul*. A top overhaul is similar to a regular overhaul, except the engine casing isn't disassembled; therefore, the condition of the main bearings and other interior parts is unchanged.

Overhaul costs might vary widely, even for the same engine. Quality also varies. Some companies specialize in precision overhauls to like-new limits. These professional shops automatically replace some parts, whether in limits or not. They produce high-quality engines, but sometimes the price approaches the cost of a factory remanufacture.

You aren't stuck with these high prices. Any A&P mechanic can overhaul an aircraft engine. There are a number of small shops performing overhauls, with quite a variation in prices. Shops range from FBOs to A&P schools to retired mechanics who like to tinker with engines.

These smaller shops usually overhaul to service limits, which can reduce the number of parts that must be replaced. You don't end up paying for the replacement of components that might have thousands of hours of life remaining. However, sometimes this can go too far. Reusing spark plugs, for instance, is an example of excessive frugality.

One other option is to rebuild the engine yourself—it's legal. However, find an experienced mechanic to watch over your shoulder and give advice when you need it. Rebuilding manuals are available through most homebuilder's supply companies.

Overhaulers don't sell engines, they sell rebuilding services. Therefore, you will have to supply a *rebuildable core*. This is an engine without major damage—generally, it means at least an intact crankshaft. Some of the larger outfits might sell you a rebuilt engine outright, but a core charge will be tacked on. This isn't unique to the aviation world, as you know if you've ever bought rebuilt car parts.

Finding the large professional shops is no problem because they advertise extensively in *Trade-A-Plane* and other publications. Or contact local FBOs; some act as agents for the large rebuilders. The price will be higher than dealing direct, but the FBO also handles crating and shipping your core.

The FBO might offer to do the job for less, or you might find an independent A&P. Ask for references, and check with other members of your EAA chapter. All rebuilds are not alike, even if they pass FAA certification.

There is one trap when having an engine rebuilt. Occasionally, the rebuilder will say, "Y'know, I can save you even more money. Since this is going on a homebuilt, we don't need all the FAA paperwork. I'll cut the price by _____ dollars if we don't make this an official overhaul."

The paperwork certifies that all parts meet the tolerances of the service limits. Eliminating this guarantee is no problem with an honest and conscientious A&P. But bad apples grow in all orchards.

A friend of mine bought a rebuilt Lycoming on this basis for $10,000. He mounted it on his homebuilt; it ran fine. However, he made a mistake with a hoist and bent a couple of pushrods. He took it to another FBO for repair.

The FBO refused to reassemble the engine. Many parts were beyond tolerance; the interior of the engine was filthy. My friend paid $3,000 more to get the engine into legal condition, besides the $1,000 for repairing the hoist damage.

The engine is the second-most-important component on the aircraft. You and your passengers are first. Rebuilding an engine is more feasible than raising the dead. No matter how much they cost, quality rebuilds are cheaper than tombstones.

Buying a Used Engine

Buying a used engine has its advantages, primarily in the cost department. Mid-time used engines can be bought for half the cost of a remanufactured engine. A used engine is ready to roar upon installation, while new engines have to be babied until they're broken in. Tight new/reman/rebuilt engines run hot; a problem compounded by the extensive ground testing before a homebuilt's first flight. The builder can easily overheat a tight engine during the static runups and taxi tests.

Used engines run cooler. In an ideal world, used engines would be for ground and flight testing and a new engine installed when the plane has proven itself. Otherwise, low-time used engines are sometimes the best compromise.

Used engines become available for a variety of reasons. An owner might be upgrading to a larger engine, or an accident leaves the engine still usable. The best way to buy a used engine is running—that is, mounted either on an airplane or a test stand.

Failing that, you'll have to take a careful look at the engine's history. Two key things to look out for are prop strikes and sudden stoppage. A prop strike might occur during a hard landing or taxiing into something medium-soft.

A sudden stoppage is the king of prop strikes because it hits something hard enough to stop the engine, right now. All that rotational energy has to go somewhere. Typically, it goes into the crankshaft, which is the single most expensive part and the most difficult to repair.

After a prop strike of any sort, the engine should be checked for internal damage. Lycoming, in fact, requires a complete teardown. Such a check should be included in the engine logs. You might need to play detective; if there is an entry stating that a damaged propeller was replaced, check into how it was damaged. Someone might have backed a pickup into it while the engine was off, or a prop strike might have been ignored.

Another factor to consider when buying any used engine is how long it's been sitting. When an engine doesn't run, rust begins to form on the inside. It doesn't do the engine any good.

Each logbook entry should include the total time on the engine and the date of service. At minimum, each annual should be indicated. If the last signoff was five years ago, there might be significant internal corrosion.

The logbook includes other information as well. Results of compression checks performed for the annual, airworthiness directives complied with, and histories of all other work on the engine. Logs aren't perfect because they indicate certain actions were taken, but not necessarily why, however, logbooks are all we have to go on.

That is, assuming they're available. You'll occasionally see engines advertised as "no logs." This is sticky. The logs could have honestly been lost. Or the engine might have some significant damage history that the owner would rather not reveal. Again, the crankshaft *is* the engine. The owner might have discovered a crack and decided to dump the engine for what he can get. The engine can't be reinstalled in a factory airplane, so the only possible market is the homebuilt.

Buy it, and turn it in as the core for a rebuilt or remanufactured? Careful, all rebuilding agreements require turning over a rebuildable engine. One with a cracked crank might not apply.

If you know the seller personally and trust him, it might not be a bad deal. Otherwise, you're taking a considerable chance. Don't pay much.

Used engines are found from a myriad of sources. Aircraft junkyards make a major portion of their income from removing and selling the engines from wrecks. Prospective homebuilders pick up engines with the intention of using them someday and eventually give up. Local FBOs and independent A&Ps often have one or two sitting around.

One of the best sources is from air-taxi operators. They cannot operate an engine beyond TBO, and the engines are babied by professional pilots and regular expert maintenance. All it's worth to them is the core charge, so you might be able to pick it up for $5,000 or so. This might seem high, but well-maintained engines can go well beyond TBO.

Check around, and let it be known which engine you need. See what the grapevine can produce. If the right one doesn't turn up locally, check *Trade-A-Plane* and other sources.

One option might be to buy a freshly-wrecked airplane. Some owners don't carry insurance, and might be more willing to sell to a private individual. You can afford to offer a bit more than the junkyards because you'd have to pay an additional markup if you end up buying it through them. In addition, the wreck might have usable radios and other parts that aren't included in your kit.

But the salvagers know how much the wreck is worth, and it's usually worth more to them. You probably don't have a need for the wings or tail feathers; they know where to find buyers. In the old days, homebuilders eagerly sought wrecks as sources of cheap bolts, pulleys, and other small hardware. But this hardware should be included with your kit.

Besides, where are you going to store something the size of a wrecked 172?

No matter how you find a used engine, have an A&P check it over. If the engine can be run, so much the better. Otherwise you'll have to trust the mechanic's judgment. But there are many problems that can be missed without a full teardown.

Know which accessories the seller is including. Any additions or deletions are a starting point for price adjustment. Buying a new alternator, starter, magnetos, or fuel pump can make a cheap used engine no bargain. After all, two new mags will cost nearly $1,000.

Used engine prices vary depending on accessories, condition, and total time since overhaul. An engine near or past its TBO is called a *run-out*, and prices are the lowest at this stage. An engine advertised as a "first run-out" has never been overhauled before, and is worth more. Common terms used in engine advertisements are shown in Table 3-3.

Just because an engine is past its TBO doesn't make it junk. The condition is more important; if it runs well, if the compression is good, and if a borescope doesn't turn up anything negative, the engine could last for years with just normal attention.

After all, what does TBO mean?

Table 3-3. Common used-engine abbreviations.

SCOH	Since chrome overhaul (chrome increases the hardness of internal components and makes them last longer)
SFREM	Since factory remanufacture
SFRM	Since factory remanufacture (same as SFREM)
SN	Since new (same as TSN)
SMOH	Since major overhaul. Usually accompanied with the engine's total time: "200 SMOH TT 3040"
SOH	Since overhaul (same as SMOH)
SNEW	Since new (same as TSN)
STOH	Since top overhaul. Usually accompanied with an indication of the engine's total time: "250 hrs STOH 1500 TT"
TSMO	Time since major overhaul (same as SMOH)
TSN	Time since new
TSOH	Time since overhaul (same as SMOH)
TT	Total time
TTSN	Total time since new (same as TSN)
TTSFRM	Total time since factory remanufacture

Other Terms of Interest

Balanced: Connecting rods and other components have been matched or adjusted to weigh the same. Results in a smoother-running engine.
First run or **First run out:** Engine has never been overhauled before.
FWF: Firewall forward. Includes all parts forward of the firewall (muffler, baffles, etc.).
Green tagged: Repairable.
Red tagged: Unairworthy.
Run out: Due for overhaul.
Yellow tagged: Ready to run.

The TBO Myth

Everyone puts great stock in the concept of the manufacturer's recommended TBO.

Here's a quote from Textron Lycoming's warranty on new aircraft engines:

> If the engine proves to be defective in material or workmanship during the period until the expiration of . . . recommended TBO, or two (2) years from the date of first operation, whichever occurs first, Textron Lycoming will reimburse you for a pro-rata.

In other words, failures are covered under the warranty for only two years. Because Lycoming engines have TBOs of up to 2,400 hours, you would need to fly one hundred hours a month.

If you fly only an hour a month, an 1,800 hour TBO doesn't mean the engine will go 150 years between overhauls. Engine manufacturers recommend a minimum usage of at least 10 hours a month. In a service bulletin, Lycoming says: ". . . because of the variations in operation and maintenance, there can be no assurance that an individual operator will achieve the recommended TBO."

When an engine doesn't reach operating temperatures on a regular basis, the moisture and acids produced by combustion and condensation collect in the engine instead of being vaporized and eliminated through the exhaust and crankcase breather. These byproducts remain and contribute to the formation of rust on the cylinder walls, camshaft, and tappets. When the engine is finally started, this rust becomes a fine abrasive. As the engine components wear, the metal scraped off attacks softer metals like piston pins.

Keep in mind, TBO isn't a regulatory number. The engine doesn't have to be overhauled at that point; and nothing guarantees it will last even that long.

Storing the Engine

It's too easy to rush out and buy the engine at the same time as the kit.

It isn't too smart, either. Building the plane might take years. Where are you going to store this 300-pound block of metal? As mentioned in the last section, sitting idle does nasty things to the inside of an engine. What's going to keep it from rusting up?

If at all possible, don't buy the engine until you actually need it. Sometimes you can't wait. The factory might be discontinuing production of the right model. The price is going up like crazy. Or a great deal might materialize, to be lost forever if not snapped up immediately.

Which leaves the problem of keeping the darn thing healthy until needed. The first order of business is to find a nice dry place to store the engine. The second is preparing the engine for storage, a practice called *pickling*.

The procedure varies between engines, but it generally begins by draining the oil and replacing it with a special pickling oil. The sparkplugs are removed and pickling oil squirted into each cylinder. Then the engine crankshaft is turned over by hand to spread the oil on the cylinder walls. Dehydrator plugs are screwed into the spark plug holes; these plugs contain silica gel to absorb moisture. Similar plugs are inserted into the exhaust pipes. Crankcase vents are sealed off, and the engine is wrapped in plastic (Fig. 3-6). The engine's overhaul manual will give exact instructions.

Pickling isn't a permanent procedure. The silica gel in the dehydrator plugs must be replaced occasionally, and the crankshaft should be turned over occasionally to keep pickling oil on the cylinder walls. Count on checking the engine at least once a month.

Fig. 3-6. A pickled Lycoming O-320.

Summary

Certified aircraft engines have many advantages to the homebuilder. Reliability has been mentioned. Maintainability is another; if you're on a trip and something breaks, most parts will be available right on the airport.

It's hard to pin down engine prices. Here's an example of what might be expected in the early 1990s when buying an engine like a Lycoming O-320:

New price: $22,000
Street price (new): $14,000
OEM price (via kitplane manufacturer): $12,000
Street price (reman): $9,000 (plus core)
Quality overhaul: $8,000 (plus core)
Average overhaul: $5,000 (plus core)
Low-time: $6,000
Mid-time: $4,500
Run-out: $3,000

These prices, especially those for the used engines, are just approximations. Shop around to find the best deal.

Determining the cost of operation is easy, after all, there are a number of production aircraft flying the same engine. The most commonly-used engines will burn fewer than 10 gallons an hour; usually in the range of 6–8 gph. As mentioned earlier, picking an engine that can run on autofuel cuts direct costs almost in half.

However, there's another "cost of operation" that most homebuilders don't realize.

Aircraft engines cost as much as they do because of their certification: guaranteed to meet the federal standards required for installation in production aircraft. To keep that certification, they must be maintained by approved individuals.

Kitplanes aren't production planes. They don't require certified engines. Kitplane builders receive "repairman certificates" that allow them to inspect and repair their aircraft. This includes the engine.

But a holder of a repairman certificate is *not authorized to maintain a certified engine*, even if it's installed on their experimental aircraft. As soon as the repairman certificate holder performs an annual inspection, or some other operation that requires an A&P, the engine loses its certification.

It makes no difference to the homebuilt aircraft because it still can be legally flown. It makes no difference to the homebuilt owner because he or she can still legally maintain the aircraft and newly-uncertificated engine.

However, the engine has just plummeted in value. Its original high price was partially due to its certified status, if the engine were a Lycoming O-320-A2B, for example, it could have been legally installed in certain models of Tri-Pacers, Cherokees, Super Cubs, Citabrias, and even the Robinson R-22 helicopter. A low-time used engine that had maintained its certification would find a ready market.

Without it, the only market is the experimental aircraft. To recertify the engine, a certified mechanic must tear it down completely for inspection. The process is the same as a complete overhaul because he or she would be checking the tolerances and replacing out-of-limit parts. Hence, the marketability of an uncertified engine is greatly reduced.

To maintain the value of your engine, all inspections and maintenance must be logged by an A&P. You can still do the work, as long as an A&P signs off the log entries. The mechanic is taking some chance at this because his signature makes him legally responsible, no matter who actually held the wrench.

But few homebuilders bother maintaining the certification. Certified or no, the engine generally stays with the homebuilt, and the issue of "market value" rarely comes up.

This problem is moot with engines that start out uncertified: the auto engine conversions and the Rotax two-strokes. Let's take a look at them.

UNCERTIFIED ENGINES

As nice as certified aircraft engines are, they also have several disadvantages.

Their basic design dates from the 1930s. High technology, in itself, is not an advantage. One can make a considerable case for the "tried and true" traditional designs. However, advances in metallurgy and electronics in the last 50 years have been considerable.

Technological advancement hasn't made inroads in small aircraft engines because development and certification costs aren't justified for available market. GM can develop a new car engine and expect to sell millions of copies. Lycoming would be lucky to sell hundreds.

A certified engine's second disadvantage is the lack of adequate engines in the lower horsepower ranges. The Continental A-50 and A-65 engines made the J-3 into the first mass-produced air recreational vehicle. Sixty-five horsepower is a pretty good value; it's adequate for spritely single-seaters and light two-seaters. Yet no in-production certified engine is available at anywhere near this power level.

The A-65 engine has been out of production for more than 35 years. It and other small engines (like the Continental O-200) have become somewhat rare and therefore expensive. At one point, the supply of small Continentals dried up due to their popularity in homebuilts like the VariEze.

The reliability advantages of obsolete certified engines can be disputed. How many times can an engine be rebuilt until it's no longer safe? Some engines and their accessories (magnetos, and generators, for instance) are kept operating through stocks of left-over original parts. Can you trust a 30-year-old fuel pump?

The final major disadvantage of the certified engine is cost. A rebuilt O-360 might cost $10,000 or more, then, if a cylinder goes bad, replacement "jugs" run around $1,000 each. Yet new auto engines go for less than $2,000; buy a reduction drive for $2,500 and you'll have a more powerful engine for half the price. Parts and services are available cheaply at the auto parts store. A set of four VW cylinders and pistons sells for less than a hundred bucks.

Let's take a look at the main contenders among uncertified engines.

The Mighty VW

Certified engine disadvantage number one, (old technology, dating from the 1930s) also applies to the Volkswagen engine. Ferdinand Porsche designed the engine for Hitler's people's car concept. Air cooling was picked in the interest of simplicity and reliability.

The people's car was magically transformed into the slabsided bucket car (Kubelwagen) for the German army. Wartime pressures refined the engine and drivetrain. The Kubelwagen continued its yeoman service after the war, obligingly being blown up, machine gunned, and driven off cliffs for the cinematic armies of Warner Brothers, MGM, and Twentieth-Century Fox.

Volkswagenwerk AG recovered quickly after the war, producing almost 2,000 civilian cars by the end of 1945. By the late '40s, humpbacked autos swarmed from the Wolfsburg factory. The Bug became the most widely recognized car in the world. While the parent factory converted to more modern designs in the '70s, the Bug and its variants are still produced in Brazil and Mexico.

Experimentation with aircraft began shortly after the war. With a continuing shortage of small aircraft engines, Europe has been the most enthusiastic proponent of VW power. Limbach of Germany even certified a VW-derived engine for use in motorgliders.

Acceptance in the U.S. was slower. Thousands of small Continental engines were still available. The first major breech was Bud Evan's Volksplane, one of the first simplified homebuilt designs. The VW engine was converted for aircraft use by removing the flywheel, starter, and generator, replacing the distributor with a dune buggy magneto, and replacing the crankshaft pulley with a propeller flange.

The VW's biggest boost came with the Q2. The original single-seat Quickie used an 18-hp Onan GPU engine. The two-seater Q2 was designed for VW power; getting almost 200 mph on the Revmaster conversion. While few kits currently list the VW as the standard engine, it's a common builder modification and they often appear on plans-built aircraft.

The usual VW-aero conversion (Fig. 3-7) moves the carburetor underneath the engine (using a Posa or other aircraft-style carb) and installs a standard aircraft magneto at the aft end of the crankshaft. Prop hubs are either the machined-to-fit or the shrink-on variety, where the hub is slightly undersized, but when heated, expands enough to slip over the crankshaft.

Horsepower can almost be selected. The standard 1,600 cc engine produces about 45 horsepower, but larger cylinders and pistons bolt on for higher performance. Only the very largest (2,100 cc and up) require machine work, and they produce about 65 hp. VW-derived engines such as the HAPI Magnum offer 75 hp and up. Ready-to-fly weight for a basic VW is about 150 pounds.

The VW has four bolt-holes on a wide flange at the rear of the engine. This is perfect for simple and cheap engine mounts; in fact, the Volksplane merely bolts the engine directly to the firewall.

Assembling your own VW is the cheapest route to an engine. A good used engine can be bought for $300-$400 or so, with another $300 budgeted to replace questionable parts. You can put together a basic aero VW for less than $1,000. Complete kits are available, starting around $1,500 for a 1,600 cc model. If you go with the build-it-yourself route, buy Rex Taylor's book, *How to Build a Reliable VW Aero Engine*. It's available through Mosler Airframes and Powerplants.

However, these prices don't include carburetor, intake manifold, accessory cases, or ignition systems. Some use automotive carbs on a stock manifold. This places the carb above the engine, which usually requires installation of a fuel pump. A below-engine mount gets enough "fuel head" via gravity. Carburetors designed for VW-aero use cost around $250.

Accessory cases make VW use far easier. The engine itself bolts directly to the case; the case, in turn, provides a shock absorbing attachment point to the mount. Cases allow direct installation of standard aircraft magnetos as well as lightweight starters and alternators (in fact, some incorporate the alternator within the case itself). They add a lot of capability for just $200.

The ignition system is another issue. Aircraft magnetos cost $400 on up; electronic ignition is a bit cheaper. Many argue that electronic ignition is more reliable. However, it does require an alternator, regulator, battery, and the rest of the

Fig. 3-7. A Revmaster VW mounted on a Sonerai. The Posa carburetor can be seen underneath. The firewall has a large indentation for the Slick magneto mounted behind the engine.

paraphernalia associated with an electrical system. If you like ignition redundancy, several companies offer dual-ignition modifications. Some even mix the modes, using a magneto for one set of plugs and an electronic ignition setup for the other.

Depending upon options, most of the machine work is normal hot rod stuff that can be done cheaply at local VW car specialists. Have the prop flange installed by a VW-aero specialist so it's done correctly.

Or you can buy a ready-to-fly engine. Prices start at $2,500 and run to $5,000 and more for dual-ignition 82-hp versions. Considering the smallest new certified engine costs over $10,000, the VW seems quite a bargain.

But disadvantages exist as well. Where certified aircraft engines get their peak horsepower at 2,800 rpm or less, VWs must run at about 3,400 for max power. To allow the engine to rev this fast, VWs must have smaller, lower-pitched, less efficient propellers. While it might have only limited affect on cruise speeds, large-diameter props are best for short field work.

These small propellers won't windmill, so a momentary glitch in flight might result in a forced landing for a starterless airplane.

The VW is also hurt by its poor power-to-weight ratio. A 50-hp VW weighs about 150 pounds, a Continental A-65 with 15 more horses weighs only 20 pounds more. And the propeller likes these horses better because the A-65 produces full power at only 2,150 rpm.

(At the opposite end of the spectrum, the Rotax 503 engine has the same horsepower, and with a gear reduction drive achieves full power at about 2,700 prop rpm. The Rotax is 60 pounds lighter.)

One curious fact appears when studying VW-powered homebuilts—except in a few cases, the VWs have been eventually replaced by aircraft engines. The Q2 turned into the Q-200, with a Continental. The VariEze went the same way. The Zenair Zodiac transitioned to the first Rotax four-cycle engine marketed for aircraft use.

VWs turn in the wrong direction, counter-clockwise as viewed from the cockpit. This can cause transition difficulties, as left rudder (instead of right) must be held on takeoff. The small prop diameter magnifies the P-factor effect, which means the plane swings more when power is applied for takeoff.

Opinions are mixed on the VW reliability issue. Single-ignition engines run their own risks, as mentioned in chapter 1. Some argue that the engine was designed for 1,600 cc and 36 hp; anything more reduces longevity. The stock VW probably operates with a narrower margin of safety than a certified engine. Then again, the VW design has stood up to auto and air racing for years, as well as pushing thousands of dune buggies.

VWs have one recurring maintenance problem. The valves have a tendency to become too tight and must be adjusted at least every 25 hours. This is a minor procedure that only takes 15 minutes or so. Some companies offer hydraulic lifters that eliminate the difficulty.

Drawbacks or no, the VW engine has remained active on the homebuilding front for one reason: It is the only proven four-cycle engine in the 36–65 hp range. Until Rotax can challenge its reliability and availability, it shall remain the frontrunner in its class.

The Liquid-Cooled Auto Engine

The VW's low cost and availability are due to its automotive background. Why not convert larger car engines?

Why not, indeed? Conversions of liquid-cooled auto engines have been popular since the 1920s. Some have even been certified, the Ford Model B engine in the Funk, for one. Quite a number of Pietenpol homebuilts have flown with Ford Model A's as well.

Traditionally, auto engines have been used to fill a gap where the properly sized aero engine isn't available. This is changing, now that the prices for new Lycomings have reached the $20,000 level.

The same arguments used in favor of the VW can be made for other auto conversions: lower cost, better price availablity. The drawbacks are also similar: single ignition and peak power achieved at inefficient propeller speeds.

But most auto engine conversions include a reduction drive to convert rpm into thrust. A 2-to-1 reduction ratio changes a 5,000 rpm horsepower peak into 2,500 rpm at the prop. With the reduction unit, the auto engine is more suited for aircraft use.

The power is usually transmitted to the prop via a cog belt drive incorporating the appropriately sized pulleys. While some design their own, most homebuilders buy one of the commercially-available units. These will cost about $2,500. The engines themselves can come from the junkyard; in fact, Dave Blanton, the designer of the Javelin Ford V-6, recommends junkyard engines over new ones. Car accidents don't produce the equivalent of prop strikes and a broken-in engine runs cooler. They can be bought for less than $1,000.

So the total for the auto engine conversion will run about $3,000. Auto engines are a little heavier than aircraft engines, but liquid cooled engines are more efficient.

There are several conversion packages on the market. The Blanton Javelin Ford (Fig. 3-8) is based on a cast-iron-block Ford V-6, and tips the scales at 440

Javelin Aircraft

Fig. 3-8. The Blanton Ford V-6 aircraft engine.

pounds, including radiator and coolant. Other converters concentrate on the Buick aluminum-block V-6, still manufactured in England.

The main drawback to selecting an auto conversion is the extra work and construction time. Converting the engine will be at least as complex as building the airplane. You can install a Lycoming with no more work than bolting it in place and connecting the cables and fuel lines. With proper preparation, it could be lifted on a rented engine hoist and installed in the mount in a single evening.

But you'll be spending a lot of time manipulating the auto engine. The camshaft might have to be reground, or coolant port locations changed. The reduction drive itself must be assembled and installed. You'll have to build a custom engine mount, design a carb heat system, install a coolant header tank, and have a special prop built.

Several times in this book, I've mentioned the extra work involved in installing an engine different than the designer recommends. Here's an excerpt from *Sport Aviation* magazine, concerning an auto-engine conversion in an RV-4:

> The first thing he had to do was remove the aft portions of the cheek cowls . . . rivet in some aluminum angle to make up, structurally, for their absence. . . . He had two long, skinny radiators made and mounted under the bottom of the fuselage . . . instead of a P-51 type scoop, he built a second bottom into the fuselage. . . . He welded-up his own bed-type engine mount, and hammered out an aluminum nose bowl. . . . [Not to mention working out routings for throttle cables, fuel and coolant lines, and more.]

All of this could have been avoided by installing the O-320 called out in the plans.

If you decide to install an auto engine, buy a commercially-available conversion. Talk to the company before deciding and determine how much they'll support your effort. After all, their only market is homebuilt aircraft. Check on the availability of standard engine mounts, and whether anyone else is installing the engine on the same type of airplane.

The Rotax Two-Strokes

Rotax single-handedly saved the ultralight industry and was a shot of fresh air to the ARVs, as well.

The ultralighters were buffeted on all sides: TV exposes on safety, noisy engines irritating the public, shoddy manufacturers trying to cash in on a craze, and cobbled-up conversions of two-stroke stationary engines that rolled their eyes up and died at the least provocation.

Along came Rotax. Engines proven in snowmobiles, not just in stationary power units. A wide range of engines. Effective tuned muffler systems. And a bolt-on gearbox that eliminated prop noise as it doubled or tripled efficiency.

Is it any wonder most ultralights and ARVs now use Rotaxes?

Traditionally, the bigger the engine, the better the horsepower-per-pound ratio. The standard VW has a 1-to-3 ratio, the O-200 1-to-2.5, the Lycoming O-360 has a 1-to-1.87 ratio, and so on.

The Rotax 532 has a 1-to-1.5 ratio; it produces 65 hp and weighs 98 pounds. Best of the lot.

The popularity of the Rotax means good support. A number of companies sell accessories. Propellers are available in a variety of styles, and at low prices. The Rotax models 503 and 532 are standard equipment on an astounding number of kitplanes, in fact, these engines are often included in the kits.

Prices aren't so bad, even if you have to buy the engine separately. The 503 runs about $1,600; the liquid-cooled 532 for a bit over $2,400. This price includes everything but the gear box, which sells for another $350 or so. Prices vary, check around.

Rotaxes are reasonably easy to install. Four studs project from the bottom of the crankcase and can be bolted to plates and angles as necessary. The engines come with a tuned muffler and exhaust system, which optimizes the performance. The length of the system is critical—don't shorten the pipes. You can bend and position the exhaust in a variety of ways, but the centerline distance should stay the same. If you wondered why so many ARVs equipped with Rotaxes have the ugly muffler and exhaust pipe hanging below the fuselage, now you know.

Two of the biggest problems of two-strokes, the need to mix oil with the gas plus a single ignition, have been solved with the introduction of the liquid-cooled Rotax 582 (Fig. 3-9). It's almost identical in weight and performance to the 532, but adds an oil injection system and dual magnetos.

Fig. 3-9. The hottest engine for the ARV-class of kitplane is the Rotax 582, which features dual ignition and oil injection.

Any other two-stroke engine needs the oil mixed with the gas. Maintaining the proper ratio is critical; too much oil, and excessive carbon deposits form. Too little and the engine might overheat and seize. If on a cross-country, you can't just taxi up to the gas pumps and fill up. Rotax recommends a separate container be used to blend the gas and oil before the mixture is added to the tank. It's difficult to carry such a container in the aircraft, so you end up carefully adding the right measure of oil after a fillup. But unless the aircraft's fuel tank can be sloshed about, the oil might not mix properly.

The oil will tend to separate if not regularly agitated. If the plane will be stored over the winter, drain the gas and use it in the snowblower instead. About a quarter of all two-stroke engine failures are caused by fuel quality problems, so extra attention to fuel mixture and quality are warranted.

This leads to the major objection to two-strokes in general: reliability. They aren't certified engines; they were designed for other purposes and adapted to aircraft. The Rotax's TBO is 400 hours; less than a quarter of a typical aircraft engine.

In the August and September 1989 issues of *EAA Experimenter* (reprinted in the June 1990 issue of *Kitplanes*), Hank Fritz of Renton, Washington, presented the results of a survey of two-stroke engine failures. The combination showing the best reliability was a Rotax 503 with dual carburetors, fan cooling, and mounted upright in the tractor configuration. Change the plugs and fuel filter every 50 hours, along with checking the timing and fan belt and cleaning the air filter.

But as discussed earlier, a 400-hour TBO means little to sport flyers. Fly 50 hours per year, and the engine might not need an overhaul for eight years. And Rotax engines are so widely used in nonaviation applications that a complete overhaul costs less than $500. Do it yourself for even less.

Two-stroke engines have other, minor disadvantages. Vibration is one. They're shaky little devils, and require careful attention to shock absorption. The Rotax gear reduction turns the prop counterclockwise, which, like the VW, requires left rudder on takeoff.

Summary

So there are the engine choices: expensive certified engines, for high reliability; the VW, cheap with a proven track record but rather mediocre performance; other converted auto engines with good performance but complex to convert and install; or the powerful little Rotax with the short TBO.

Actually, there are even more. In addition to marketing Rex Taylor's HAPI VW conversion, Mosler sells its own 35-hp four-stroke engine, and includes it with the kits for its various N-3 Pup variants. Other kitplane manufacturers specify Cuyuna, KFM, or other engines.

Engine selection depends upon a number of factors: the kitplane manufacturer's recommendation, your engine budget, and your mission.

Further information on aero-VWs and other auto-engine conversions can be found in TAB's *Automobile Engines for Homebuilts* by Joe Christy.

PROPELLER SELECTION

Prop selection isn't as easy as you might think. Not only must it be matched to the engine, but to the aircraft as well. A Kitfox and a Dragonfly might both use the same VW engine, but the Dragonfly's mission of high speed cruising requires a different prop.

Props are specified by *diameter* and *pitch*. A 52 × 46 prop is 52 inches in diameter, and, ignoring slippage, would pull itself 46 inches forward in a single revolution.

Three items affect selection: mission, aircraft, and engine.

The mission decides your choice of a prop optimized for climb, cruise, or a compromise between the two. A cruise prop will maximize speed at the cost of short field performance. A constant-speed prop is the best solution, but the dollar and construction-time cost is incredible.

The aircraft has already been optimized for both the engine and the mission, how much can it affect propeller selection? Simple: What ground clearance does it leave for the prop? Small aircraft have short gear legs. A standard prop might be too long; even if it doesn't actually touch the ground, a hard landing might make the prop hit. So some planes use a shorter prop for clearance reasons, and compensate with increased pitch.

The engine affects prop selection in two ways: First, there's no single standard prop hub. The bolt circle diameter varies; and between four and eight bolts might be used. Second, the pitch and diameter must be selected based upon the mission. The horsepower output of the engine is directly related to its rpm. But the rpm isn't just controlled by the throttle—forward speed decreases the blade's effective angle of attack and allows the engine to turn faster.

Your objective is to select a prop based on the desired aircraft speed at which engine power must be maximum. If you need short field performance, the engine should be running at its maximum torque point at full throttle and 60-80 mph airspeed. This requires less pitch and a larger diameter.

How much difference does it make? Cessna 150s and early Lancairs both mount Continental O-200 engines. A typical cruise prop for the 150 is 69 inches in diameter with 50-inch pitch, and gives about 125 mph. The same engine on the Lancair must be propped to allow 200 mph; one prop maker recommends a 58-inch diameter and a pitch of *71 inches*. The Lancair has limited ground clearance as well. Both factors serve to increase pitch.

At some point, you'll have to chose between props made of wood or metal. There aren't any metal props made specifically for homebuilts; hence metal props will be expensive certified models. If you must have a new metal fixed-pitch prop, count on spending about $2,000. However, used props can be found for less than $1,000. Propeller shops can often make a good homebuilt prop by cutting-down a damaged certified model.

Wood props are cheap in comparison; prices run about $500 or less for most engines. A Rotax prop might only cost $200 or so. There are a number of custom-propmaking companies aiming specifically at the homebuilt markets.

If you're going for a constant-speed prop, expect to pay out a good bit of change. A minimum of $3,000 for a used one, and many times more for selected new models.

Should you pick metal or wood? Like wood airplanes, wood props seem to run smoother. Because they are in production at low cost, you can order a prop specially for your aircraft and mission. Wood props are classier, as well.

However, metal props are durable. Bent props can be repaired to a certain extent; wooden props don't bend, they break (although by breaking, they can prevent engine damage in a prop strike). If you buy a wood prop, you're stuck with the diameter and pitch. Metal props can be shortened and repitched by certified propeller shops. Of course, for the price of a single new metal airscrew, you could buy a wood prop for every season.

Metal can pick up some nicks and dings from gravel and debris, but these can be easily repaired. Flying rocks can break wooden blades. Finally, a metal propeller is impervious to weather. Wood props should be protected from moisture; small cracks in the varnish or epoxy allow water to enter and cause an unbalanced condition. Use a cover or move the blades to horizontal when the plane must be kept outside. Also, flying in the rain can damage the wood's finish.

Wooden props need careful attention to the prop bolt tension. The wood expands and contracts with the weather. After a long dry spell, the hub shrinks and makes the prop bolts relax their grip. Always install a crush plate in front of the prop (Fig. 3-10) and periodically retorque the bolts. If you buy a wooden prop well in advance, store it in the stereotypical cool, dry, place. The wall of your den might be a good spot. . . .

Fig. 3-10. Wood propellers must include a crush plate to equalize the pressure of the prop bolts.

If the kitplane manufacturer recommends a particular engine, they'll specify the propeller as well. If they don't, picking the right prop can be difficult. If you've decided to go with metal, consider the same propeller as the production aircraft closest to your kitplane with the same engine.

When hunting for propellers and other aviation accessories, you often hear the term *tagged*, as in yellow tagged and green tagged. When a component is checked by an approved maintenance facility, a tag grades the component based upon the condition of the part. In the case of propellers, the facility checks various dimensions of the propeller against the minimum standards published by the propeller manufacturer.

A prop that meets all standards is given a yellow tag, which means "This component is airworthy and operational." If the prop is damaged, the facility compares the amount of damage to the manufacturer's repair allowances. A prop missing a little from a tip can often be shortened slightly, or bent blades straightened to some extent.

If the discrepancies are within specified limits, the prop gets a green tag: "Unairworthy, but repairable." Once repairs are completed the tag is replaced by a yellow one. The rest are given red tags, which prohibit use in certified aircraft.

But kitplanes aren't certified. Propeller shops can rework some red-tagged units into perfectly acceptable homebuilt props. These sell for between $500 and $1,000.

Some of the bigger-engined RV-4s and -6s are being built with constant-speed propellers: new cost around $4,000. One common Hartzel model is used on selected Mooneys and Arrows. The RVs can take a slightly cut-down version of this prop; prop shops start with a red-tagged Hartzel, cut the blades down a bit, and sell it for about $1,000 less than a new one. Be warned, though, because the Hartzel has an AD note out, requiring a teardown and inspection every five years.

If you buy a green-tagged fixed-pitch prop, count on spending about $250 to get it rebuilt. If you need it repitched, add around $100. Check with the local propeller facilities before buying a new propeller. You might be able to save 50 percent with a used, refurbished model.

But metal can't do everything. There's a limit to how much a prop can be repitched and shortened. Sometimes it's easier just to buy a custom wooden prop. Talk to the propmaker; see if he's made props for your kitplane before. Props for the more common kitplane/engine combinations can often be bought right off the shelf. It doesn't guarantee a perfect solution. A friend has bought three props for his VariEze and still isn't satisfied.

However, you aren't buying from a factory. Wooden prop shops are generally one-man operations. Many don't make certified props, and therefore aren't necessarily familiar with federal standards. Ask around and look at some examples. A local wooden prop owner found little pieces of epoxy varnish littering the ground one day; flakes were peeling off his brand-new prop.

Selecting and finding the prop isn't a quarter of the battle involved in picking the right engine. But it has almost as great an affect on performance and satisfaction.

INSTRUMENTS

Your instrument decisions are fairly easy: Are you building for VFR or IFR?

If VFR, your airplane must include the following:

- Airspeed
- Altimeter
- Magnetic compass
- Tachometer
- Fuel gauge

Other instruments are required depending on what type of engine is installed. For four-stroke engines, you must have an oil temperature and oil pressure gauge. If you're using a liquid-cooled auto engine conversion, add a coolant-temperature gauge. Similarly, add gear lights if the plane is equipped with retractable gear, and a manifold pressure gauge if the engine is turbocharged or has a constant-speed prop.

What's interesting about this list (from FAR 91.205) is what *isn't* required for VFR flight:

- Vertical Speed Indicator (VSI)
- Directional Gyro (DG)
- Turn and bank/turn coordinator
- Clock
- Artificial horizon

Fig. 3-11. A well-instrumented VFR homebuilt.

These are required for IFR flight, but not for VFR. Many builders install more (Fig. 3-11), but if you're never going to fly into the clouds, why spend the money and waste time on their installation?

Prices are interesting; each of the basic instruments costs about the same. A new sensitive altimeter, airspeed, or VSI will cost around $200. The compass is cheaper; about $60. Overhauled gauges run about 75 percent of the new price. If you want special features, like a 2-inch gauge rather than the standard 3-inch, you'll pay extra for it.

Or buy used instruments, and save 50 percent or more. You'd like to find a gauge directly removed from another aircraft, or one that's yellow tagged.

Keep in mind that you're building an experimental aircraft, and don't have to use TSO'd instruments. I've seen automobile compasses installed in homebuilts. The airspeed indicator requirement could be met by a spring-loaded, calibrated windvane.

They're legal solutions; but don't get too unusual. Stick with what you're used to. Besides, the aircraft will eventually be inspected by an FAA inspector who might take a dim view of such shenanigans. He or she can't legally down-check the aircraft on the basis of nonstandard instruments, but they can always find some other excuse to reject it.

That doesn't mean to only use TSO instruments. The homebuilder's supply outfits sell many good-quality non-certified gauges meant for homebuilts. These are perfectly acceptable. But if you're building an IFR airplane, use TSO'd only.

ELECTRICAL SYSTEM

Do you want an electrical system? Do you *need* an electrical system?

There are only two valid reasons to install one: You plan on operating at night or in IFR, and your location requires transiting TCAs or ARSAs.

By not installing an electrical system, you'll save 50 to 100 pounds in weight, and tens of hours in construction time. Your plane won't have to carry a battery, an alternator, a regulator, a starter, an ammeter or voltmeter, fuses/circuit breakers, switches, dozens of feet of wire, etc. All of which you'd otherwise have to buy and install.

Astute readers might have noticed the inclusion of the word "starter" in the previous paragraph; as in "your plane won't have to carry a starter. . . ." Hand-propping isn't a dead art, and with care and proper training, it's reasonably safe. Rotax operators have it easy because a manual pull-starter is standard with the engine. The plane will already carry one starter: you. Why add a heavy mechanical device?

This mostly applies to the simple knock-around planes. I'm not telling anyone to hand-prop a Glasair III.

If you don't install an engine-driven electrical system, then you aren't required to have a transponder to operate within the 30-nm radius TCA veil. This isn't much help to those in the California or East Coast beehives, where a transponder is practically required to operate a pogo stick.

Fig. 3-12. A VariEze with a solar array. The solar cells trickle-charge an onboard battery that powers a small radio. A considerably larger array would be necessary to eliminate the battery. Also, if the aircraft were kept in a hangar, the array could not keep the battery charged.

In my case, I live near two uncontrolled strips within 10 miles of the primary airport of the local TCA. The TCA is unconvoluted and reasonably small. If I just stay away from it, the aircraft does not need a transponder; with an electrical-systemless airplane, a transponder is not required. Or I can install an electrical system that isn't powered by an engine-driven device: a wind generator or solar cell array (Fig. 3-12).

Of course, it doesn't work for everybody. An auto-engine conversion probably won't have a magneto, and hence must drive a generator. Eliminating the electrical system on the heavier airplanes won't add much payload, and would result in ridiculous mission limitations.

Night Operations

If you're planning to fly at night, a few more items will be necessary: *position lights* (left and right wingtip and tail) and an *anticollision light*.

The position light requirement can't be met just by having bulbs stick out at the various locations. The lights must be visible within certain angular limits. For example, both wingtip lights must be visible from in front of the aircraft. The individual lights must be visible through 120 degrees in the fore-and-aft direction. If you want the FAA inspector to approve your kitplane for night VFR, you'll have to prove compliance. The easiest way is by using approved position light assemblies.

The same is true of the anticollision light—angular coverage requirements must be met. Either strobe lights or rotating beacons are acceptable. The usual practice is to install either a single strobe at the top of the horizontal stabilizer, or install one in each wingtip.

In the cockpit, *post-mounted* instrument lights are the easiest to install, but the small spotlight-style are more flexible. Getting the light full and even is the biggest problem. Proper instrument lighting is another one of those arts where solutions are generally found only by trial and error.

The only time a landing light is legally required is when the aircraft is being operated for hire, which a homebuilt can't do anyway. So you don't have to put one in. If you do, keep in mind that hot filaments don't like vibration. Wing-mounted lights have a lower rate of bulb failure than cowling lights. But then, they don't provide even coverage, either. Your choice.

Electrical System Voltage

The choice is between 12-volt and 24-volt systems. Except there *is* no choice, use 12 volts.

There might be a few extenuating circumstances. If you already own a stack of 24-volt avionics, for example. Yes, there are some advantages to 24 volts, for example, the wiring can be lighter because same power can be passed at half the current. And yes, many production craft now use 24 volts.

But the preponderance of evidence is solidly on the side of 12 volts. Components cost half as much. Battery chargers can be bought at many hardware and automotive stores for $20 or so. If the battery goes flat, you can jump-start using the battery in your car.

Forget 24 and go 12.

AVIONICS

The primary reason to install electrical systems is to power the avionics stack. Again, avionics selection should be determined by your mission.

If your mission is to transit the airways with more electronics than James T. Kirk, boldly go ahead. My friend Bret is building a Questair Venture. He spent almost $60,000 for the kit and engine. He's spending another $50,000 on avionics. Everything from an autopilot to a moving-map display.

But he's got everything figured out. The alternator will provide enough power; the panel enough space. The weight of all that equipment causes other problems. He'll have enough payload capacity to carry a toothbrush, but will have to buy toothpaste at his destination.

He knows what he's giving up. A lot of homebuilders don't. They go hog-wild at the avionics store, and willy-nilly mount radios and gadgets in every nook and cranny. Then they wonder why their pride-and-joy climbs slowly and stalls viciously. Their wives wonder why the credit-card bill is so high.

Avionics cost the homebuilder in three ways: pocketbook, panel, payload. You want to minimize the first two and maximize the last.

Pocketbook is obvious. As a rule of thumb, count on spending at least $1,200 for each gadget (comm radio, VOR receiver, Loran, etc.). Some are cheaper, most aren't. These tiny little boxes "burn" a lot of money.

Panel space is somewhat ignored until too late. If the plane is tandem seating, there might not be much room for anything but a compact navcom and a transponder. Side-by-side airplanes have wide panels, and should have sufficient room for all but the most extravagant layouts.

Avionics weight can add up in a hurry, chewing into useful load. For modern avionics, count on about seven pounds for each box, including tray, cables, and antennas. Two nav-Coms, a transponder with blind encoder, a Loran, and a control panel will weigh about 35 pounds. For a simple two-seater, that weight can reduce full-fuel useful load by 10 percent. Used, older-model avionics might be 50 percent heavier.

Let's take a more specific look at the type of avionics you might install.

Communications Tranceivers

A *nordo* (no-radio) airplane in these days is a real problem. Even when you operate from uncontrolled fields, the world is full of ninnies who can't see an airplane until they hear it on the radio. I've been flying nordo for three years; I've learned to operate as if I were invisible. Too many pilots haven't seen my red-and-yellow airplane in the pattern.

My plane is old, with an unshielded ignition system that makes hash of any attempt at using a radio. Modern kitplanes and engines should support communications with few problems.

With a simple knock-around aircraft, just a hand-held communications transceiver (Fig. 3-13) will do. There's nothing wrong with mounting an antenna on the aircraft, hooking up a power plug to the electrical system, and installing a small holster to hold the radio. The radio itself will cost about $400, and you can pay as little as $30 for a simple whip antenna or $100+ for a fancy blade type. External antennas work much better than the little "rubber duck" included with the radio. One nice hand-held feature is its built-in theftproofing. You don't have to leave it in the airplane as a target for thieves, and you can walk around and listen-in to the flying chatter at airshows.

But hand-helds aren't the perfect solution. They're low in transmited power; hence shorter-ranged and less likely to catch the listener's attention. The displays and controls aren't designed for easy in-flight use.

Fig. 3-13. Most ARVs can get by with just a hand-held transceiver like this Narco HT 870.

Avia-Tech Marketing

More conventional comm radios (Fig. 3-14) will cost around $800 on up. Navcoms (that include VOR navigation as well as the communications tranceiver) start only a little higher.

Whether you buy two comm radios depends on the type of airspace you fly through, whether you're flying IFR, and whether you're used to dual radios. An audio control panel (Fig. 3-15) is necessary to allow a single microphone and speaker to be used with both radios. Control panels are light (three pounds or fewer) but cost up to $1,000. And, like the radios themselves, they need to be installed and hooked up.

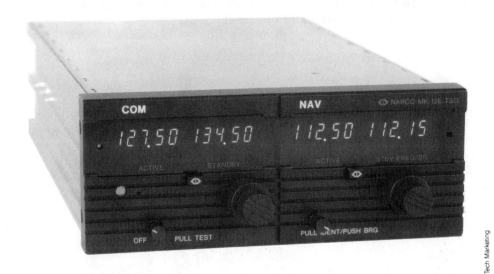

Fig. 3-14. A typical navcom, the Narco Mark 12E. A navigation indicator would also be required to display the VOR navigational information.

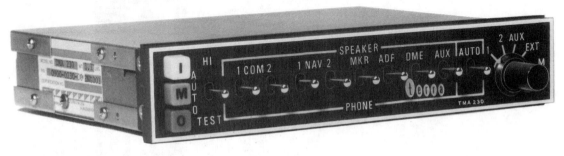

Fig. 3-15. Multiple communication radios require a control panel to route inputs and outputs.

Transponders

Transponders aren't required everywhere, but they eventually might be. Might as well save a spot in the panel now.

If you do install a transponder, you might as well install one with Mode C. The transponder itself will cost about $1,000, and a blind encoder another $300-$500.

Considering that the blind encoder costs $500 and a sensitive altimeter runs an additional $200 or so, consider spending a few hundred dollars more for an encoding altimeter instead.

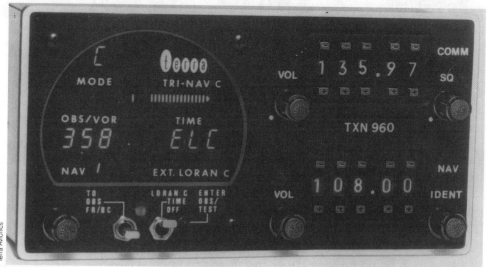

Fig. 3-16. Some navcoms include the navigation indicator as a digital display.

Navigation

Combination navcom prices start around $1,000. However, some nav units include the indicator (in the form of a digital display, as shown in Fig. 3-16) and others don't. For these, you'll have to buy a separate indicator for about $800. The ability to receive the localizer for instrument approaches adds about $600 to the navcom price and about $100 to the indicator. If building an IFR aircraft, you'll need two VORs and probably will want DME, ADF, and other units.

Of course, the hottest navigation item is Loran-C (Fig. 3-17). There are hand-held Loran receivers; a little bulkier than a comm radio and a bit higher price. Otherwise, count on spending $1,000 on up, depending upon options.

Emergency Locator Transmitters

ELTs are fairly cheap and light: around $250 and two pounds. One nice feature is that installation consists of bolting it to the structure, little or no wiring is needed.

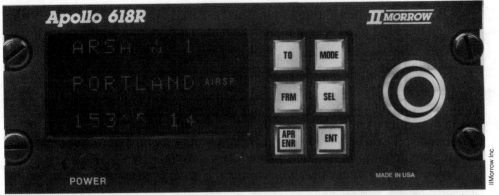

Fig. 3-17. LORAN has practically become the standard navigation instrument of homebuilt aviation.

They aren't required in single-seat aircraft, but is $200 cheaper than spending a week (or eternity) in a crashed airplane.

Intercoms

Most homebuilts are noisy. I recommend wearing a headset, so why not buy two and install an intercom? An intercom is light, and costs less than $250.

Then again, they can be easily added later. Anyway, you won't be able to carry passengers until the test flying period is completed.

Tunes

Why ride bored on long trips? Install an automotive cassette stereo. Don't bother with speakers because converter boxes allow connection with the standard aviation headset. Don't hook it up directly, because aviation headsets have 600-ohm impedances, and the stereo requires 8 ohms. Pick out a cheap rugged unit. A $35 stereo from a discount store weighs less than a pound and has survived six years of operation in my plane.

Antennas

One point to remember about antennas: They're just wires of particular lengths. You don't get a longer one for more powerful radios, the wire length is set by the transmit and/or receive frequency. Because the precise length is critical, and commerical models are cheap, buy one rather than build. Those with composite or wood airplanes can install custom antennas internally, drag and leaving the exterior uncluttered. Several companies offer kits.

However, check with the kitplane maker first. Several composite kitplanes contain graphite, which will block radio signals almost as well as metal.

Saving Money on Avionics

The prices listed above are approximate, based on typical advertisements. Careful shopping can find even greater savings. Avionics list prices are the maximum you should pay; most places sell for less.

And like everything else, you can save on avionics cost by buying used. Have an avionics shop test prospective purchases. Sometimes the equipment is still installed in an aircraft, and you can verify operation yourself. Solid-state units are hardy because if a unit does not fail in the first 10 hours of operation, it'll last forever. Used avionics are a good deal.

But be especially careful when buying used communications transceivers. The FCC has proposed tighter frequency tolerance standards that will make thousands of older radios illegal. These are listed in Table 3-4. Besides, these are mostly 360-channel radios. To operate in the modern ATC system requires at least 720 channels. You can quickly tell how many channels a radio has by checking the frequency interval. For example, if the radio tunes from 118.00 to 118.05 to 188.10, it's a 360 channel unit (118 MHz to 136 MHz in steps of .05 MHz). A 720-channel radio will tune 118.00 to 118.02 to 118.05 to 118.07 to 118.10. (The 118.02 is actually 118.025 MHz; and 118.07 is actually 188.075. The ".005" is dropped for simplification.)

Another option for those handy with a soldering iron is a radio kit. RST sells kits for common avionics, with prices equal to used equipment. It'll add to your construction time, but is a cheap way to upgrade your panel.

Table 3-4. Used radios to avoid due to threatened FCC action.

Genave	A-10	A-100	A-190
	A-200	A-200A	A-360
Narco	Mark 2	Mark 3	Mark 4
	Mark 5	Mark 6	Mark 7
	Mark 10	Mark 12	Mark 12A
	Mark 12B	Mark 16	Mark 24
	E-110A	Comm 10	Comm 10A
	Comm 11		
King	KY-90	KY-95	KX-120
	KX-130	KX-150	KX-150A
	KX-150B	KX-160	KX-170
	KX-170A		
ARC	1E	515	516
	115R	517R	540
	528A		

Installation

The standard width for regular aircraft radios is 6.25 inches. Allow at least 12 inches behind the panel. The height of the units varies, but typically runs around 1.5 to 2 inches.

The components are not solidly mounted to the panel; instead, a tray or rack allows the unit to be easily removed for service. Some trays have a set of electrical contacts that automatically connect the unit to power and antenna; others merely allow the unit to slide out once it's manually disconnected.

Space-saver units are designed to fit into a standard 3.25-inch instrument hole (Fig. 3-18), and must be mounted from behind. Keep in mind that the trays might not be included with the purchase of used avionics.

Don't neglect cooling. The traditional mounting procedure is to stack the radios atop each other, which concentrates the heat. Solid-state electronic gear runs far cooler than the old tube radios, but is also more sensitive to temperature variations.

One last point about installation: Carefully study the warranty when buying new avionics. Some companies only give warranty coverage if factory-certified technicians perform the installation. Other companies give triple the warranty if certified techs do the job. Solid-state avionics are wonderful, but the little chips aren't as tolerant of mistakes as vacuum tubes. One little snap of static electricity, and they're gone.

There are certain functions, like antenna tuning, that take specialized equipment. But the wiring harnesses aren't any different from handling any other electrical wiring. Discuss your situation with the avionics manufacturer. They might allow you to install the wiring, and only require an outside check-out before the boxes are connected.

You may wish to leave expensive or complex installations to a professional. Then if it blows up, you don't have to pay for it.

By now, you should have decided on your kitplane's equipment and other factors. You know which engine will be installed, what instruments, and know right avionics for your mission. It's time to get ready to build.

Fig. 3-18. Many avionics units are available in space-saving packages that mount in standard instrument holes. This Terra TRT 250 transponder is an example.

4

Preparations

THE KITPLANE AND ITS EQUIPMENT HAVE BEEN SELECTED. Delivery times vary from weeks to months. No need to sit around moping because there are plenty of things to do in the meantime.

SHIPPING

At the time the kit is ordered, you normally select the shipment mode. It usually boils down to having it shipped commercially or picking it up in person. In either case, getting the kitplane from the factory to the workshop is not trivial.

All kit prices are F.O.B. (*free on board*) the manufacturer, which means that shipping costs are extra. These can vary from a few hundred to a couple thousand dollars, depending upon the size and weight of the kit and the distance you live from the manufacturer. You'd like to save the money, but, like fabricating a part instead of buying a completed one, the hassle is counterbalanced by the savings.

Because it might be weeks before the kit's shipped, you have plenty of time to arrange for the trailer and other necessary equipment. And if you can't get it together by the time the kit is ready, just send a check for commercial shipment.

Factory Pick-Up

Materials kits and/or multiple subkits can often be carried in the back of a pickup. If the manufacturer is local, you can make several trips, if necessary.

If a pickup is too small, a rental truck might be adequate. These cost $60 a day, plus 10 cents or more per mile. Rent the truck locally or fly to the factory's town, rent a truck, pick up the kit, and drive it home. Calculate the total expense for round-trip or one-way, considering time and distance, and choose lower cost or better convenience to your preference.

The truck will be much wider than the crate, so tie it down to prevent shifting. Check with the manufacturer and find out if anyone else in your area has a kit on order. You might be able to carry both and share expenses.

Such trucks might handle a variety of kits. But some composite and tube-and-fabric kits include components 15 or more feet long. The Glasair, for instance, comes in a single crate about 25 feet long. In these cases, a trailer is the best solution.

Get the crate size and weight from the kit maker, then check around with your friends. Simple flat trailers are best, but the addition of a few boards might make a boat trailer suitable. If you can't borrow a trailer, you'll have to rent one; they go for about $30–$40 a day.

Carry lots of rope and bungee cords to secure the kit to the trailer. It wouldn't hurt to take along some boards and a saw, in case extra support is needed. Generally, the crates aren't waterproof, so take some plastic and a staple gun.

Depending upon the kit, the trailer and crate will weigh 1,000 to 2,000 pounds. Small pickup trucks can handle lighter kits (Fig. 4-1), but heavier loads and/or steep terrain might require a full-size pickup. A truck could be rented, but if you've got to rent both the truck and the trailer, you're better off having the kit shipped.

Fig. 4-1. Most kitplanes come in large crates. Hugh Sincock's Avid Flyer was no exception. He saved several hundred dollars by picking the kit up at the factory and carrying it home on a boat trailer towed by a small pickup.

How much will you save by hauling it yourself? Assume the factory is 500 miles away, you own an adequate truck, and a rental trailer costs $40 a day.

Fuel mileage is going to be moderately low on the way down and worse on the return. Speed is the same way because you'll hold the speed limit without much problem going to the factory, but the weight of the kit will reduce your average speed on the return leg.

Count on getting to the factory in a single day, and we'll assume your fuel consumption is about 15 mpg. You'll need a hotel room overnight. Pick up the kit at the factory the next morning. The weight of the crate reduces speed enough to require a day and a half on the return leg. Mileage drops to 8 mpg. For simplicity's sake, we'll assume gasoline costs a buck a gallon. (Apply the current cost of gasoline for an accurate estimate.)

You drove 500 miles at 15 mpg and 500 at 8 mpg. Gas cost is about $96. Two nights in motels, nothing fancy, at $25 a night. Meals and incidental costs probably come to another $25. Total rent on the trailer (three days) is $120.

For about $290, you've delivered the kit to your door. In the case above, a kit manufacturer quoted $500 for truck freight, hence, a savings of $210.

There are drawbacks, though. You'll probably have to take at least one day off from work. The trailer can get flat tires or even break free from the tow vehicle. Any damage to the kit on the way back comes out of your pocket. And how well are you going to sleep, while your $20,000 kit lies unattended in the motel's parking lot?

There's usually a break-even point between commercial shipment and factory pick up. One hundred miles is ideal for hauling it yourself; five hundred is iffy; across country, it might be better to just bite the bullet and write the check.

The Delivery Problem

Having the kit shipped can be far less stressful. But there's one aspect many overlook.

Truck drivers are paid to drive, not be stevedores. They knock on the addressee's door, point out the package, and watch the crate get taken off the truck. And they work normal working hours, when you and I aren't home.

In plain English: They don't do the unloading. They won't slide it into the garage for you. And you're *still* going to have to take a day off from work.

Unloading is a serious problem because remember that the crated kit might weigh 1,000 pounds and be 20 feet long and 5 feet square. The driver expects a forklift. How are you going to unload it?

One way uses a set of 10-foot-long 2 × 6 boards. Pick up about five of them. Have the manufacturer call when the kit is shipped, and have them instruct the shipper to call on arrival (COA) before attempting delivery.

When they tell you the kit has arrived, set up a day and an approximate delivery time. Cajole a couple of friends into helping. When the truck arrives, have the driver back into the driveway. Arrange the three boards into a ramp, and carefully slide the crate down to the pavement.

Or you can use the 2 × 6s to build three sturdy sawhorses. Have the truck back into the driveway, and tie a rope from the crate to something sturdy inside the garage. Have the trucker drive forward a few feet until the slack disappears and the crate slides partially out of the truck. Set a sawhorse right behind the rear bumper, directly under the crate.

Have the driver pull forward a couple more feet. Set a second sawhorse behind the bumper, again. As the truck goes forward again, the crate will be pulled past its center of gravity. The end closest to the garage will tip down; you and a friend should be ready to make sure it lands atop the first sawhorse.

When the crate is almost all the way out of the truck, set the last sawhorse behind the bumper. Inch the truck forward. As the last edge of the crate comes

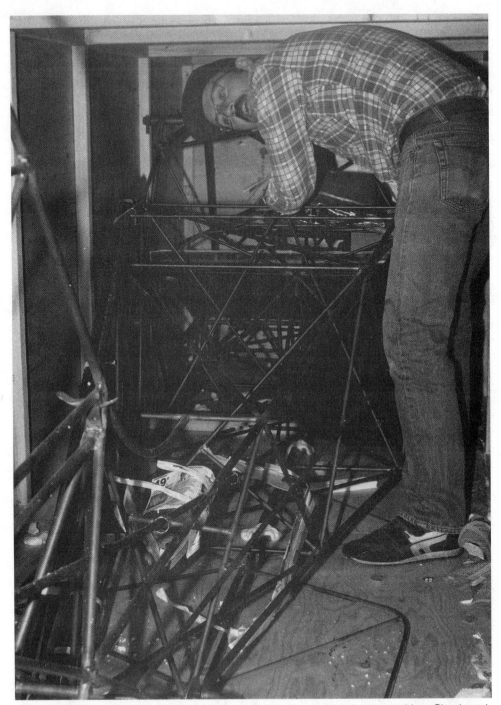

Fig. 4-2. Hauling everything out of the crate at once is a good way to lose something. Check each item off the packing list as soon as it comes out. Many of the smaller items for Sincock's Avid Flyer were secured in the cockpit area of the prewelded fuselage. Rather than unload the 800-pound crate from the trailer, Sincock elected to remove the end of the crate and unload it first.

free, you and a friend should guide it safely atop the sawhorse. This method works best with kits in long crates.

Another solution is to pick it up at the freight office. Borrow a trailer and get a couple of buddies to leave work a bit early. The freight company will already have the crate off the truck, and will use their own forklift to load it onto your trailer. Because the shipping charges call for the crate to be delivered to your home address, the freight company will reimburse the difference.

If the shipping company doesn't have a local office, arrange to meet the trucker at a convenient loading ramp. Often, industrial districts have a public ramps for this purpose, which is literally a ramp that ends abruptly at truck-bed height. You can slide the kit off the truck, and you and your buddies can carefully load it onto the trailer.

Inspection

No matter how the kit comes into your hands, inspect it immediately. If there is external damage to the crate, open it and inspect *before* you accept delivery. Damages will have to be noted on the acceptance paperwork. Signing the shipment papers doesn't get the trucker off the hook, it merely certifies there was no obvious external damage at time of delivery.

Damages that weren't apparent on delivery should be reported to shipper and the kit manufacturer. Any shortages aren't the shipper's fault, unless a hole was knocked in the crate and components might have fallen out.

Inventory the contents of the crate as soon as possible. The kit should include a complete parts list. You don't have to count all the bolts, but be sure nothing expensive is missing. Probably the easiest way is to check off each part as it comes out of the crate (Fig. 4-2). Some parts (Fig. 4-3) should be checked for fit right away, of course. . .

Other Components

Anything not included with the kit can usually be shipped by a low cost means like parcel post or UPS. Instruments, paint, tires, and other items take only a week or so to arrive, so don't order them too much in advance to save storage space.

Engines are too heavy for anything but truck freight. Lycomings come in a box about 3 feet square by 2 feet high, which means you can carry the engine in a pickup if necessary. Rotaxes are much smaller and lighter.

Remember, there's little reason to buy the engine, instruments, or other components immediately. As mentioned last chapter, the engine must be pickled if it won't be run for a while. Even when pickled, the engine crankshaft has to be turned over a few times every month, so why do it more months than you have to? Find out the typical delivery time, and order it when you're ready.

Fig. 4-3. Go ahead, have your picture taken in the cockpit. Everybody does it . . . .

WORKSHOP REQUIREMENTS

The workshop is like a budget because no matter what you have, more is needed: more tools, more space, more light, more power. The following are some general requirements for the homebuilder's workshop. Specific requirements, including tools, will be covered in later sections.

Space Required

As mentioned in chapter 2, the kitplane manufacturer usually will indicate the space required to build the aircraft. Most specify a two-car garage.

Is that a garage big enough to hold two mid-60s Cadillacs, or sized for two Hyundais? Garage sizes have changed drastically in the last 20 years.

Suffice it to say, the more room you have, the better. Sure, it's not absolutely vital. I've seen a Q2 built in a two-bedroom apartment. I watched an Osprey (a large amphibian) being built in a one-car garage.

But it's much, *much* easier with sufficient room.

It really doesn't make much difference, as we have to use what's available. We have a basement, a garage, or spare bedroom. Most of us can't afford to build a workshop in the backyard (Fig. 4-4). At best, we might rent a hangar. But the majority of kits are assembled at home, in whatever space is available.

If you have enough room for the single largest complete component (usually the fuselage, except for one-piece wings), you have a chance. Building in such limited quarters will be difficult. Moving components around will be a chore—imagine trying to arrange a hoist to lift an engine in place when you barely have enough room to turn around.

Fig. 4-4. The ideal solution for construction space is to erect a separate building in the backyard. Most of us have neither the money nor the room.

Some of the oldest jokes in homebuilding are about guys who have to disassemble the house to get the completed airplane out of the basement. These stories are true to varying degrees. A 35-inch-wide fuselage won't necessarily fit through a 36-inch door. Wing attach plates might project out; the hallway beyond the door might take a sharp turn. If aircraft egress room is questionable, slap together a full-sized plywood mockup and experiment.

Clever solutions can be found. The Osprey builder had a garage door eight feet across, with a minimum fuselage width of eight and a half feet. He developed a slick little system to give him extra clearance (Fig. 4-5).

If building room is especially tight, you'll eventually have to complete construction somewhere else. You'll probably end up paying rent. But the longer the move can be delayed, the better.

Power

There are two main questions regarding your shop's electric power: Is it adequate and is it convenient?

If you plan on using any 220-volt power tools, of course you'll make sure the shop is wired for it. Most industrial-grade floor tools require 220-volt *three-phase* power. While the right voltage is usually available (electric clothes dryers) three-phase power often isn't.

Fig. 4-5. The opposite extreme. Kirk McCarty built his Osprey II in a single-car garage. The fuselage was wider than the door, so he cut away part of the door pillar and used a jack to temporarily support the building. Once the fuselage was through the door, he replaced the missing section and removed the jack.

How many circuits are available for standard 120-volt tools? A $^1/_3$-hp motor takes between 5 and 10 amps. Standard house wiring is 20 amp; so two small motors is all you'd want on any single circuit. Of course, you're not likely to use more than one tool at a time, unless someone else is helping. At least two 120V circuits should be available.

Modern garages and basements shouldn't have any problem, but watch older homes. Some 10-amp circuits are still out there; a moderate-sized table saw will blow the fuse. I've a friend who lives in one of these houses, his motto is, "A penny saved is a fuse." He's just kidding, but if you're stuck with low-current wiring, you might have to pay an electrician to upgrade the workshop area.

Convenience is usually of greater impact to the homebuilder. You'll be roaming all over the workspace with cord-type electric hand tools. Where are the outlets? In my garage, all the outlets are on one wall. Extension cords are a fact of life.

They are hard to escape in any case. Outlets are usually on the walls, while the aircraft sits in the middle of the floor. One or two 25-foot cords should suffice.

Outlet strips are useful accessories. These plug into an outlet and provide six or so additional outlets controlled by a single switch. Workbenches and tools limit access to the wall outlets themselves, and the outlet strips bring power closer to where it's needed.

Outlet strips sell for around $10, and can be found on sale for half that. Most come with their own circuit breaker, which, unfortunately, are usually rated at fewer than 20 amps. Try not to use an outlet strip with several large tools running at the same time.

One feature to ignore is those that have *power protection* or are *surge limiting*. This feature is intended to protect delicate electronics (like computers) from power fluctuations. Power tools aren't sensitive, they *cause* surges.

Lighting

For shop lighting, nothing can beat ceiling-mounted fluorescents. The fixtures are cheap, $10 or less (without tubes) if you watch for sales. Shop carefully for tubes, prices vary from $1 to $3, depending if it's marketed toward home use (individually wrapped, found beside the rest of the bulbs at the grocer's) or shop use (no box, piled in a rack in a hardware store). I buy a box of 10 for $8.

The fixtures can be either screwed to the ceiling or hung from hooks. I hang mine; in fact, I've set up groups of hooks in various areas and move lights as needed. An electrician could add ceiling outlets, but I just tack up extension cords using cable holders.

Because of the shadows cast by ceiling lights, you'll need portable lighting as well. The standard shop trouble light works fine and acts as an extension cord, too. Bulbs specifically designed for trouble lights cost around three dollars each. Regular light bulbs work, but their filaments are a bit sensitive to shaking and bumping. I'm a klutz—I use the special bulbs.

There are also portable fluorescent bulb holders. While more awkward to handle, the long tubes cast fewer shadows.

Heat

If you live in the cold country, you'll want some heat for your workshop. You don't want to spend 30 hours a week shivering in the garage and handling frosty aluminum. The only alternative is to stop work during the coldest months, but these are months you'd rather work, because you can't fly.

Just because you live in warm climes doesn't excuse the need for occasional workshop heat. A composite aircraft requires approximately 70-degree temperatures, as do wooden kitplanes. Tube-and-fabric models are fairly impervious to temperature, until it comes time to apply the fabric and dope. And you'd like to rivet skins on aluminum kits while the skins are warm; otherwise they might get a bit wrinkled in warm weather.

Comfort heating might be taken care of by a portable electric heater costing $20 or so. But building aircraft isn't a stationary enterprise. You'll be moving around too much to benefit much from a small heater.

Options for full heat abound. Attached garages often contain the house's furnace, and the ductwork is accessible. Install a few junctions and the furnace will heat the garage as well. You won't get the temperature up too far because the thermostat is still in the house. But any little bit helps.

Large kerosene-burning space heaters sell for $200 and up. Wood stoves, multiple electric heaters, and waste-oil-burning furnaces might fit your particular situation. You might even buy a used house furnace.

However, pay careful attention to local codes. For example, kerosene heaters are banned in my town but legal in the surrounding county. Your options when using a basement as a workshop (inside the house) are more limited than might be allowed for a separate garage.

Finally, don't forget insulation. Adding batting between the joists can do wonders. Garage doors are a particular nuisance because they don't seal worth a hoot and are a major source of drafts. Try a curtain of plastic, stapled to either side of the door, with the foot weighted down by boards or sandbags. Also, lay some scrap carpet on the floor of the work area. Cold cement sucks heat from feet and legs, and a little bit of insulation goes a long way. It also keeps dropped parts from rolling too far.

Worktable

One thing necessary for many homebuilts is a worktable big enough to build the major components on. Generally speaking, this table is about 3 feet wide and 12 feet long. A height of about 3 feet seems the most comfortable.

Not all airplanes need a worktable. Composite airplanes often use a special support system, as shown in Figs. 4-6 and 4-7. Templates are cut from plywood and attached to sawhorses. These templates match the shape of the fuselage or wing, and hold it solidly in place.

The worktable might have a solid top, or might be an open trusswork to allow access to the components from underneath. *Both* modes might be necessary; for instance, the fuselage might be built inverted on an open truss, and the wings built on a solid tabletop.

You can build the worktable out of nearly any material. The top must be wood because you'll probably be attaching temporary supports to it. Construction must be sturdy. Not only is the airplane going to come together on it, but

B. WING JIG CONSTRUCTION

The dihedral and airfoil of the wing are molded in at the factory. Until the upper wing panel is bonded to the lower panel, however, the wing panels are free to twist slightly. A wing jig is needed, therefore, to support the wing in an untwisted condition until wing fabrication is complete. The wing jig consists of a sturdy support table to which a series of plywood supports is fastened. The idea in fabricating the wing jig is to position the supports on the jig stand so that they most closely match the contour and dihedral of the molded lower wing panel while supporting it in an untwisted condition.

STEP B-1

Using the supplied full size support templates, cut out the (9) wing supports from 1/2" plywood. Make (1) "A" support and (2) each of supports "B", "C", "D", and "E", as shown in FIGURES (C-4) and (C-5). To duplicate the B, C, D, and E supports, use sheet rock screws or the equivalent to fasten two pieces of plywood together, glue the support template to one of the sheets, and cut both supports at once.

A level line is marked on each of the wing jig support templates. Transfer these lines to all of the plywood supports.

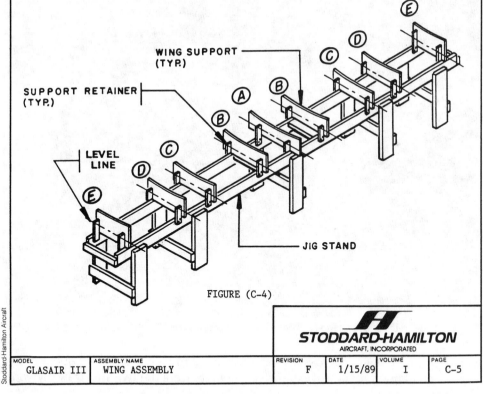

FIGURE (C-4)

MODEL	ASSEMBLY NAME	REVISION	DATE	VOLUME	PAGE
GLASAIR III	WING ASSEMBLY	F	1/15/89	I	C-5

STODDARD-HAMILTON
AIRCRAFT, INCORPORATED

Stoddard-Hamilton Aircraft

Fig. 4-6. To ensure a straight and true wing, the Glasair plans include the directions and templates to build a combination jig and worktable for wing construction.

Fig. 4-7. A close-up of a fuselage jig. The full-size template is glued to the plywood and cut out with a saber or bandsaw.

you'll end up crawling across it to access a hard-to-reach portion. I spent about $150 to build mine; the most expensive elements were the two sheets of plywood for the top.

The biggest requirement of the worktable is its precision. The table is the basis for accuracy when building the aircraft. You'll build your wing on the table, and if the table is warped, the wing will be warped, too.

The correct top material can cut down on warping. Plywood is most common, but has a lower resistance to warping than particle board. But particle board tends to crumble. Hollow-core doors or drafting-table tops are great at resisting warpage, but can't take the abuse as well. Some builders add a plywood overlay over doors. Whatever you choose, get the thickest you can afford, 5/8 for an absolute minimum, 3/4 inch or greater preferred.

The table must be precisely horizontal. This allows checking accurate placement of components by using a carpenter's level. This complicates workbench construction, as your shop's floor probably isn't level. The workbench must be able to compensate.

Sample workbench construction is shown in Figs. 4-8 and 4-9. Your aircraft plans might also include some workbench ideas.

When it comes time to adjust the table, three tools are important: carpenter's level, ruler, and ball of string. Designate one corner of the table as a reference. Attach one end of the string to a block of wood, and stretch it above one long

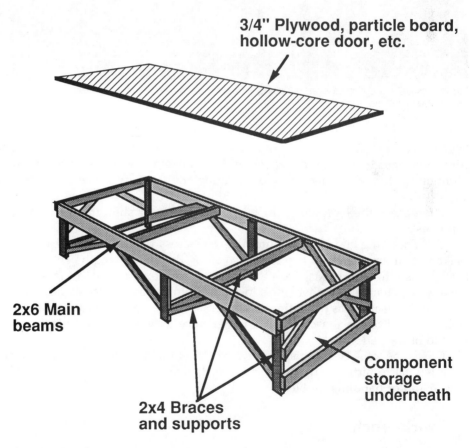

3/4" Plywood, particle board, hollow-core door, etc.

2x6 Main beams

2x4 Braces and supports

Component storage underneath

Fig. 4-8. The traditional homebuilder's worktable, constructed from dimensioned lumber and plywood. Diagonal braces ensure rigidity yet allow storage underneath.

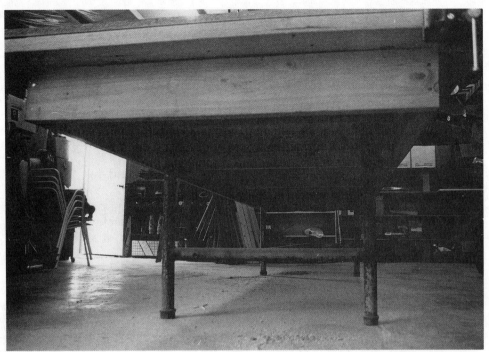

Fig. 4-9. A table of similar design, but using pipe for the legs. Note the lateral support boards and the unobstructed storage space.

edge of the table. Attach it to a same-sized block of wood in the far corner. Measure the distance between the string and the tabletop at points between ends. Shim up the top where necessary, eliminating any sag. This process is shown in Fig. 4-10.

Then check the level along the edge, and adjust or shim the table's legs to compensate.

Use the string and ruler to eliminate sag across the short axis of the table. Make measurements every two feet or so along the table's length. Then use the level across the short axis. Don't touch the adjustments on the legs you've already set; change the other side's instead. Shim the table edge as necessary.

Repeat the above as many times as necessary to get the errors as small as possible. If I had to make a choice, I'd rather have a table that's a little off level, instead of warped.

Check the table occasionally afterwards, or if the table is moved. Again, the table is vital to airframe trueness; a mere degree of error could cause performance and/or handling problems.

Workbench

You can use the worktable for all the activities involved with building the aircraft, but you don't really want to. Parts with sharp edges can cut up the surface,

Slide square along tabletop, measuring height of string

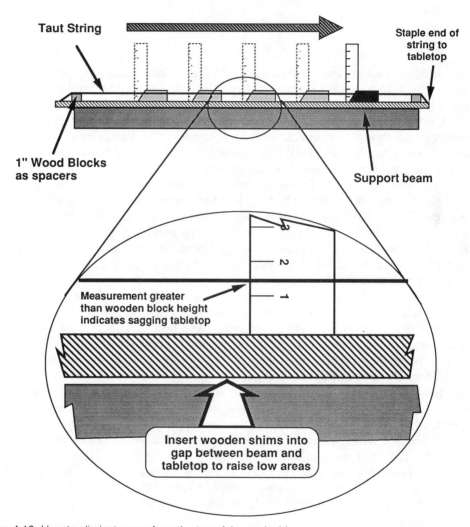

Taut String

Staple end of string to tabletop

1" Wood Blocks as spacers

Support beam

Measurement greater than wooden block height indicates sagging tabletop

Insert wooden shims into gap between beam and tabletop to raise low areas

Fig. 4-10. How to eliminate sags from the top of the worktable.

or heavy components can throw the table out of alignment. It's a good idea to have a separate workbench for the rough, heavy work.

There aren't really any specific requirements for the workbench. Make it the same height as the worktable. Don't sweat alignment and warpage too much; however, you might use the bench as practice for the full-size worktable.

Or buy a commercial model, they come with drawers, electrical outlets, and other things that make building easier. I bought a used drafting table for $35; a used metal office desk would work just as well, with a sheet of plywood as a work surface.

If you're very limited in space, don't bother with a workbench. But fit one in, if possible.

Other Items

Building an airplane generates a lot of dust. Shop vacuums are nice, but I prefer the little handheld rechargeable models. A simple pushbroom and dust-pan suffices for the floor.

Don't plan on building the entire airplane standing on your feet. A lot of jobs involve assembling components at the workbench. Get a chair or stool or two. I keep a lawn chair handy, and if I have to polish a lot of small parts, I set the chair in the sun and use my chest as a workbench.

One thing you'd like in the workshop is a telephone. You are going to get dirty, and there's no reason to be tracking dust into the house and getting greasy handprints on the phone. If you watch the sales, you can buy a cheapie phone for $10 or less. Wiring the garage for a telephone is relatively easy. The garage is the typical entry point for phone wiring nowadays. Phone service is supplied though a small four-conductor cable, similar to the extension cable that is used from the wall jack to the phone.

First, go to Radio Shack and buy a four-position terminal strip, a modular phone jack, and 10 feet or so of four-conductor phone wire. When you get home, find the point where the phone line enters the house. This should be a small multiconductor wire traveling on its own. Find a good spot along the wire run inside the house to mount the terminal strip, then carefully pull on the wire to gain as much slack as possible.

Cut the wire and strip back the outer shell to reveal the four conductors: red, black, green, and yellow. Strip a little insulation away from the wires on both sides. Mount the terminal strip on the wall or joist. Connect the red, black, green, and red wires on one cable to separate terminals on one side of the terminal strip. On the corresponding terminals on the other side, connect the four wires of the other cable. Also, strip back one end of the phone cable you bought and connect them to the same colors on either side of the terminal strip. Run the cable where you'd like to mount the phone, and install the modular jack per instructions.

There are other little comfort details you could attend to. As mentioned, an old piece of scrap carpet in the primary working area prevents dropped parts from rolling too far, as well as keeping your feet warmer. A radio makes the time pass pleasantly. Hang up a few pictures of the type of plane you're building, both for inspiration and to answer the questions of visitors. You'll be spending a lot of time there; make it as homey as possible.

Two last points regarding the workshop: never turn down anything free and be inventive. One local builder found an old rollaway typing table. He clamped a plywood top to it, and had an instant portable workbench (Fig. 4-11). A veterinarian gave me a broken safe, which I use as a filing cabinet for receipts and magazines. Grab up old carpet and blankets for padding, scrap lumber for braces, glass jars to store parts and leftover liquids. The motors for many homebuilder's bench sanders have been liberated from junked appliances. Muffin pans work great for storing small parts—surely the wife would love to have a new set while the old set disappears into the garage.

It's one thing to specify what items are necessary to the workshop. But if you don't arrange them properly, every little task becomes a chore.

Fig. 4-11. One person's junk is a kitbuilder's treasure. A castoff typing table converted into a nifty portable workstand.

WORKSHOP ARRANGEMENTS

Let's count up the major components of the homebuilder's workshop. Worktables and workbenches. The aircraft's fuselage. Two wings. A heavy engine. Freestanding power tools, like drill presses, bandsaw, and table saw. Racks of waiting parts and components. Cupboards of bolts and fasteners. Not to mention enough room to assemble the entire aircraft, an object 20 feet long with a wingspan approaching 30 feet.

Workshop space is not unlimited, hence the need for efficient layout.

Hand Tool Storage

Most of us keep our hand tools in a plastic or metal box. That works out OK most of the time. We can scoop up the box and carry it to the car, the broken shelf, or the unassembled bicycle.

But building an airplane means using the same tools continually in the same location. How much time do you want to waste digging through the mess in an old toolbox looking for the $7/16$-inch wrench?

Make it easier on yourself and set up a tool board, as shown in Fig. 4-12. A tool board consists of a piece of plywood in plain sight, with often-used tools hanging from nails. The outline of each tool is traced onto the wood with a marker or crayon.

A four-foot-square piece of plywood or particle board is adequate. Don't overdo the number of tools on the board, and don't cram them too tightly. For example, don't hang your entire set of wrenches on the board. You'll probably most often use the $3/8$ inch and $7/16$ inch; hang these only and keep the rest somewhere else. Better yet, buy extras of these wrenches and keep the stored set intact.

Tool boards are a marvelous way to keep track of tools. They work even better if you put each tool back after you use it. Every good homebuilder should develop this habit. Let me know if you figure out how because I certainly haven't picked up the knack.

Fig. 4-12. A tool board is one way to store often-used tools. Outlines are drawn around each to mark the tool location.

The tool board can be seen from anywhere in the shop. If the tool is on the board, fetch it. If it's not there, then start searching the worktable and workbench. At least you didn't have to dig through a toolbox first.

A broomstick with a bent nail in the end is a useful gadget. If you spot the errant tool on the other side of the table, use the broomstick to grab it and slide it within reach.

While the most-used tools occupy the tool board, the rest must wait elsewhere until the call to arms. Your old toolbox would be fine, but if possible, pick up a rollaway unit with a multidrawered box. These units consist of a toolbox atop a wheeled cabinet. They normally are fairly expensive, around $100 for the most basic unit. Cheaply made bargain units can be found for half that, and for the amount of use they're likely to see, they'll do just fine.

Access

In kitchen design, the distance from the sink to the stove to the refrigerator and back to the sink should be no more than 10 feet; the same can be said about aircraft building. Minimize the distance between the worktable, workbench, the tool board, and the free-standing tools.

This rosy view of workshop design is discolored by real-world problems. The first is the worktable itself. The table is 3 feet wide and 12 long. Because you'll probably need to gain access to all sides, at least 2 feet of clearance must be maintained on all four sides. This makes the table workspace 7 feet wide and 16 long, or about the size of a typical car.

This not only restricts where the table can be placed, its very size ensures that perfection can't be approached. There'll be plenty of times when the tool you want is on the opposite side of the table. So you'll have to walk all the way around and all the way back.

If possible, place your tools at the ends of the table, rather than the sides. It seems shorter when you don't have to turn so many corners.

The layout doesn't have to be fixed. My worktable occupied the center of the garage during fuselage construction. Once the gear was installed, the worktable was shoved off into one corner, ready for later use building the wings. (Don't forget to relevel and check the warpage whenever the table is moved.)

Storage of Kit Parts and Components

An aircraft kit has hundreds or thousands of pieces that must be carefully stored until ready for use. Problems caused by bad storage can range from ruined material to chronically barked shins. While some general hints are given here, each type of material has specific handling requirements that must be followed. These are given in the appropriate construction chapters, under **Material, Fasteners, and Safety**.

Sometimes the kit manufacturer makes it easy for you. Flat crates can be used at temporary tabletops, as shown in Fig. 4-13.

The hardest pieces to store are large sheets of aluminum or plywood. They're

Fig. 4-13. Can you spot the RV-4 wing kit? Storage of kit components until needed can be difficult.

awkward to handle, for one thing. For another, wood must be protected from rot and warpage, and aluminum sheets are easily scratched or dented. At least easy access isn't a requirement, because you won't use a sheet every day. Some suggestions:

- Ceiling racks. There's usually more headroom than you need, so ceiling racks move components and materials out of the way. Getting a sheet down requires special care.
- Rolled. Aluminum sheets up to .032″ or so can easily be stored in a roll. In fact, it's often delivered this way, so just leave it packed until needed. Fiberglass comes in rolls, which can be set up on racks on the end of the worktable, as shown in Fig. 4-14.
- Wall racks. Ten sheets of .040 aluminum are only a half-inch thick. With a wall rack, they can be stored with minimal impact on working area. Wall racks for plywood must clamp the sheets tightly or warping will result.
- Off-the-wall racks. Flat sheets of aluminum can be stored in the darndest places. Why not under the bed, between sheets of scrap plywood? Curve a sheet around a corner. Tack them to walls. They'll snuggle down into ceiling trusses. Protect them from scratches with scrap plywood, leftover Christmas wrap, old blankets, mattress pads, whatever.

Other components are smaller and easier to store. Racks for tubing and board lumber are easy to build, but must support the load evenly. Underneath the worktable is a dandy place for tubing and aluminum extrusions, and for larger kit components like cowlings and canopies.

Smaller parts are even easier; however, access is more important. Nuts and bolts could be stashed anywhere, but you'll be needing them all the time. Plastic

Fig. 4-14. For composite airplanes, the rolls of fiberglass cloth can be set up on racks similar to paper towels. However, the rolls must be protected against contamination when not in use.

Fig. 4-15. Stoddard-Hamilton supplies the Glasair's small hardware in see-through plastic boxes.

Fig. 4-16. Milk cartons and old boards make a cheap and adequate storage cabinet for small hardware.

multidrawered storage cabinets are cheap and can be mounted anywhere, or not permanently installed. A small easel-like mounting lets you bring all the hardware close to the job.

They aren't necessarily the best solution. The cheaper ones have awkwardly small drawers, and those with larger drawers are more expensive. The flat type, with the large top-hinged door, is better for access. A no- or low-cost alternative is to store the hardware in old cans, milk cartons, or what-have-you. Label the exterior with felt-tip pen on masking tape, and build some cheap shelves to keep them on. Some kits even ship small parts in storage boxes (Fig. 4-15).

A cheap alternative is to mount some shelves and cut the bottoms off milk cartons to hold the parts (Fig. 4-16). It costs next to nothing, and the cartons are wide enough to make access easy.

Storing Finished Components

Safe storage of finished components is slightly different. A ruined sheet of plywood merely drains your pocketbook; a ruined component wastes building time. Similar, but more careful procedures must be used.

If allowed by the kitplane's plans, build the control surfaces first. They're easy to store on the walls or ceiling (Fig. 4-17). You'd like to build the wings next, as they can be similarly stored (Fig. 4-18).

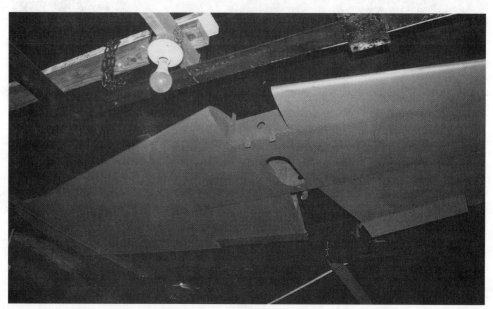

Fig. 4-17. Control surfaces and wings can be suspended from the ceiling to get them out of the way.

Another good way to store wings is shown in Fig. 4-19. A board is bolted to the root fittings, and a pivot bolt slides down into the vertical part of a wheeled L-shaped trolley. This allows the wing to be turned horizontally for work (with the tip supported by a sawhorse), then turned vertically and rolled into a storage area.

The fuselage is the hardest item to shunt aside temporarily. The wings are thin, but the fuselage is long and wide. One option, if your workshop is located with an open truss ceiling, is to delay engine mount installation and set the fuselage vertical.

Fig. 4-18. Two RANS S-10 wings stored on the wall of a workshop.

Watch for mice. They love the little compartments of metal wings; they enjoy chewing on wood and insulation. They'll happily build their nests in your composite speedster, and gleefully "piddle" on anything within reach. Ceiling and wall storage is a good first step toward prevention.

TOOLS

The kitplane plans should indicate what tools are required. But even an official tool list might not include everything. Minimal tools is a selling point; practically everybody has a power drill, so it looks better if that's the only drill the plans call for. However, a *drill press* will make your life far easier.

Where to Get Them

Two kinds of tools are actually needed for building a kitplane: *specialty tools* and *shop tools*.

Specialty tools are those that are used mostly in professional, specific applications. For example, most hardware stores sell tin snips, even aviation tin snips. But you'll have to go to a professional supplier to buy a hole duplicator or cable swager. These companies advertise in the homebuilder's magazines. The prices are uniformly high for the specialty tools, and don't vary much from dealer to dealer.

Shop tools can be found at any hardware or discount store. Pliers, screwdrivers, electric drills, and the like can be bought almost anywhere. Prices vary drastically. Shop around.

Fig. 4-19. Kirk McCarty built rollaway workstands for his Osprey II's wings. When rotated to the vertical position, a pin would lock the wing securely in place. The wings could then be stored in a small area.

There are several ways to lower tool costs. If the kit won't be delivered for awhile, keep an eye on local garage sales. Good deals in both hand and power tools can often be found. Similarly, selected companies have a public outlet for selling used tools. While buying a moderate-quality new power tool is preferable to the clapped-out tools offered at these outlets, certain smaller items offer significant savings. One local outlet sells hole cutters in sizes from $7/8$ inch to 3 inches for $3 (Fig. 4-20). That's a third the cost of a new hole saw, and the used cutters are heavy-duty models instead of light hobbyist tools.

Fig. 4-20. Surplus heavy-duty hole cutter.

Another option is the discount tool merchants. Some are willing to bargain—mention you saw the same tool at a lower price somewhere else, and they might offer a better deal. Or pick out several tools and make a single-price offer. I got a good deal on a drill press, vise, and some hand tools this way.

The quality of their tools varies widely, making careful evaluation necessary.

Evaluating Quality

Prices and quality are generally related. The more you spend, the higher the tool's quality is likely to be.

Admiral Gorshkov, the commander of the Soviet Navy during the 1970s, had a favorite saying: "'Better' is the enemy of 'good enough'." In other words, there's no reason to buy higher quality than is necessary.

How good is good enough? For obvious reasons, you don't want to buy the cheapest. But you don't have to pay top price for tools of sufficient quality, either. Stick with name brands where possible, but better prices and adequate quality are available in some off-brands as well.

Some signs of good-quality tools:

Polished Surfaces. Cheap metal tools have an overall dull or matte finish. It takes extra effort to make 'em shine. Tool-mating surfaces (the area where the tool actually makes contact with the work, such as the inside of a socket) *must* be polished.

Rounded Corners on Nontool-Mating Surfaces. Cheap ones are sharp everywhere, and you'll probably feel left-over ridges as well.

Fine Machining on Grip Areas. The places where the hand holds the tool often have a crosshatch pattern in the metal to give a nonskid grip. The deeper, smoother, and finer the pattern is, the better it works.

Chamfered or Radiused Tool-Mating Surfaces. Sharp corners chip easily; the edges of the mating surfaces of good-quality tools should have a very slight curvature or be cut at a 45-degree angle to reduce chipping.

Straight Edges. Look inside a cheap socket and you can actually see a little reverse curve on the socket edge. This causes a sloppy fit, and can damage parts.

Heft. Good tools are made of good steel; they're heavy, and produce a satisfying ring when dropped.

Materials. Table-model tools are built of a variety of substances. Steel is better than cast iron, cast iron is better than plastic. Steel has smoothly machined surfaces, cast iron has a rougher surface (especially in areas not in plain sight), and plastic is, well, plastic. This isn't a hard and fast rule; for example, there's nothing against a nonstructural motor cover or safety guard made from plastic.

There's nothing *against* buying the best-quality tools available. Some people use kitbuilding as an excuse to assemble a top-quality workshop.

Having cheap tools is no reason to rush out and replace the lot, either. If it works, fine. But if the wrenches keep slipping or the Phillips screwdriver doesn't mesh in the slot very well anymore, it's time to consider replacement.

Sometimes you'll come across a do-everything tool, one that's supposed to do the work of several tools. While these often work out for odd jobs around the house, they usually can't hack intensive use.

I own a small screwdriver within which nest several smaller drivers. It's a handy little gadget; it's good quality steel and brass, and works quite well. However, the grip area is poor quality. I've raised blisters after removing a single tight screw. The handles on the smaller screwdrivers are too small and awkward to use.

Despite this, I own about five of these. They're great for keeping in the desk or car's glove compartment, but I'd never consider building an airplane with one. Individual screwdrivers with large, comfortable grips are preferred.

More Ways to Reduce Tool Costs

Check out the plans, and find out when particular tools might be needed. For example, a good-quality Nicopress sleeve swager costs around $150, yet they generally aren't needed until the controls are being rigged. You might not need one for a year or more. In that case, wait. Keep watching, and try to find one for less. The kit manufacturer might even start selling prefabricated cables. It'll cost more but will reduce building time.

Or you might find someone to lend you the tool. I'm all for "neither a borrower nor a lender be." But the specialty tools necessary are just that, special, and can't be used for much else. Someone who has completed his aircraft has little use for a hole matcher or bending brake. He or she might not be willing to sell the tool (anticipating upgrades or maintenance), but might be quite willing to lend it for a few months, or let you come over and use it on occasion.

Check with your EAA chapter. Some own a variety of specialty tools for checkout to members.

The Basic Tool List

The following are the basic tools required for just about any homebuilt. These are the general-purpose tools; those that aren't construction-specific. Chapters 6 through 9 list the additional tools needed depending upon the mode of construction.

You can get by with just a single example of most tools, but that gets fairly irritating when the only screwdriver gets misplaced and work must halt until it's found. And *get* the right tools, and *use* the right tool. Remember the old saying: "When all you have is a hammer, everything looks like a nail." By having the right tools on hand, you won't be tempted to use a hacksaw as a file, or whatever.

You won't need metric tools. U.S. measurements rule in homebuilding. (Please pardon the pun.) There might be one or two Canadian kits using metric, but I'm not aware of any.

Wrenches

Have at least one set of *combination* (open end/box end) wrenches, in sizes from 1/4 inch to at least 7/8 inch. The box end should be at a slight angle to the handle of the wrench. Acceptable-quality sets can be found for $20 or less.

The *adjustable* wrench (crescent wrench) is occasionally useful, but should be limited to those situations where a regular wrench doesn't fit or isn't available. Get one small one, and consider picking up a really big one eventually.

You'll eventually need a *torque* wrench. There are two types. The first has a dial indicator that indicates the torque level. The second is the breakaway type, where the desired torque is selected, and the tool clicks when that level is reached. This type is preferred, of course, because you don't have to keep your eyes on the dial and have less chance of overshooting the desired amount.

Torque wrenches start around $25 for the dial type, but check around; you probably can borrow one from a friend.

Socket Sets

The minimum is a *3/8-inch drive* (the little square hole in the socket is 3/8 of an inch across) set, with a *ratchet* wrench and sockets the same size range as the wrenches. The smaller sockets are probably *1/4-inch drive*, so an adaptor should be part of the set.

An easier way would be to pick up a 1/4-inch drive set as well. Its ratchet wrench will be smaller and easier to swing in tight corners.

Eventually, you might need a *1/2-inch drive*. As you might expect, this size is more expensive. Count on picking up the individual components as you need them, rather than buying a full set.

Screwdrivers

Screwdrivers come in two flavors, *standard* and *Phillips*. The standard type is the conventional flat blade, while Phillips is the cross-shaped model. Standard screwdrivers come in combinations of length and blade size; get a variety. One of those items a builder always needs eventually is a short wide-bladed standard screwdriver, so be sure to include one on your list.

Phillips screwdrivers come in three sizes. Get one of each to start; by-and-by, you'll probably need a couple extra in the two larger sizes (#2 and #3).

Rechargeable electric screwdrivers are cheap enough, but I use mine around the house more than on the airplane.

Pliers

Another tool with limited use on aircraft is pliers. Yet, when you need one, there's no substitute. Keep one around and dust it occasionally.

Locking pliers (*Vise Grips*) are more useful. Pick up one or two, they're pretty cheap.

The main problem with most pliers is that they are designed to grip; they don't care what damage they do to the material in the jaws. You don't want to use a pliers on an aircraft part because it is going to chew up the part or leave marks. Slipping a sacrificial piece of wood between the jaws and the part is an option.

A variation of the pliers is *side cutters*, or *dykes*. Get a medium sized pair for cutting safety wire and electrical cable.

Snips

You'll need a pair of tin snips to cut thin aluminum. The aviation style are comfortable and easy to use, and aren't all that expensive. They come in *straight-cut*, *left*, and *right* varieties. If you're just buying one pair, get the straight. It'll cut curves, but not as tightly as either of the turning varieties. If you're cutting a lot of sheet aluminum with curves, by all means buy a left-cut and/or right-cut model.

If you've got a couple of extra bucks, buy *offset* snips. These are shaped more like a set of garden clippers; both handles are on the same side of the jaws. They reduce the cuts and nicks from the freshly-cut metal edge.

Hammer

A standard claw hammer is required, for no other reason than to nail together the worktable and workbench. A useful addition is a rubber- or plastic-tipped hammer. These can be used to tap tight-fitting parts into place without fear of marring the part or the surrounding structure.

Automatic Center Punch

The automatic center punch is used to place a small dimple on metal prior to drilling. It prevents the bit from walking.

The center punch works by a small amount of trickery. Place the point at the center of the desired hole. Push down with the palm of your hand, and the tool resists for a moment, then snaps downward.

The punch can be set to make various dimple sizes, and costs fewer than $10.

Vise

A bench vise is a necessity. They're sold by the width of the gripping area; a 4-inch model is probably about the smallest you should get. Bench vises are designed to be bolted to a tabletop. Clamp-base models can be moved easily, but they're more expensive, and larger sizes are rare. Forget the vacuum-clamp models. They just won't stick well to a wooden tabletop.

If portability is a requirement, get a standard bench model, place it on the corner of the table, and use C-clamps to hold it.

Clamps

Buy two types: *C-clamp* and *spring*. The first is the conventional C-clamp. Get a number of them in assorted sizes. You'll never have enough, anyway. I'm just scrimping by with six, and I'd definitely need more if I were building almost any other type of aircraft.

The other is the spring clamp; the kind that looks like enormous clothespins. Again, the number you'll need depends upon the way your airplane is built. These are sized based upon approximate opening width. Buy a few, each between one and three inches, and eventually get more depending upon which you use the most.

Saws

Numerous handsaws exist, but there's one you'll need for sure: a *hacksaw*. Most take 10- or 12-inch blades. Twelve-inch blades give the longest wear and fastest cut, but 10-inch blades are easier to use in tight quarters. Pick a hacksaw with long blade-mounting posts because the blade's less likely to pop off under pressure. Remember, when mounting the blades, the teeth point away from the handle, towards the end of the hacksaw. The saw cuts on the forward stroke.

Buy other types of handsaws as the plans call for, or upon recommendation by another builder.

Hand-held *power* saws come in two basic flavors. The first is the *jig* or *saber* saw. These are useful for cutting a variety of material, from wood to aluminum. They generally cost around $40 or so.

The second type is the *circular* saw. While not used for cutting aircraft materials, it makes workbench and worktable-building a breeze.

If you can afford only a single floor-model tool, buy a *bandsaw*. A bandsaw drives a continuous-loop blade through a small slot in a steel worktable. It's the best all-around tool for any power cutting. While hand tools like hacksaws or saber saws can be used, the bandsaw gives the easiest and most exact cuts. Saber saws drive their blade back and forth; this jiggles the piece being cut and requires a steadying hand. All while trying to steer an awkward saw along a precise path.

Bandsaws, on the other hand, run the blade in one direction: down through the worktable. The piece stays steady, and both hands can be used to guide the cut. Bandsaws can cut almost any material, if the right blade and speed are selected. Bandsaw speed is stated in blade feet per minute. A speed of 1,000 fpm is needed for wood. Aluminum can be cut at the same speed, or preferably slower, but steel requires no more than 150 feet per minute.

Bandsaw prices start at $150. Blades cost around $5 – 10. One option to consider is the combination horizontal-vertical bandsaw. These can be pivoted downward for precise cutoffs of lumber or tubing. They feature removable worktables for regular cutting. Their main drawback is their working height is close to the ground; you'll need a low stool when using it as a regular bandsaw.

Table saws use circular saw blades in a fixed mounting. They're more important on plans-built airplanes than kitplanes because the kitplane should include precut materials.

Drills

As an utter minimum, you'll need a hand-held *power* drill with at least a 1/4" chuck (can grip and use bits 1/4" diameter and smaller). Reversability is nice, but variable speed helps start holes in metal. I keep a cordless drill around for tight spots, but the cord models have more power.

An *air-powered* drill is a possibility, especially if you're building an RV or other riveted metal airplane. Air lines are more awkward to move around than electrical cords, though.

Like a bandsaw, a *drill press* is vital. And they're not that much more expensive than hand-held drills, small tabletop units can be found for about $80. Hand drills can't be held steady enough; the drill wavers slightly, and the hole gets oblong. They're well suited for drilling or cutting large diameter holes, and even the cheapest models have at least a 1/2-inch chuck. Drill presses let you position the hole very carefully, too.

Tabletop drill presses (Fig. 4-21) can't make very deep cuts. Not only is the chuck-to-worktable space limited, the very length of the drill bit cuts down on the available room. *Floor models* are more flexible because they allow the work to be properly positioned no matter its thickness, or the length of the bit.

A tabletop drill press worktable is clamped in place, while a floor model can be cranked up and down. The crank allows more accurate placement, and more importantly, repeated placement. After the tabletop worktable is swung away, it's hard to reposition it exactly in the same spot under the drill point.

The usual next step is to buy a nice set of drill *bits*, typically to 1/4-inch diameter. However, it's unlikely you'll need all the values in between, and you'll

Fig. 4-21. A floor drill press is better, but a bench model is adequate for many kitplanes. Note the grinder and bandsaw.

undoubtedly need more of the ¹/8, ³/16, and ¹/4 inch sizes. So don't buy a huge set of bits. Get a set of eight, from ¹/16 to ¹/4 inch with ¹/16-inch increments. Replace them as necessary. In fact, buy extras of oft-used sizes.

Sizing drill bits in fractions of an inch is a concession to the home hobbyist. Machinists use a different system, assigning a number or letter to given sizes, as shown in Table 4-1. If the plans call for using a #30 bit, use that size. There are good reasons not to specify the nearly-identical ¹/8-inch bit. These numbered bits can be found at well-equipped hardware stores or at professional outlets.

Reams are special drill bits. Conventional bits are designed to quickly open a hole of approximately a given diameter. Reams concentrate on exactly the correct size. They won't enlarge a ¹/8-inch pilot hole to ¹/4 inch, but if you drill a hole to ⁷/32 inch and then use a ¹/4-inch ream, the hole will be perfectly circular with exactly the indicated diameter.

Files

Files are cheap; stock up on an assortment. Files are described by their length, cross-sectional shape, cut, and grade. Length is obvious. The cross-sectional shape can vary from flat, to half-round, to round, and other geometrics. *Mill* and *flat* files are almost identical, but the mill file has one smooth edge. The smooth edge protects the opposite side when filing in tight places.

Table 4-1. Decimal equivalents of drill sizes.

Most builders are familiar with drill bits available in fractional sizes 1/8 inch, 11/16 inch, and so on. However, the fractionals are just a subset of the bits available for precision drilling. Here are the decimal equivalents of drill bits from 1/16" to 1/4".

Size	Decimal Equivalent	Size	Decimal Equivalent	Size	Decimal Equivalent
1/16	.0625	32	.1160	3/16	.1875
52	.0635	31	.1200	12	.1890
51	.0670	1/8	.1250	11	.1910
50	.0700	30	.1285	10	.1935
49	.0730	29	.1360	9	.1960
48	.0760	28	.1405	8	.1990
5/64	.0781	9/64	.1406	7	.2010
47	.0785	27	.1440	13/64	.2031
46	.0810	26	.1470	6	.2040
45	.0820	25	.1495	5	.2055
44	.0860	24	.1520	4	.2090
43	.0890	23	.1540	3	.2130
42	.0935	5/32	.1562	7/32	.2187
3/32	.0937	22	.1570	2	.2210
41	.0960	21	.1590	1	.2280
40	.0980	20	.1610	A	.2340
39	.0995	19	.1660	15/64	.2344
38	.1015	18	.1695	B	.2380
37	.1040	11/64	.1719	C	.2420
36	.1065	17	.1730	D	.2460
7/64	.1094	16	.1770	1/4	.2500
35	.1100	15	.1800	E	.2500
34	.1110	14	.1820		
33	.1130	13	.1850		

The *cut* describes whether the file's teeth are long and parallel or small and diamond-shaped: *single cut* and *double cut*, respectively; the double cut is more effective at removing metal.

Finally, the grade determines the true roughness of a file. There are four grades: *smooth, second cut, bastard cut,* and *coarse.* The coarse file removes a lot of metal quickly, while the smooth cut would be used for finish filing.

However, the grade doesn't tell the whole story. The grade is for comparison between files *of the same length.* A 12-inch second-cut file will be rougher than a 6-inch model of the same grade.

If you're facing a ragged edge of steel, the best file would be a 12-inch double-cut coarse or bastard file. For taking the last little notches out of a piece of aluminum, select a small single-cut smooth one.

Now you see the importance of picking up a variety of files. The *half-rounds* are the most useful, followed by the *flat* files.

You'll also need other aids to smoothing wood and metal, such as *sandpaper* and *emery cloth.*

Bench Grinder

Another cheap power tool. Hand filing is slow, especially when working with steel. The bench grinder does the job much faster.

Rulers and References

A variety of rulers will be necessary, including devices that can measure angles and levelness. A wide variety is easy to get because each is pretty cheap to buy.

A 12-Foot Tape Measure. You'll notice the little metal tip appears loose. Don't try to tighten it. It slides inward to compensate for the width of the tip during inside measurements.

A Metal Ruler, 12 Inches Long, Calibrated to 1/32 of an Inch. Buy a *combination square* and you'll get both the ruler and a handy reference tool.

A Micrometer or Vernier Slide Caliper. These take extremely accurate measurements, depending upon the skill of the operator. Study the instructions and learn how to read the tool quickly and correctly. Slide calipers come in either decimal or fraction varieties (reads in .001 inch increments or 1/128 inch). I lean toward the decimal variety. Buy one than can take up to one-inch measurements. Good-quality micrometers and vernier calipers are expensive, so plan on spending at least $20.

A Carpenter's Square. These are nothing more than a big L-shaped piece of steel with ruler markings on all edges. They are used to help make good 90-degree angles. You'll need at least one. They are pretty cheap.

Clecos and Cleco Pliers

Clecos are small spring-loaded temporary fasteners with prongs sticking out one end, and a button on top of the other. Insert a cleco in the *cleco pliers*, squeeze the pliers and the prongs get closer together. Insert the prongs into a hole, release pressure on the pliers, and the prongs spread to hold the cleco fast.

If you're trying to bolt two pieces of metal together, you can drill one hole through both pieces, insert a cleco, then drill a second hole and insert another cleco. From that point on, each additional hole will maintain proper alignment. Figure 4-22 shows an example of their use and a further description is given in chapter 5.

Clecos are color-coded depending upon fractional inch size. The code is:

3/32:	Silver
1/8:	Copper
5/32:	Black
3/16:	Brass
7/32:	Silver
1/4:	Copper
9/32:	Black
5/16:	Brass

Fig. 4-22. Clecos in action. Note the cleco pliers on the floor.

Note that the color code repeats after ³/16 inch. You shouldn't have any trouble telling a ¹/8-inch copper-colored cleco from a ¹/4-inch.

Cleco pliers sell for about $8; clecos for around 50 cents each. I've bought used ones for as little as 18 cents. It pays to find the lowest cleco prices, especially if building a metal airplane because you might need hundreds.

To begin, just buy a few. Two dozen each, for example, of the ¹/8-inch and ³/16-inch models.

Safety Items

No matter what type of airplane you're building, it involves working with fast-turning machinery and/or toxic chemicals.

The first line of defense is *safety goggles*. Using any sort of power tool puts you at risk to eye damage because they all fling small particles of material. You need two eyes to fly an airplane. Wouldn't you feel real silly spending three years building your dream plane, yet losing your medical for not wearing $3 worth of protection?

Don't count on being safe because you already wear normal eyeglasses. They offer little protection in other directions. Get safety goggles, instead. Keep them in a convenient place; and get used to wearing them.

I did not wear safety glasses. One day I was drilling plastic and a piece about 1/16 inch across flew up and stuck in one eye. A careful bit of tweezer work removed it, with no permanent damage. I wear goggles faithfully now, and so should you.

A set of ear plugs isn't a bad idea, especially when riveting. The earmuff type can be found for about $10.

One thing for sure about working with homebuilts: The toxic chemical risk is not going to get any better. Many consider the biggest advantage of metal airplanes over composites is that metal aircraft construction avoids hazardous materials like epoxies.

But zinc-chromate primer, the most popular corrosion preventative, causes cancer. And most, if not all, aluminum on a homebuilt should be treated before assembly.

In any case, there's no such thing as a safe paint. If you plan on painting the aircraft yourself, protection is vital. At the very minimum, wear a mask specifically designed for spray-painting use. Wear an old coat, gloves, and hat, as well.

Chemical problems also come in the hyperallergic department. Back when VariEzes were under construction, a few builders would work contentedly for months without problems. Then, one day, they'd arrive in the workshop and their skin would break out into an itchy rash.

Regular exposure to the epoxy had resulted in an allergic reaction. It's permanent; it never gets better, it never goes away. Not everyone is susceptible. But if it hits, forget completing the aircraft.

Modern epoxies drastically reduce this problem. But it can still happen. Don't touch the chemicals and don't apply epoxy with your bare hands. Buy rubber or plastic gloves, don't work in bare sleeves. Incidental contact shouldn't cause a reaction, so don't worry about a few drops here and there. Several how-to articles show people applying epoxy or related mixtures barehanded, and that's bad, bad, bad.

Finally, toxic dust can be produced when working with certain composite materials. A simple dust mask will suffice when cutting foam, or sanding a completed structure. But more elaborate protection is needed when painting.

The best advice I can give is *read the label*. Don't assume the company lawyers forced Chicken Little statements down the product manufacturer's throat. Follow the safety precautions.

While there are a number of smaller items you'll need, the above tools, summarized in Table 4-2, will suffice to start. Again, this is the *basic* list. More tools, and possibly more expensive tools, shall be necessary, depending upon what type of aircraft is being built. The individual construction chapters include an applicable list of additional tools.

Table 4-2. Basic tool list.

Tool	Approximate* Cost	Comments
Hand Drill	$ 40	Cord-type preferred
Bench drill press	100	
Floor drill press	250	Optional but preferred
Tabletop bandsaw	150	
Floor bandsaw	250	Optional but preferred
Jig (saber) saw	40	Get variable speed model
Bench grinder	75	
Combination Wrench Set	20	
Crescent Wrench	10	
Torque Wrench	25	Borrow one
3/8″ Socket Set	20	
Screwdriver Set	10	
Pliers	5	
Snips	10	
Hammer	5	
Automatic Center Punch	10	
Files (set)	15	
C-Clamps	20	Some A/C need much more
Spring clamps	20	Some A/C need much more
Tape measure	5	
Micrometer	20	
Carpenter's square	15	
Combination square	10	
Vise	30	
Cleco pliers	10	
Set of various clecos	25	Some A/C need much more
Safety glasses	5	
Dust Masks (box of 50)	10	
Respirators w/filter	15	

*Your actual cost might vary by plus or minus 20 percent, depending upon local prices and sales.

OTHER SHOP ITEMS

There are a number of items that fall into the netherland between tools and aircraft parts. The best description for them might be *shop supplies*.

When building the airplane on the worktable, the parts must occasionally be temporarily held in place. Hence, a supply of scrap 1 × 2 or 2 × 4 lumber is a good start. The 1 × 2 is commonly called a *furring strip*. Set aside a convenient location to store these scraps. Cut as necessary, then reuse them. Similarly, you'll need nails to hold the pieces in place. *Six-* or *eight-penny (6d or 8d)* nails are about right.

Scrap plywood comes in handy. You can build forms for bending tubing, or make templates out of it.

In fact, much of the preceding should be left over from building your work-table and workbench. As you build, you'll find it comes in handy, buy more as needed.

Hopefully, sufficient nuts, washers, and bolts came with your kit. Even so, pick up some extras. They always roll away into the dark corners, or a different size must be installed for some reason. They aren't that expensive, and it beats stopping construction on a Saturday evening because you've run out of bolts. Buy them from an aircraft supply outlet, though, not the corner lumber store. Chapter 5 explains the differences between aircraft-quality and commercial hardware.

One handy thing from the hardware store is threaded rod, sometimes called *readi-bolt*. It's like a very long (typically 3 feet) bolt without a head. Combined with a couple of blocks of wood or some scrap aluminum, it can be used to apply significant tension during the building process. Pick up a length of 3/16-inch readi-bolt, just to have around.

If you can find a low-cost source of cheap aluminum, pick up some .016 or .025 sheet for making patterns and templates. This doesn't have to be aircraft-grade aluminum because it is not going to be installed on the aircraft. You'll use it to make templates, or to practice a particular technique on cheaper materials. Similarly, various sizes of aluminum angles are useful.

Hardware stores sell nonaircraft aluminum, but their prices are generally equal to what the homebuilder's catalogs charge. Check in the Yellow Pages under surplus and call around. Often, the catalog companies offer a grab bag of aluminum parts.

If you do end up with some nonaircraft grade aluminum, make sure it doesn't end up incorporated into the aircraft. Spray it with orange paint, scratch it with a chisel; do something to make sure it isn't mistaken for good aluminum.

Set up a scrap pile to dump your mistakes. There are two reasons for hanging onto scrap. First, there might be enough good material for another, smaller part. I've often taken tubing from the scrap pile, cut off the ruined portions, and had sufficient material for a different item. Second, the aluminum scraps are recyclable. You can get a small return when your project is finished. Selected recycling centers won't take aluminum alloy, though, so check before taking it in.

You'll also need pens, pencils and markers to draw lines, sharpeners for the pencils, and the like. Garbage cans. Brooms. Dustpans, and a dozen other small items that meet individual needs.

Paperwork Preparations

You don't have to notify the FAA about your project until it's ready for inspection. However, at that time, the inspector will want proof that *you* built the aircraft.

The first item of proof is the builder's log. These are available on the market, or you can just buy a spiral-bound notebook. Make a log entry every time you work on the aircraft. Include the date, the number of hours worked, and the tasks performed. Especially note any variation from the plans. For example, my aircraft's plans call for the installation of a tail skid instead of a tailwheel. I designed and built a steerable tailwheel assembly, and included drawings and descriptions in the log.

The second proof is pictures. Take photos of various stages of construction, including the changes. These don't have to be magazine quality photos, just keep an old camera in a drawer and take a picture every once in awhile.

One person you will want to speak to before you start is an EAA technical counselor. Have him inspect your project on a regular basis. The sooner that problems are found, a correction will be easier. The counselor's suggestions have no legal bearing, but the counselors have vast experience, and their recommendations should be followed.

READY TO BUILD

You're all set now: picked out the kit, figured out engine and other options, workshop is equipped, and tools are ready. The kit lies on the worktable like an anesthetized patient awaiting the first touch of the surgeon's blade.

But it's not enough to know where the kit's appendix is located; or for that matter, how to rivet two pieces of aluminum together. The surgeon can't just hack away to the target, nor can the homebuilder be blind to the other requirements of aircraft manufacture.

There exists a body of standards, that together are defined as *aircraft-quality workmanship*. Before the individual construction methods are detailed, let's look at some of the background details of aircraft construction.

5

Basics

THE INTRODUCTION OF KITPLANES changed the homebuilt industry forever.

Not totally for the better, though.

In the early days of homebuilding, $20 would buy a 10-page plan set. The average person couldn't even attempt to start a project of this size on his own with such limited information. Prospective builders had to work closely with an experienced mechanic or builder to complete the project. The builder would learn craftsmanship, not just how to build an airplane.

Nowadays, kits arrive with prefab parts. The pieces are cut to size and holes computer-drilled. All you add is the battery and the paint. Construction manuals are marvels of exactitude. Anyone who can keep a lawn mower running can build an airplane with little or no outside help.

But there's one crucial difference between keeping the Lawn-Boy spinning versus building an airplane: aircraft-quality workmanship. You can't drill just any hole when building an airplane, and you can't make parts out of just any material. In earlier days, the "old hands" taught the basic practices. Violations of accepted practice are subtle and they might not be detected once the part is installed.

New builders typically worry about whether they'll pick up the skills necessary to construct the aircraft's structure. The tasks might be complicated, but rarely difficult. And in most cases, kitplanes are designed to be tolerant.

Homebuilts are strong. Most if not all in-flight structural failures are due to either serious deviations from the plans or overstress during aerobatics. If you take reasonable care, your airplane won't break up.

However, that's not the same thing as saying the airplane is safe to fly. Poor workmanship can kill you just as fast, through jammed controls, warped surfaces, and parts that fail prematurely.

As pilots, we occasionally deride the FAA's rulebook mentality, and the minutiae contained in the FARs. However, there's one unfortunate truth: nearly every regulation came about after someone flew an airplane into the ground. Overbearing as they might occasionally be, the air regulations are a good guide to safe flying.

So it is with aircraft workmanship. Every requirement, from self-locking nuts, to safety-wiring turnbuckles, to edge margin, grew out of an aircraft crash.

The ultimate source of workmanship standards is FAA Advisory Circular AC 43.13, *Acceptable Methods, Techniques, and Practices*. It's available through many sources, including the EAA.

This chapter reviews the basics of aircraft-quality workmanship. Read, learn, and remember.

THE NUTS AND BOLTS

Bolting two pieces together is simple. Run a bolt through matching holes, place a washer over the end, then add a nut to hold the bolt in place. Just about the same as working on the car. However, there are several rules to follow.

Probably about the biggest one is the requirement to use aircraft-standard hardware. Of course, anytime the word "aircraft" is associated with anything, you know the price is going up. Why not buy bolts at the local hardware store?

Over the years, the government has established a number of standards for the construction and testing of aircraft hardware. When the kitplane plans call for the installation of an AN4-32A bolt, it means something more than a mere manufacturer's part number. The number refers to a federal standard specification; the bolt you install must meet this spec.

The major specs are:

- MS (Military Standard)
- NAS (National Aerospace Standard)
- AN (The granddaddy of them all; originally the Army-Navy standard, it's been renamed Air Force-Navy instead.)

The specifications will list the minimum and, in some cases, maximum requirements for approval. These requirements include dimensions, tolerances, strengths, and finishes.

For example, bolt specifications will list minimum strengths in both *shear* and *tension* applications. The specification guarantees the bolt can withstand the strain up to a certain load. Shear loads are those that try to make opposite ends go in different directions. A tractor trying to pull a stump out of the ground applies a shear load; if the roots are too strongly set, the stump will break just above the ground. Tension loads are those that try to stretch the bolt.

Most AN hardware is made from steel, then cadmium-plated. The cadmium gives the metal a goldish color and provides corrosion resistance. Bolts and washers are also available in aluminum. They're half as strong as the steel hardware, so do not use them unless specified.

Similar parts in the different systems meet different requirements, complicating substitution. An AN3 bolt and a NAS623-3 machine screw are similar in appearance; the machine screw's head is Phillips instead of hex. But if your plans specify the NAS bolt, don't substitute the AN variety. The National Aerospace Standard specifies a higher degree of heat treating. The NAS bolt is 33 percent stronger.

That doesn't mean you should blindly substitute NAS bolts for AN ones. There are good reasons to use AN bolts, cost and availability are big ones. A competent designer will make allowances for the type of hardware. In any case, the limiting load factor is usually set by the aircraft's structural design and material, not the hardware that bolts it together.

Because you're building an experimental aircraft, hardware-store bolts and nuts are perfectly legal. But the use of approved aircraft hardware gives you a guarantee, of sorts. You know an AN3 bolt can withstand more than 2,200 pounds of tension. At what point will the hardware store bolt break?

The man at the hardware store will give you a guarantee: "If they don't work out, just return them for a full refund." It's kind of like the old joke about parachutes: "If it doesn't open, we'll give you your money back." Hardware store bolts are made of cheap steel with limited heat treating. They're about a third as strong as approved hardware.

Bolt Basics

The standard bolts used in homebuilt aircraft are the AN3 through AN20 series. A bolt consists of two parts, the *head* and the *shank*. The head is the hexagonal portion that fits the wrench. The shank is the portion designed to slip into the hole. It's threaded at one end, with the unthreaded portion called the *grip*.

Whether a bolt is aircraft quality can often be determined by looking at the head. The usual marking is a four- or six-pointed star, possibly accompanied by a couple of letters. If the bolt has no markings on the head, it isn't an approved bolt. Typical head markings are shown in Fig. 5-1.

A typical bolt specification is *AN3-14A*:

- *AN* indicates the bolt meets the AN standards.
- *3* is the diameter of the bolt, in $1/16$ of an inch.
- *14* is the approximate shank length.
- *A* indicates the thread area isn't drilled for a cotter pin.

The *diameter* is the width of the grip. There is a small amount of allowed tolerance, so don't be surprised if one bolt fits more tightly than another.

The *length* description is rather tricky. The first digit indicates the number of whole inches; the second digit is the number of $1/8$ inches. So an AN3-14 bolt is about $1 4/8$ ($1/2$) inches long from the bottom of the head to the tip of the shank (Fig. 5-2).

The drilled/undrilled status of the threads is important depending upon how the nut will be secured. The undrilled type uses self-locking nuts. Most of your installations will use self-locking nuts, so buy undrilled bolts.

Castle nuts require drilled bolts in order to pass the cotter pin. Although there are special jigs available to make it easier, avoid drilling the bolts yourself. Not only won't your work match the exact specifications of the standard, but the new hole won't be cadmium-plated. It's acceptable for non-structural purposes.

Drilled-head bolts are available for special cases; for instance, if the bolt is installed into a *tapped hole*. Safety wire is then passed through the hole to secure

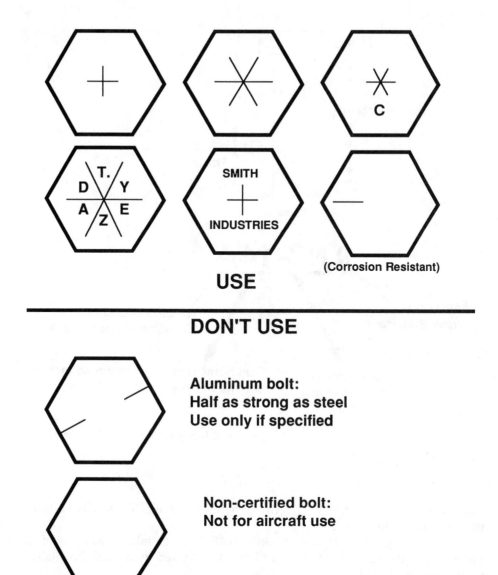

USE

(Corrosion Resistant)

DON'T USE

Aluminum bolt:
Half as strong as steel
Use only if specified

Non-certified bolt:
Not for aircraft use

(No markings on head)

Fig. 5-1. Typical bolt-head markings.

the bolt. Drilled-head bolts are identified by the addition of an H after the code for bolt diameter: AN3H-10A, for example. The AN76 series has a larger head with multiple holes.

As mentioned, the AN3 through AN20 series of bolts have some small amount of shank diameter tolerance. This amount of variance is acceptable in some applications, but not in others. For example, the designer wants to minimize slop in the wing attachments, and so might specify a close-tolerance bolt.

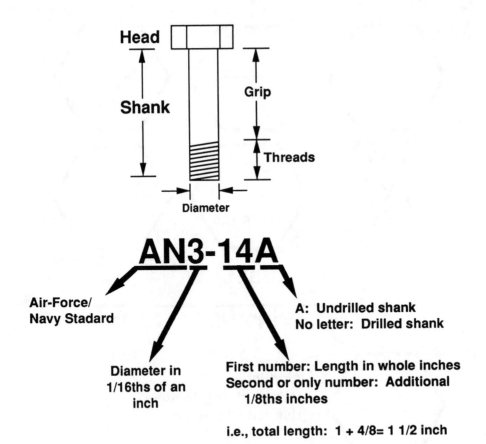

Head

Shank

Grip

Threads

Diameter

AN3-14A

**Air-Force/
Navy Stadard**

**Diameter in
1/16ths of an
inch**

**A: Undrilled shank
No letter: Drilled shank**

**First number: Length in whole inches
Second or only number: Additional
1/8ths inches**

i.e., total length: 1 + 4/8= 1 1/2 inch

Fig. 5-2. AN bolt specifications.

These are generally of the AN173 through AN186 series. These have a triangle marking on the head.

Occasionally, the designer specifies NAS bolts to take advantage of their higher strength. These high-strength bolts are marked by a small bowl-shaped depression in the bolt head.

The cadmium plating holds corrosion in check for most applications, but it isn't perfect. Over a long period, the moisture retained in wood can overcome the plating of a bolt installed through it. To make matters worse, the surface corrosion caused by the contact with the wood tends to lock bolts in place, complicating removal and inspection.

Stainless steel bolts are available, but are expensive. Cheaper alternatives are to protect the bolt (by applying zinc chromate primer) or to varnish the hole to isolate the moisture from the bolt.

Sometimes, you'll find bolts that have no grip length; the entire shank is threaded. These might be useful in lightly-loaded applications, but cannot be used in lieu of standard AN3-AN20 bolts.

Nuts

Aircraft nuts must not come off accidentally. They undergo a lot of vibration that tends to loosen plain nuts. The consequences are obvious. There are three ways to secure them: *lock washers, self-locking nuts (stop nuts),* or *external safetying*.

AN-standard lock washers are rarely used. Weight is one reason because every bolt must have a lock washer in addition to the nut and a regular washer. Longer bolts are necessary, and installation is more complex.

Self-locking nuts are the most common method. The self-locking feature is implemented in one of two ways. First, the nut might incorporate an elastic fiber collar slightly smaller than the bolt. The pressure of the fiber on the threads then keeps the nut from turning. Second, the threaded portion of the nut can be manufactured with a slight distortion, adding sufficient friction to resist vibration.

The AN365 elastic stop is the most popular self-locking nut. It is cheap, less than a dime each. They can be reused as long as the fiber still holds the nut in place. If the nut can be turned by hand, junk the nut. The AN364 is a low-profile version, but it can only be used in shear applications. Use the AN365 unless your plans say otherwise.

Both nuts have their equivalents in other standards; the AN365 is paralleled by NAS1021N and MS20365 nuts, and the MS20364 and NAS1022N nuts can be used in lieu of the AN364.

Dash numbers after the main specification are used to differentiate sizes: AN365-428. The dash number indicates the thread size. The system is rather awkward to memorize, but here's a short cut, for AN4 and larger bolts, the first digit of the dash number is the same as the AN bolt diameter: AN4 takes AN365-428, AN5 takes AN365-524, AN6 takes AN365-624, and so forth. The short cut doesn't work for AN3, so you'll just have to remember that it uses an AN365-1032 nut.

Elastic stop nuts are the most popular, but they have one drawback; the fiber insert is certified to only 250°F. This is adequate for most of the airframe, but they can't be used forward of the firewall. An all-metal stop nut is necessary.

The most common of these is the AN363. By using threaded metal fingers, the hole diameter narrows slightly at one end and applies enough friction to prevent self-rotation. The AN363 uses the same dash number scheme as the AN365, and is about 20 percent more expensive.

Self-locking nuts aren't the universal solution, though. If the bolt is intended to connect moving parts, the motion might overcome the nut's self-locking ability. An example would be a clevis attached to a control horn, hardly the place you'd want a nut to disappear from.

Castle nuts are designed for these applications. They look like conventional nuts with notches cut out of one side like a castle's ramparts. The notches pass a cotter pin through the hole in a drilled bolt. The AN310 castle nut is for all uses, or the low-profile AN320 for shear-only applications. Just to add to the confusion, the dash numbers of the AN310/320 series actually correspond to the AN bolt size the nut goes with: AN310-3 for AN3 bolt, AN320-6 for AN6 bolts. AN310/320 nuts cost about twice that of the equivalent AN365.

For the belt-and-suspenders types, there's the MS17825 self-locking castle nut. Nice concept, but you can buy 10 AN365 nuts for one of these fancy varieties.

Washers

Washers act as shims, spread the compressive forces over a wider area, and protect the surface from the rotation of the nut during installation.

The AN960 is aviation's standard washer. It comes in *regular* and *thin* (one-half the thickness of the regular washer). The light washer is used as a shim, especially when using castle nuts. Stock up on the regular variety, and get just a few thin ones.

The dash number of the AN960 washers gives the size bolt they're designed for. And guess what? It's yet *another* system.

A typical washer spec is AN960-416. The three-digit dash number gives the diameter of the center hole in $1/16$ of an inch; in this case, $4/16$ or $1/4$ inch. Light washers will have an L suffix, as in AN960-416L.

However, dash numbers of one or two digits indicate the bolt or screw size. An AN960-8 washer goes with a #8 screw, while an AN960-816 is for a $1/2$-inch bolt. The right washer for AN3 bolts is the AN960-10.

As mentioned, one function of the washer is to spread the load over a wider area. When the bolt goes through a softer material like wood, you'd like to spread the load even more. Hence the AN970 flat washer, which has at least twice the diameter of equivalent-sized AN960. There is no thin model of AN970. Just to make your day, the AN970 uses a different scheme for the dash number; the dash number on the AN970 is the same as the bolt it goes with. In other words, for an AN6 bolt, you could use either an AN960-616 or an AN970-6. But don't try an AN960-6 because it's too small.

Table 5-1 shows which nuts and washers go with which bolts.

Table 5-1. Matching bolts, nuts, and washers.

Bolt Size	Elastic Stop Nut		Metal Stop Nut	Castle Nuts		Cotter Pin
	Regular	Shear		Regular	Shear	
AN3	AN365-1032A	AN364-1032A	AN363-1032	AN310-3	AN320-3	AN380-2-2
AN4	AN365-428A	AN364-428A	AN363-428	AN310-4	AN320-4	AN380-2-2
AN5	AN365-524A	AN364-524A	AN363-524	AN310-5	AN320-5	AN380-2-2
AN6	AN365-624A	AN364-624A	AN363-624	AN310-6	AN320-6	AN380-3-3
AN7	AN365-720A	AN364-720A	AN363-720	AN310-7	AN320-7	AN380-3-3
AN8	AN365-820A	AN364-820A	AN363-820	AN310-8	AN320-8	AN380-3-3

Bolt Size	Steel Flat Washers		Wide
	Regular	Thin	
AN3	AN960-10	AN960-10L	AN970-3
AN4	AN960-416	AN960-416L	AN970-4
AN5	AN960-516	AN960-516L	AN970-5
AN6	AN960-616	AN960-616L	AN970-6
AN7	AN960-716	AN960-716L	AN970-7
AN8	AN960-816	AN960-816L	AN970-8

Bushings

Bolts are sometimes used as a pivot for a moving part. If the part moves over a broad angle, or is highly loaded, the plans will often require installation of a bearing.

But if a part is lightly loaded, any bearing is overkill. In the short term, there'd be little wrong with just passing the bolt through a hole drilled in the part. But motion causes wear. Aviation bolts are hard and strong; much more durable than aluminum or wood. So the hole begins to enlarge and distort: *ovaling*. The pivot becomes sloppy, and eventually the part doesn't move right.

To avoid this problem, a *bushing* is installed between the bolt and the part. Bushings are tubular pieces of moderately-hard metal with a hole drilled down the middle. They're pressed into a slightly-undersized hole in the part, and the pivot bolt runs through the hole in the bushing.

Typically, bushings are aluminum, bronze, or a grade of steel softer than the bolt. The bushing wears at a slower rate than aluminum or wood. But its tubular shape makes it easy to replace: when fit gets too sloppy, push out the old bushing and insert another.

There isn't really an AN standard for bushings because they aren't really under load. They just pass the loads to the components designed for them.

Other Common Hardware

Clevis pins are like bolts without thread, used solely in shear applications. A typical application is in cable shackles and forks. The most common ones are the AN393 (^{3}/$_{16}$ inch diameter) and the AN394 (^{1}/$_4$ inch diameter). A dash number divides the number of ^{1}/$_{32}$ of an inch between the head to the cotter pin hole. In other words, an AN393-39 clevis pin in ^{3}/$_{16}$ inch in diameter and ^{39}/$_{32}$ of an inch long; or 1 and ^{7}/$_{32}$ inches.

Clevis pins and castle nuts are locked by the AN380/MS24665 *cotter pin*. The diameter and length are given by a bizarre little dash-number code. One big thing to remember about cotter pins is that they cannot be reused. Don't bend them straight and reinstall them.

Anchor nuts are lock nuts that can be solidly attached to the surrounding structure by rivets or screws. With the nut fixed in place, you don't have to hold it in place with a wrench whenever the component is removed or installed. If the plans call for an anchor nut, don't use the regular variety. You'll hate yourself if the part has to be removed later.

Typical Installation

Let's assume we're bolting a piece of ^{1}/$_8$-inch angle to a 1-inch diameter steel tube. A good set of plans wouldn't leave us in suspense, they'd say "bolt the angle to the tube using an AN3-14A bolt, an AN365-1032 stop nut, and an AN960-10 washer."

More typically, they'd say, "bolt the pieces together with an AN3 bolt," or

". . .with a 3/16 bolt." Again, the best kits reduce your building time by telling you exactly what to do.

Let's operate on the assumption that the bolt diameter is the only thing specified. Selecting the right nut should be easy. We know the size of the bolt. If the bolt were meant to act as a pivot, we'd know to use a castle nut and cotter pin. Instead, the simpler self-locking nut will do. Because the assembly doesn't mount near the engine, the cheap elastic stop nut is acceptable. Should you use the AN364 shear-only style or the AN365 full-tension one? Without any guidance from the plans, better use the AN365.

Because neither of the items being joined are made of wood, we won't need the large diameter AN970 washer. That leaves the AN960-10 washer; for right now, let's assume we won't need the thin variety.

Which leaves the bolt. Because we're using a self-locking nut, whatever bolt we select should have the "A" (undrilled) suffix. The bolt's grip length must be approximately equal to the total thickness of the pieces being joined. Selecting the right bolt is awkward, as the AN system refers to total length, not grip length. On AN3 bolts, the threaded area is about 7/16 of an inch long. For AN4, the threads are 15/32 inch, while AN5 and AN6 is about 1/2 inch.

The threaded area develops the bolt's full strength in tension, but is not designed for shear loading. So the threaded length of the shank should start just where the bolt emerges from the material. A little bit (say, one full turn of threads or so) can still be inside. But you don't want too much coming out, either. If the bolt is too long, the nut will reach the end of the threads before it applies pressure to the work. Up to three washers can be used to compensate for overlong bolts, but a shorter bolt would be lighter.

Sometimes, when all the bolts are too long you might be tempted to use a die to add more threads. Don't. The AN specification states exactly the characteristics of the threads, and your home-threading attempt can't match it. In addition, cutting new threads removes the cadmium plating, giving corrosion a place to start.

The absolute limits to bolt length are no more than one turn of the threads should be left within the hole and no more than three washers can be used to allow the bolt to apply its proper tension.

Select the correct grip length one of two ways: intellectual or practical. By adding the thicknesses of the material together, you derive the proper grip length. In other words, a 1/8-inch thickness angle and a 1-inch diameter steel tube needs a bolt with a 1- and-1/8-inch grip length. Look at Table 5-2 and you'll see that means an AN3-14A bolt.

Or, hold the pieces together, take a wire or piece of wood, and slide it into the hole to measure its depth. Then measure the length or compare it to several bolts until the correct one is found.

Slide the bolt into the hole. It's strongly preferred that the head faces up or forward, so if the nut comes off, gravity or the slipstream will hold the bolt in place. But if there's some reason you can't install the bolt that way, don't worry about it.

Table 5-2. Approximate grip length for various bolt sizes.

Dash Number	AN3	AN4	AN5	AN6	AN7	AN8
3	1/16	1/16	-	-	-	-
4	1/8	1/16	1/16	-	-	-
5	1/4	3/16	3/16	1/16	1/16	-
6	3/8	5/16	5/16	3/16	3/16	1/16
7	1/2	7/16	7/16	5/16	5/16	3/16
10	5/8	9/16	9/16	7/16	7/16	5/16
11	3/4	11/16	11/16	9/16	9/16	7/16
12	7/8	13/16	13/16	11/16	11/16	9/16
13	1	15/16	15/16	13/16	13/16	11/16
14	1 1/8	1 1/16	1 1/16	15/16	15/16	13/16
15	1 1/4	1 3/16	1 3/16	1 1/16	1 1/16	15/16
16	1 3/8	1 5/16	1 5/16	1 3/16	1 3/16	1 1/16
17	1 1/2	1 7/16	1 7/16	1 5/16	1 5/16	1 3/16
20	1 5/8	1 9/16	1 9/16	1 7/16	1 7/16	1 5/16
21	1 3/4	1 11/16	1 11/16	1 9/16	1 9/16	1 7/16
22	1 7/8	1 13/16	1 13/16	1 11/16	1 11/16	1 9/16
23	2	1 15/16	1 15/16	1 13/16	1 13/16	1 11/16
24	2 1/8	2 1/16	2 1/16	1 15/16	1 15/16	1 13/16
25	2 1/4	2 3/16	2 3/16	2 1/16	2 1/16	1 15/16
26	2 3/8	2 5/16	2 5/16	2 3/16	2 3/16	2 1/16
27	2 1/2	2 7/16	2 7/16	2 5/16	2 5/16	2 3/16
30	2 5/8	2 9/16	2 9/16	2 7/16	2 7/16	2 5/16

Slide the washer over the end, then start the nut by hand. The fiber locking material inside the nut should make it impossible to turn by hand after the first turn or so. If the nut can be tightened all the way by hand, throw the nut away and use another.

When the nut gets hard to turn, take a combination wrench and socket wrench (3/8 inch for AN3 bolts) and tighten it the rest of the way. You'd prefer to use the socket on the nut and the combination wrench on the head. Otherwise, the cadmium gets scraped off the bottom of the bolt head. But do it the other way if necessary; space constraints might dictate the situation.

How tight? AC 43-13 gives the numbers, but few homebuilders use torque wrenches for the ordinary bolts. One guy told me, "Tighten it until the head starts to twist off, then back off half a turn." It took me a couple of months to realize he was kidding.

Anyway, massive amounts of torque aren't necessary. The maximum torque of the AN3 bolt is 25 inch-pounds. If you're using a ratchet wrench with a 6-inch handle, that's only 4 *pounds* (4 pounds × 6 inches = 24 inch-pounds) of pressure on the end of the handle. AN4 bolts specify a maximum of 11 pounds at the end of the wrench; AN5 gets all the way to 23. Less if your wrench is longer.

Tighten them up until they're snug, and don't apply insane amounts of torque. Use a torque wrench occasionally and compare the readings with the figures given in AC 43-13.

Once the bolt is in place, look at the end of the nut. You should see at least one thread projecting past the fiber. If not, the bolt is too short and should be replaced with a longer one.

If there are more than three threads showing, the bolt might be too long. Try to rotate the bolt with the wrench. If it turns, the nut bottomed out before proper pressure could be applied. Remove the nut and add another washer. Retighten, recheck, and repeat if necessary. If the bolt still isn't tight in the hole with three washers on it, replace with a shorter bolt.

If installing a drilled bolt and using a castle nut, the nut must be in a position to allow the cotter pin to pass through the drilled hole in the bolt. The thinner

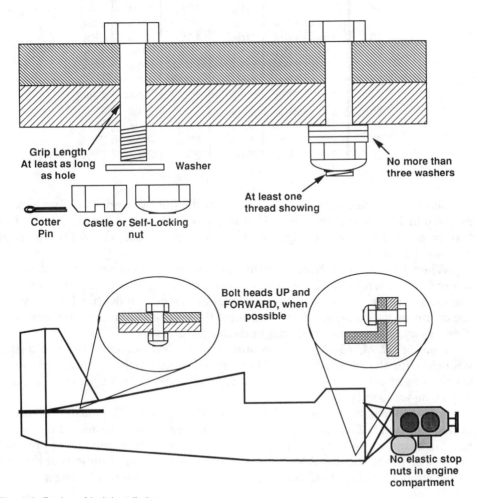

Fig. 5-3. Basics of bolt installation.

washer (AN960-10L for AN3) can be used to ensure both proper torque and positioning. Bend both prongs of the cotter pin after insertion. Remember, never reuse one.

Let's summarize the nuts-and-bolts of aircraft-quality workmanship:

- Either a self-locking nut or a castle nut/cotter pin combination must be used (Fig. 5-3).
- No self-locking nuts on moving components.
- No more than three washers can be used per bolt.
- Little or no portion of the bolt's threads should remain inside the drilled hole.
- If possible, the bolt's head shouldn't be mounted down or backwards. This is a preference; not a hard-and-fast rule. There'll be instances where it isn't possible.
- Use aircraft standard hardware in all structural applications. Commercial-grade hardware is suited for non-structural duties like attaching placards, upholstery, etc.

The preceding won't fit every situation that comes up while building your airplane, but should give you a good start. Ask a local A&P or your FAA technical counselor for more information.

BASIC METAL CONSTRUCTION

Don't go skimming by this section just because you're building a composite airplane. You'll be making a lot of fixtures from aluminum and steel, and you should know the basic rules of metal construction. Let's start off by looking at the metals themselves.

Aluminum Alloys

Pure aluminum is great stuff, but must be alloyed with other metals for maximum strength. The type of alloy is given by a four-digit number, like 2024 or 6061. The first digit identifies the major alloying elements. Alloy 6061, for example, consists of aluminum, magnesium, and silicon. The next digit indicates major modifications in the basic process, such as changes in percentages. The last two are essentially serial numbers for an exact alloy.

The alloy number is always followed by a dash, and a *temper designation*, like 6061-T3. Temper designations range from zero (-0) to -T9, with the first digit after the T indicating the actual temper. Other numbers after this first digit are modifiers that don't really concern us. The temper indicates the degree of workability of the metal. T3 might crack if excessively worked, but you can fold a zero-temper sheet in half and it probably won't break. However, it's only half as strong as -T3 in the same alloy.

Alloy 2024 is the most commonly seen. Other popular ones on homebuilts include 6061 and 7075. What's the difference? Strength, mostly. Of the common aircraft structural alloys, 7075 is the strongest, 6061 is the weakest. Why not use 7075 exclusively? Well, the extra strength comes at the expense of poor corrosion resistance and brittleness; 7075 is great for flat plates away from moisture, but has a higher tendency to crack if flexed; 2024 is a good compromise, not quite as strong as 7075, but less brittle.

It's still susceptible to corrosion, though. Pure aluminum won't corrode, so aircraft grade 2024 and 7075 alloys are often clad with a thin layer of pure aluminum, hence, Alclad.

Most if not all applications using 2024 or 7075 will specify Alclad. The Alclad status, as well as the alloy and temper, will be printed on the sheet. The term *BARE* is a positive indication that the sheet is not cladded. An example of corroded bare aluminum is shown as Fig. 5-4.

Most aluminum-monocoque aircraft use 2024-T3 throughout. Aluminum-tube aircraft (Murphy Renegade and CIRCA Nieuport) use 6061-T6 tubing. Alloy 7075 is fairly rare, but often finds its way into special fittings.

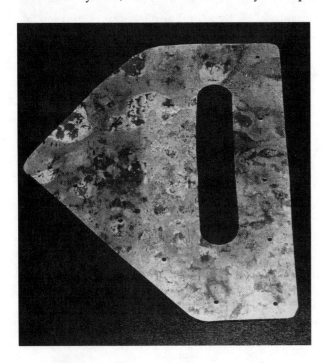

Fig. 5-4. Corroded unclad 2024 T-3 aluminum. While this piece can't be installed on an aircraft, it can be used for a template.

Steel

Steel has several advantages over aluminum. It's three times as strong, and can be easily welded into complex shapes that are far more durable than a bolted or riveted aluminum structure.

Steel is quite a bit heavier, but sometimes it's worth it. Most of the lightest homebuilts feature fuselages made from welded steel tubing.

It's a "magic metal" that we see too often to fully appreciate. The characteristics of a steel part can be tweaked by processes like normalization, annealing, quenching, or tempering to produce exactly the required performance. It's not a modern wonder-metal governed by iron-clad (or steel-clad) patents. Quenching, for example, increases strength and hardness when the red-hot metal is suddenly plunged into a cooler liquid. A thousand years ago, swords made of Damascus steel were quenched in blood.

Shops specializing in more modern steel treatments are found in most cities. If you need some sort of special work done, you shouldn't have trouble finding a place to do it. But it's a rare kitplane that needs any steel work, especially custom treatment.

Like aluminum, steel alloys are identified by a four-digit code that specifies the alloying components. Unlike aluminum, one steel alloy predominates for aircraft: 4130, which incorporates chromium and molybdenum, called *chromoly* steel, for short.

Alloy 4130 is usually sold *normalized*, that is, it has been heated to its *critical temperature* and allowed to cool at room temperature. It's required after any sort of heat treating (including welding) to equalize stresses and maximize the hardness.

Occasionally, you might find 4130 in the annealed condition. Don't use it on your aircraft without additional treatment. It's soft and easily formed, but has no real strength. You can use it to build complex parts, but the parts must be heat treated afterwards. Look in the Yellow Pages under "Heat Treatment – Metal."

The steel alloy and condition is printed directly onto the metal. The condition is often abbreviated: 4130 N for normalized; 4130 A for annealed.

Other types of steels are *stainless* steels. These are rarely used for structural purposes. However, aircraft firewalls must be able to withstand a 1,200°F flame for five minutes, and stainless steel is one of the few materials that can take it. Alloy 304 is commonly used for firewalls.

The preceding discussion of aluminum and steel alloys and hardening is simplified, the matter is actually a bit more complex. For example, aluminum tempers aren't always single digits; further numbers indicate modifications. For example, a piece of aluminum angle might be labeled 6061-T652; indicating the alloy was compressed to produce a permanent set as part of the heat-treating process. As far as the average homebuilder is concerned, only the first digit of the temper is of interest. Greater detail is given in AC 43.13.

Buying Aluminum and Steel

Hopefully, every speck of metal necessary came with the kit. In most cases, however, this is the exact quantity the aircraft requires. There's no obligation to supply extra to allow for occasional flubs. And you're going to make mistakes, especially at the beginning.

One kit manufacturer supplies templates for the most economical cutting of sheet aluminum for the aircraft. Unfortunately, several cuts are shared between two pieces. One little mistake over a cut several feet long, and both pieces are ruined.

It is best to develop an additional supply source. The kit manufacturer will sell you additional steel or aluminum, but if the part is made from *sheet stock* (raw sheet metal), you can probably get it, faster and cheaper, other ways.

One way is by mail order through catalog companies like Aircraft Spruce and Specialty or Wicks Aircraft Supply. Homebuilders are their primary customers, so you can buy small or large quantities.

But for large orders where you can't wait, check the Yellow Pages under "Aluminum" or "Steel Distributors." There's nothing specifically aircrafty about 2024 aluminum or 4130 steel. As long as they can supply exactly the alloy and temper/condition required, there's nothing wrong with buying from public outlets. And you can pick it up the same day.

One glitch: these companies often have a minimum order stipulation. Locally, several local outlets won't sell less than $100 worth of metal at a time. In other words, if you need just a two-foot piece, you're out of luck. Also, their prices operate on a sliding scale—the more you buy, the less per-pound you pay.

If you need just a single sheet or so, try to combine your order with another homebuilder, to exceed the minimum and to get the lowest possible price.

An option for those in larger cities is surplus outlets. Companies that use aluminum generally buy it by the roll or sheet. Occasionally, odd pieces or the end of a roll are left over. Large companies find it more economical to sell the excess than to waste time trying to use every square inch. These companies often sell the scrap at public outlets, at scrap metal prices. A sheet that costs $30 new might sell for around $6. It's not just theory, I've made several good buys this way.

One point to remember: never buy ungraded and unmarked metal for your airplane. If you buy from a scrap dealer, make sure the piece is marked with the original manufacturer's printing. Sure, you can tell whether it's aluminum or steel. But no one can determine the alloy or temper just by looking at it. Sheet metal has the alloy and temper/condition printed in repeating blocks, so only the smallest pieces will end up unlabeled. The printing often indicates the thickness, too.

The printing also indicates the *grain* of the metal. The grain is the long axis of the metal before it's cut and rolled. This becomes important when bending the material.

Cutting

OK, you're working on your airplane and you come across an instruction to cut a shape from sheet stock. The choice between a hand or power tool will depend on the thickness of the material and the physical size of the part. Selection criteria include speed, control, cleanness of cut, and ease of use.

Let's look at cutting aluminum first.

How about a pair of scissors? Don't laugh because on aluminum that is .016 inch or less, they work quite well. Don't steal the wife's dressmaking pair because cutting aluminum quickly dulls the edge. Similarly, a paper cutter is fantastic for long, straight cuts in thin metal.

The most common tool for thin sheet aluminum is metal *shears*. The maximum thickness is limited mostly by your hand strength, about .040 inch is as thick as I care to try. Depending upon the thickness, shears make a reasonably clean cut. But they can leave some pockmarks on the edges of thicker material. Long, straight cuts are somewhat of a bother because the two edges tend to get in the way of the handles and slice fingers. *Offset snips* alleviate this problem, but they cost more.

Another hand tool is the sheet-metal *nibbler*, which punches out sections of metal. They leave beautiful edges, and tight curves can be easily negotiated. Like most hand tools, they are limited in the thickness they'll cut. I overdrove a nibbler once, and the head broke off and whizzed past my ear. One major problem is slowness because the nibbler cuts, at most, an $1/8$ or $1/4$ inch per bite, and each bite has to be carefully set up. But they work great for cutting instrument holes.

Hacksaws are pretty awkward for thin sheet, but will cut away at plate as long as you're willing to saw. Blade selection is a compromise between speed and a clean edge. I prefer clean cuts over fast ones, so I get blades with 32 teeth per inch. But it still leaves a pretty rough edge.

As mentioned in the previous chapter, a bandsaw is usually your best solution. It's really your only choice for thick aluminum. Set the speed to about 1,000 fpm, 400 fpm is even better, and use a metal-cutting blade.

Otherwise, you can opt for the lower cost of the humble *saber* or *jig* saw. They will handle aluminum to .125 inch and thicker, which is quite adequate for most homebuilts. Get fine-toothed blades for smooth cuts.

Cutting steel is a bigger problem. Most hand tools just aren't suited for anything larger than the thinnest steel. Hand tools cut steel eventually and your hands won't be worth much afterwards. A bandsaw is a necessity for steel. Use a very slow blade speed; about 120 fpm.

Your first step is to mark the shape of the part on the sheet. The most accurate method is to paint an area with blue machinists' ink, and scribe the outline of the part into the dried ink. This works great when using a bandsaw, but a saber saw's shoe (the skid plates on either side of the blade) will leave long scratches on unprotected aluminum.

Don't scribe any lines over the part itself. Scratches disrupt the anticorrosion effect of Alclad and initiate cracks.

An alternative would be to apply masking tape to the sheet and draw the pattern on the tape (Fig. 5-5). If using a saber saw, stick the tape over a wide enough area so the shoe won't contact unprotected metal. Contact paper, instead of tape, is another alternative. Or draw the shape on a piece of paper and glue it to the metal with rubber cement. The cement will easily peel off when you're done. Clean up tape residue with MEK or acetone.

If your plans include a full-scale template, make a photocopy and glue it down. This cuts down on a lot of work, as the templates generally include the locations of drilled holes, too. Again, if using a saber saw, include enough paper to keep the shoe from touching bare metal.

Because the bandsaw blade travels downward, friction should hold the piece flat on the table. Start and finish all cuts away from the outline of the part, if possible. When you've gained enough skill, you can try to shave the line closely. Otherwise, keep $1/16$ inch to $1/8$ inch clear of the line and count on trimming the

Fig. 5-5. Covering aluminum with masking tape provides a surface for drawing the part outline. The tape also protects the metal from scratches during cutting.

piece down with files. Otherwise, a minor jiggle could ruin the part. Don't try to cut corners too sharply because the blade might bind. Back off and come in at a tangent, if necessary.

Saber saws work differently. The blade goes back and forth, and if the metal isn't clamped or otherwise supported, it flaps (Fig. 5-6). Support is especially important when cutting thin metal. If you *must* use a saber saw on very thin aluminum, make a sandwich out of plywood with the metal in the middle. Use rubber cement, instead of catsup, to keep the aluminum from sliding around in the sandwich. Otherwise, use the same technique as with the bandsaw.

Smoothing the Blank

With a little experience, it takes very little time to cut the *blank* out from the sheet stock. It isn't a part yet. Feel the edges and realize that no matter what kind

Fig. 5-6. When using a saber saw, the metal must be securely clamped to the table. A full-size template of the part has been glued to the aluminum.

of tool you used, it left some burrs, ridges, and rough spots. It's vital to get the edges smooth with generous curves on inside corners. Premature failure of the part might result otherwise.

To illustrate the problem, take a cardboard rectangle about the size of a playing card. Cut a V-shaped notch in the side and bend the card from top to bottom. The part will fail at the notch, right? Obviously, because there's less material there.

But take another piece and trim one edge, leaving a sharp-edged triangle in the middle, like a mountain on a flat plain. This piece fails at the sharp angles where the plains meet the slopes. Why?

Notches concentrate the stresses induced in the part. Inside corners, or the rough edges the saw leaves, cause unequal stress distribution. By smoothing the edges and making inside corners with wide radii, you essentially confuse incipient cracks. They don't have an obvious place to start. Try the cardboard mountain again, but make the transition from plain to slope gradual. Quite a difference, huh?

To translate this example to an aircraft component, all inside curves should be gradual, and all marks left by the saw must be smoothed with files and emery cloth. Outside corners don't concentrate stress, but we'll knock the point off just for safety's sake. Besides, because those sharp corners don't contribute to the strength of the piece, they're just wasted weight (Fig. 5-7).

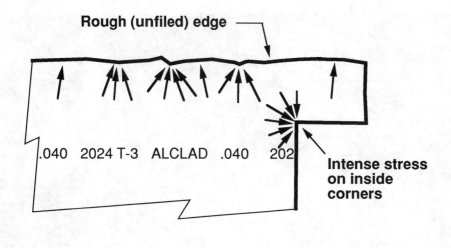

Rough (unfiled) edge

.040 2024 T-3 ALCLAD .040 202

Intense stress on inside corners

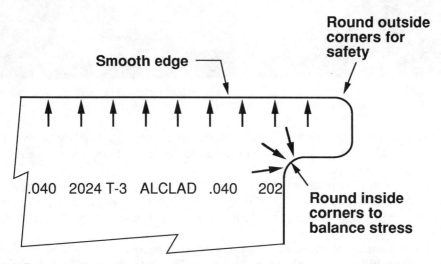

Round outside corners for safety

Smooth edge

.040 2024 T-3 ALCLAD .040 202

Round inside corners to balance stress

Fig. 5-7. Edges must be filed smooth and inside corners must be well-rounded, or premature failure might occur.

Prepare to file the blank by clamping it in a vise between two pieces of wood. The wood keeps the jaws from scarring the metal. One edge of the blank should stick just above the top of the vise, and just beyond the edge of the wood. This allows you to work on one edge and a corner.

Metal files cut only in one direction: from the handle towards the top. Don't saw back and forth with a file. Pulling the file back on the metal only dulls the file and makes machinists cringe.

The best method to smooth the edges of the blank is to draw-file, as shown in Fig. 5-8. Grasp the handle of the file in your left hand, and the top in your right hand. Stand at the end of the vise and place the file flat on the end of the blank's

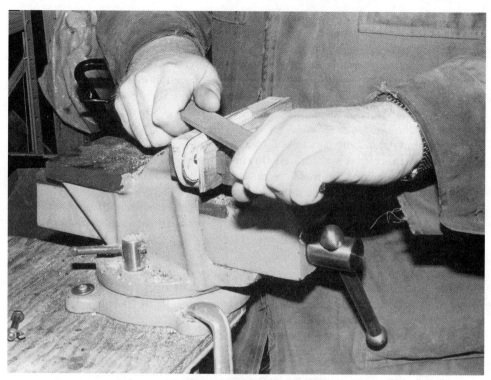

Fig. 5-8. Smooth out metal edges by placing the file as shown and pulling it towards you. The file is at a 45-degree angle to the edge and the handle end is farthest from the body.

edge farthest away. Point the top of the file slightly toward you. Apply a little pressure and pull the file along the edge. When you reach the end, lift the file, set it back, and repeat. When the saw marks are almost faded, switch to a finer file and continue.

Smooth the curves with short strokes of the file. A good file can shape curves and corners with ease. Occasionally tilt the file at 45 degrees to apply a slight bevel to the edge. Reposition the blank in the vise as necessary.

Use the fine file to remove the remaining tool marks. Remove the blank from the vise and polish the edges with small pieces of emery cloth. When finished, the edges have a shiny gleam with no tool marks, and should be silky smooth to the touch.

Those inside curves might require a half-round or rattail file. A bench grinder makes quick work of straight edges, or use a Moto-Tool with a grinder wheel for complex curves. Be careful not to eat away too much metal with power tools, though. In any case, finish up with the emery cloth.

Bending

Simple, straight-line bends are discussed here and chapter 7 includes information on the more complex bends involved in making metal aircraft.

Before undertaking a bending job, take a look at the metal extrusions available. If you're just going to make a 90-degree bend, why not see if Wicks or Aircraft Spruce carries the correct size of angle aluminum? U-shaped fittings can often be made out of U-channel, or square tubing with one side cut away. I was able to avoid making a thick aluminum U when I found the correct-size T section at a local surplus yard.

If a stock fitting isn't available, you'll have to make the bend(s) yourself. Follow three basic rules: make the bends across the grain, observe the minimum bend radius, and allow for springback.

We all understand the concept of grain as it applies to wood. The grain in metal refers to the direction it was drawn through the rollers while being formed. The grain is the long axis of the metal before it was cut and rolled. The printing (alloy, hardness, and temper) is applied along this axis.

When the metal is bent, the bend line should travel vertically through the printing. As Fig. 5-9 shows, the labeling's relationship to the crease should be like the fold in this book, not like a matchbook cover. It'll be 20 percent weaker if bent in the wrong orientation.

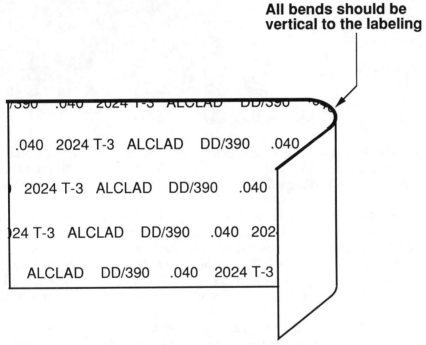

Fig. 5-9. All bends should go vertically through the metal's labeling.

Every metal has a particular amount that it can be bent before it loses strength. Going back to our analogy, take a piece of cardboard (not corrugated) and carefully roll it into a wide cylinder. Release it, and you'll find it's taken a set; it now includes a bit of a bend. But the cardboard is still whole, just as stiff as it was before. Now grab both ends and bend the cardboard until it folds. It's not stiff anymore. All the strength has gone out of it.

It works the same way with metals. Each alloy, each temper, each thickness has a minimum bend radius. Bent beyond that point, the metal might crack. The approved limits should be approached with caution, and it's better if more generous radii can be used.

Normalized 4130 steel must be bent with a minimum radius equal to three times the thickness of the material. That's pretty fair—a .080-inch thick piece (and that's thick, for steel) must use a .25-inch radius. Steel that is .090-inch thick, or less, can be bent through a 180-degree angle, as long as this minimum bend radius is followed. Thicker than .187 ($3/16$ inch), the steel can be bent no farther than 90 degrees, and between .090 and .187, 135 degrees is the limit.

Aluminum is slightly more complex. The minimum radius depends on many factors. Table 5-3 gives the values for commonly used alloys.

Table 5-3. Bend radius table for 2024 T-3 and 6061 T-6 aluminum.

Metal Thickness	2024 T3		6061 T6	
	Normal	Minimum*	Normal	Minimum*
.016	.050	.025	.016	0
.020	.064 ($1/16$)	.032	.022	.002
.025	.094 ($3/32$)	.044	.031	.006
.032	.128 ($1/8$)	.064	.048	.016
.040	.168 ($3/16$)	.088	.064 ($1/16$)	.024
.050	.227 ($1/4$)	.128	.089 ($3/32$)	.039
.063	.315 ($5/16$)	.189	.126 ($1/8$)	.063
.071	.362 ($3/8$)	.220	.150 ($5/32$)	.075
.080	.420 ($7/16$)	.257	.178 ($3/16$)	.090
.090	.486 ($1/2$)	.306	.216 ($1/4$)	.108
.125	.750 ($3/4$)	.500	.375 ($3/8$)	.188

Fractions shown are approximations of the indicated decimal value.

*Absolute minimum bend radius. Difficult to achieve without cracking or tearing the metal. Use the normal value unless tighter radius strongly justified.

One item of interest is that annealed steel, and non-tempered aluminum (2024-0, for instance) can be bent with very small radii. This is useful in some cases. However, such pieces cannot be used structurally unless they are heat-treated prior to installation. It makes sense in some cases, like building dozens of complex aluminum ribs. Make them quite easily out of 2024-0, then send them to be treated to T3. The treating process does include some warping, so you'll have to straighten them out a bit before installation.

The last factor to be concerned with is *springback*. Not be confused with the South African deer, springback refers to the simple fact that metals don't stay where you bend them. Take a piece of aluminum and bend it exactly 90 degrees. When released, it'll flex back to some lesser angle. You actually have to bend the metal past your target angle.

How you set up for the bend depends upon the material, its thickness, and the required radius. Thin, small pieces can generally be bent by hand when accuracy isn't an issue. The CIRCA Nieuport's aluminum tube fuselage uses pop rivets and .025 inch aluminum gussets to hold the longerons. One end of the guest is riveted in place, then wrapped around the longeron to hold the traverse truss members. See Fig. 5-10.

But most applications require more accuracy and better leverage. One end of the sheet or plate must be held firmly while the other end is bent over past the proper position to allow for springback.

A hand tool called a *seamer* (Fig. 5-11) makes tight curves in sheet. The tool is like a pair of pliers with very wide jaws. Clamp it across the desired bend axis, and bend the metal.

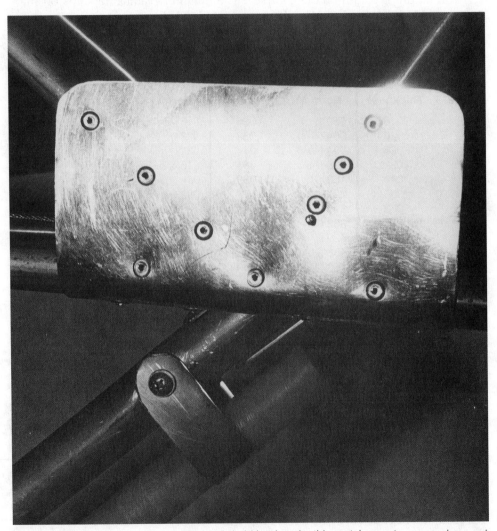

Fig. 5-10. CIRCA Nieuports' fuselage truss is held in place by thin metal gussets wrapped around the longerons and pop-riveted in place.

Fig. 5-11. The hand seamer in action.

The shop vise can be used for many pieces. Start by making a bending block out of scrap wood. The main feature of this block is a curved edge equal to or greater than the minimum bend radius you need. Whittle the wood with a jackknife, rasp, plane, or any other convenient tool. Sand the radius smooth when the approximate radius is reached.

If you're making a 90-degree bend, the radiused area will have to continue past the right angle. In other words, the top of the block will have to either be semi-circular in cross section, or beveled. This is for springback allowance.

Clamp the metal in the vise between two pieces of wood, like you did for draw-filling the edges. Only make the bending block one of the pieces, with the curved radius at the position of the bend.

There are formulas to tell you exactly how much the metal is stretched and compressed during the bend, and where exactly to place the bend line. Hopefully, the kitplane's plans will tell you. Otherwise, don't cut your blank too close to the final size. Leave significant overlap, and trim it to fit when it's bent.

Now it's time to bend the free end of the metal over the bending block. A seamer works well here, otherwise, set a scrap of wood across the end of the metal and push on it with the palm of your hands. Don't just grab the end and push it towards the vise. This results in a wide curve instead of the desired radius. Think of it as pushing the metal sideways, not down. If bending thin metal, you can put the pushing block right on top of the vise. Otherwise, you'll have to get closer to the end.

Sometimes, the metal is too thick or the bend is too awkward. It's acceptable to bash it with a plastic or rubber hammer to force it into place, but bend the free end evenly. In other words, don't try to force the left corner down and leave the right corner standing. Force one corner slightly down, slide a bit over, bend that section down, then slide farther over. Don't make any dents or crimps.

When you get the correct angle, remove the piece from the vise and trim the sides as necessary.

The preceding assumes you don't have a *bending brake*. The brake automates the previous process; about all you have to do is clamp the metal in place and lower a lever. Brakes are fairly cheap, and can be built for even less.

Drilling

Now you have a part, not just a blank. But to get useful work out of it, you'll have to drill some holes so it can be bolted or riveted in place.

Obviously, if a hole is drilled too close to the edge, the little bit of metal between the hole and the edge might tear away under stress. You must maintain an minimum *edge margin*, or distance, from the center of a hole to the nearest edge, to maintain full strength.

The standard edge margin is twice the diameter of the hole. For example, the centerpoint of a 3/16-inch (AN3) hole shouldn't be closer than 3/8 inch (2 × 3/16 inch) to any edge. This can be easily measured where only a single edge is concerned, but is complicated when you must drill a hole near several edges. By happy coincidence, the wide-diameter series of aircraft washers (AN970) has an outside diameter just slightly larger than the correct edge margin for the appropriate center hole. Align the washer so its edges just come to the part's edges, and mark the center of the hole for drilling (Fig. 5-12).

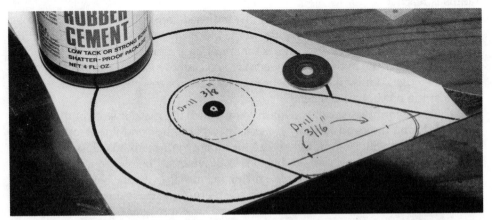

Fig. 5-12. The AN 970 series large-diameter flat washer is of slightly larger diameter than the required edge margin for its center hole. It can be used to quickly check for proper edge margin.

In a similar vein, the minimum separation between hole centers is three times the diameter. Remember, the edge margin ensures full strength, and applies to all holes: bolt holes, rivet holes, lightening holes. However, if the application isn't under high stress, you can shave the edge margin. A good example is lightening holes in plywood or metal ribs. They violate the edge margin constraints, but there's very little force trying to crush them flat. Instrument holes also need not follow the edge margin rules unless the panel is a structural member.

The edge margin parameters and a handy table are given in Fig. 5-13.

When you drill, you want the hole exactly where you place the bit. To keep the bit from skittering, use a punch to dimple the metal at the drill point. An automatic centerpunch is best because it can be worked with one hand, as shown in Fig. 5-14.

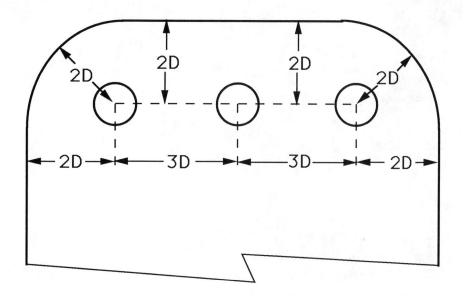

Diameter (D)	2D	3D
3/32	3/16	9/32
1/8	1/4	3/8
3/16 (AN3)	3/8	9/16
1/4 (AN4)	1/2	3/4
5/16 (AN5)	5/8	15/16
3/8 (AN6)	3/4	1 1/8

Fig. 5-13. Edge margin basics.

Fig. 5-14. Use of the automatic center punch.

But wait one minute. Where do you want that hole? Sure, the plans might show a precise location. But how precisely can you drill? If you drill two 3/16-inch holes 2 inches apart on plate A, and plate B has the holes 2^1/32 inches apart, they won't bolt together. Factories with numerically controlled machines can drill holes exactly, as can selected craftsmen. I certainly can't.

The only way to ensure the parts will match is to drill them simultaneously. Drill all the holes on plate A with a 1/8-inch bit. On plate B, drill only one hole. Cleco through this hole to the corresponding hole on plate A. As explained last chapter, clecos are little spring-loaded temporary rivets that hold the parts in alignment during drilling. You could use small bolts, but clecos are much easier to install and remove.

Position both parts, then clamp in position. Hopefully, you've been able to move the pieces to the drill press; otherwise use a hand drill. If you look through one of the holes in plate A, you'll see an undrilled section of plate B. Set the 1/8-inch bit in place, and spin the drill for a moment. This will mark the correct spot on plate B. Separate the pieces, drill 1/8 inch all the way through on plate B, then clamp them together using two clecos. You will drill the rest of the holes now, because the clecos will maintain alignment. Insert a cleco as each hole is completed.

Sometimes you will want to drill a hole to match up with a hole underneath. In this case, an inexpensive hole matcher is a lifesaver. The hole matcher is two strips of metal attached at one end; at the other end, one strip has a post the size of the holes being drilled. The other strip has a bushing with that exact inside diameter. Slip the post into the existing hole and slide the sheet to be drilled between the strips. Drill the sheet through the bushing, and the resulting hole will be lined with the existing one. Insert a cleco, and go on to the next hole.

When all holes are drilled, separate the parts. Drill holes to one size less than the final size (for instance, drill 11/64 inch for a 3/16-inch hole). Then drill one hole in each part to the final size (a matching hole, of course), cleco them together, and final-drill and cleco all of the remaining holes.

Drilling directly to the final size is sloppy because the cut bits of metal rattle around and enlarge the hole. By first drilling slightly undersized, only a slight amount of metal is actually cut during the final drilling. This ensures the hole is exactly the desired size.

What size *did* you want that hole? It's not as simple as you might think. After all, there has to be some small amount of clearance to allow the bolt or rivet to pass.

Bolts are slightly undersized . . . for instance, an AN3 bolt, nominally 3/16 inch (.1875 inch) is actually .185 inch. But rivets are exactly the indicated size, so the hole must be a bit bigger. For 3/32-inch rivets, use a #41 bit, 1/8 inch takes a #30, 5/32 inch goes to #21, and a #12 bit is used for 3/16-inch rivets.

The drilling process has likely left a ridge-shaped burr around each hole. You must eliminate these burrs for the same reason as you smoothed down the edges of the part, to eliminate stress concentrations. The best way is to take a deburring bit or countersink and twirl it by hand on each side of the hole. This cuts the ridge off and bevels the edge.

A major cause of burrs is dull drill bits. To sharpen a bit, hold it in your right hand with just the tip showing. With an underhand motion, loft it towards the nearest trash can. Then go buy another. Bits are cheap enough that you should maintain a stock of the smaller ones.

Some folks deburr a hole using a large-diameter drill bit. But the bit tends to chatter and leave small nicks, even when turned by hand. While better than not deburring at all, this method isn't recommended.

Once the holes are deburred, the part is finished, though you might want to go over the edges with emery cloth to clean up last-minute dings. A small piece of metal has been magically transformed into an aircraft-quality part.

Depending on the final use of the part, it might require corrosion-proofing and painting. Guides for this process are included in chapter 7.

AIRCRAFT CABLE AND ACCESSORIES

No matter what type of kitplane you're building, you'll have to work with aircraft cable at some point. Even composite homebuilts with *cantilever wings* and *pushrod-*

actuated controls usually have a *cable-controlled rudder*. Many designs use cables for aileron and elevator as well, and biplanes and other ragwings still use them for external and internal bracing.

Aviation cable is rarely used for a trivial purpose. It moves control surfaces, keeps the wings attached to the airplane, and maintains the rigidity of the structure. Cable failure still kills homebuilders. But today, it's rarely the cable's fault; most often, it's just plain poor workmanship. Let's look at building aircraft-quality cable assemblies.

Cable Construction

The basic unit of aircraft cable is a single wire formed of *galvanized* (zinc-coated) *steel* or *stainless steel*. The cable could consist of this single wire; if the wire were 1/8 inch thick, it'd be as strong as 1/8-inch steel rod. And that's exactly what it would be, a solid, strong, and stiff steel rod. Unfortunately for those wanting to move control surfaces, a 1/8-inch rod won't go around a pulley. It has no flexibility. We could use a smaller-diameter wire, but we'd lose strength. But use a bundle of these tiny wires, and the resulting cable is almost as strong and far more pliant. You get both strength and flexibility.

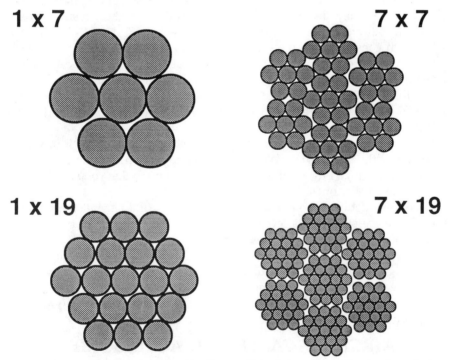

Fig. 5-15. Cable construction is defined by the number of strands and the number of wires per strand.

Aircraft cable consists of a number of wires (usually seven or 19) twisted around each other to form a strand. If more flexibility is needed for a given outside diameter, smaller wires are used in multiple strands, which, in turn, are also twisted together.

Aircraft cable is specified by diameter, construction, and material. While all three factors affect the cable's strength, diameter has the greatest effect. One-sixteenth-inch cable can withstand about 500 pounds, and its strength approximately doubles for every 1/16-inch increase in diameter.

Construction affects strength to a lesser degree. A cable's construction is described by the number of strands and the number of wires per strand (Fig. 5-15). Hence, 1 × 19 cable has one strand consisting of 19 wires, while 7 × 19 has seven strands, each of which has 19 wires. If both cables have the same outside diameter, the individual wires of the 1 × 19 cable are thicker, and some sizes can be 30 percent stronger than the same diameter in 7 × 19. But the thick wires reduce flexibility.

Single-strand cable is best for bracing and other applications that don't pass through fairleads or pulleys. The so-called *flexible cables* (7 × 7 or 7 × 19) are best for control systems. The 7 × 19 construction is sometimes called *extra-flexible*. For a given cable diameter, the wires of 7 × 19 are smaller, and hence more easily damaged. Go with 7 × 19 construction if your design uses small pulleys. Otherwise, 7 × 7 gives reasonable flexibility and wears better.

The relationship between material and strength is interesting. For single-strand cable, galvanized or stainless steel cables are equal. However, galvanized is stronger in multistrand construction.

Here's a comparison of the wire strengths:

Diameter	Material	One Strand	Seven Strands
1/16 inch	Galvanized	500 lbs.	480 lbs.
	Stainless	500	480
3/32 inch	Galvanized	1,200	920
	Stainless	1,200	920
1/8 inch	Galvanized	2,100	2,000
	Stainless	2,100	1,760
3/16 inch	Galvanized	4,700	4,200
	Stainless	4,700	3,700

Cable Selection

If you're building a kitplane, you shouldn't have to select the cable to use because it should be spelled out in the plans. But sometimes the plans just say, "Run a cable between the fitting and the rudder horn."

For certified aircraft, the FAA requires control cables to be at least 1/8 inch in diameter. This doesn't apply to kitplanes, of course. Certain kitplane manufacturers voluntarily comply. But current practice tends toward 3/32-inch cable, for a number of reasons. Attaching a fitting to the smaller cable requires only one compression with a Nicopress tool, for example. And the lighter cable is more flexible. It has sufficient strength for nonbracing applications, but the smaller size is less tolerant to damage.

Should you use galvanized or stainless cables? Galvanized cables cost half as much as stainless. But stainless is more resistant to corrosion and its shiny appearance is more attractive. When the aircraft will be stored in salty or wet climes, or the cable will be placed so that inspection is difficult, opt for stainless. Whenever galvanized cable is used, the annual inspection should include lubing with graphite grease or oil.

Cutting Aircraft Cable

Cutting galvanized or stainless cable is easy. Don't use a cutting torch because the heat will anneal and weaken the cable. It must be cut mechanically. Cable cutters are available for $15 and up, but a $3 cold chisel and a hammer work quite well. Wrap a piece of tape around the cable. Position the cable on a hard surface. An anvil works best, or the flat top of a vise. Set the chisel on the taped cable, and rap it sharply with the hammer (Fig. 5-16). This might seem crude, but tools for cutting thick steel cable work on the same principle.

Cable Terminations

Whichever type and size of cable is selected, we have to be able to attach the cable to the structure. A terminal must be added; either an appropriate fitting must be *swaged* directly onto the cable, or an eye must be formed to allow connection via other hardware. This termination must develop the same mechanical strength as the cable itself.

Swaged terminals are strong, fast, and attractive (Fig. 5-17). The terminals include a tubular section with the inside diameter equal to the cable diameter. The cable is inserted, then the tubular section is compressed with a swaging tool. The steel of the fitting is forced into the crevices of the cable, ensuring a strong, permanent connection. It also produces an uncluttered appearance with no exposed cable ends.

Complete kits with extensive predrilled holes and jigging are apt to supply completed cable assemblies with swaged fittings. Other kits might include at least one end of the cable swaged, with the other end free to set the proper length upon installation. The remaining kits require the builder to construct the entire assembly.

Unfortunately, swaging tools are expensive. They *start* at around $3,000. The usual practice is to have assemblies made for around $5 a fitting, plus parts.

But that can add up in a hurry. A typical biplane might have 10 bracing wires per side; each with two fittings. That's $200 for both sides, plus cabane braces and controls. Not to mention redoing any mismeasured cables.

The alternative is to wrap the cable upon itself and make a small loop, or eye. Pass a shackle through the loop, and the cable can be easily bolted in place. This eye can be formed in three ways: *splicing, soldering,* and *Nicopressing.*

Unless you're a masochist or antiquer, don't even *think* about splicing. The cable's end is looped around a bushing or thimble, then spliced into itself. It's cheap, a marlinspike is the only tool required, but you'll need cast-iron fingertips and infinite patience. If you feel you must try splicing, consult AC 43.13 and read your C.S. Forester. Don't say I didn't warn you.

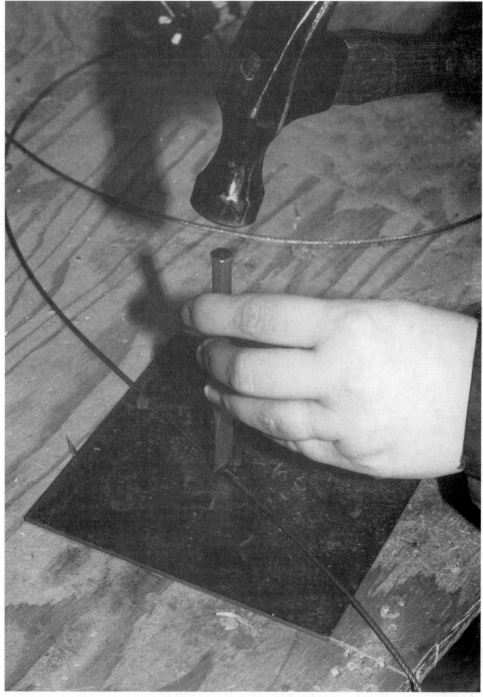

Fig. 5-16. Aircraft cable can be cut with a cold chisel and a hammer. A solid surface is also required. Note the steel plate atop the workbench.

Fig. 5-17. Swaged fittings are strong, attractive, and expensive. These are on the rudder cables of a Sonerai II.

Similarly, soldering is a lot of bother. After the cable is looped around a thimble, steel wire is wrapped around the cable and the free end. This wrapped section is then dipped in a molten-solder bath. It's too much effort, and the terminal ends up only 90 percent as strong as the cable.

Which leads us to the Nicopress system, where the loop is held by a swaged copper sleeve. It's ugly, but perfectly serviceable. This has become the standard homebuilder's terminal because it's fast, easy, and costs only about 40 cents per end. It was developed by the National Telephone Supply Company.

The cost difference between Nicopressing and traditional swaging is due to the fact that copper is softer than steel. The copper Nicopress sleeves can be easily compressed by low-cost hand tools. The standard tool costs about $150. This bolt-cutter-like device can compress a sleeve in seconds. The mini-swager applies pressure by tightening two bolts. It takes a lot longer, but the price is right, fewer than $20. These tools are shown in Fig. 5-18.

You *must* use a Nicopress tool to compress the sleeves. Don't use a vise or a locking pliers. It doesn't work. One homebuilder used locking pliers on the flying wires of his Fly Baby. He died. Surely your life is worth $20.

Fig. 5-18. Tools and accessories for applying Nicopress fittings.

Making a Nicopress Termination

Let's assume we're installing a cable terminal using the Nicopress system. To begin, pass a 1-inch piece of rubber hose over the end of the cable. Its inside diameter must be large enough to pass two cables. Automotive vacuum system hose is adequate and cheap.

Next, slide a Nicopress sleeve over the end of the cable. Make sure to use the correct sleeve; one intended for larger cable won't make good contact. Plain copper sleeves are used on galvanized cable, but zinc-plated ones must be used on stainless steel.

Bend the cable back upon itself and insert the end through the other hole in the sleeve, then through the rubber hose. If you're forming the eye inside one end of a turnbuckle, pass the cable through the turnbuckle before inserting it back into the sleeve and hose.

A loop has been formed, but it won't work on its own. Tension will flatten it out and concentrate the stress at one point, resulting in quick failure. A steel thimble placed inside the loop will equalize the forces. Like the sleeves, different sized cables take different thimbles. The exception is $3/32$-inch and $1/8$-inch cable, which take the same size.

Before inserting the thimble, clip off the four points at the ends with a pair of sidecutters. This gives a tighter assembly. If using a turnbuckle, the ends of the thimble might have to be spread a bit to get it over the turnbuckle's end. Or, if an AN115 shackle is going to be used, insert it now. The shackle *can* be added once the terminal is complete, but not very easily.

Now the *terminal* (Fig. 5-19) is ready for compression. Make sure the cable is lying flat in the thimble. Slide the rubber hose toward the loop, shoving the copper sleeve as close as possible to the thimble. Clamp the whole assembly in position using a hardware-store cable clamp around the hose.

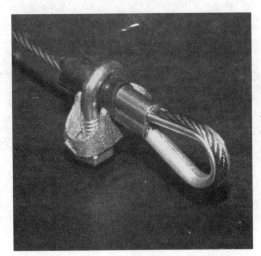

Fig. 5-19. A sleeve ready for swaging. A short piece of rubber hose slid around the cable protects it from the clamp that temporarily holds the assembly together. Note the cut-off ends of the thimble.

Position the Nicopress tool around the major axis of the sleeve. The intent is to squeeze down the largest dimension to make the sleeve more circular, not flatter. The standard tools have multiple grooves with a letter code to indicate cable size. The most common codes are:

C ($1/16$"), G ($3/32$"), M ($1/8$"), and P ($5/32$").

Incidentally, the same code is used when ordering sleeves. The miniswagers are easier to figure out; they're just marked with cable sizes.

Center the tool between the sleeve ends and begin compression. Use a slow, steady motion with the conventional tool (Fig. 5-20). If using a mini-swager, tighten the bolts evenly and equally (Fig. 5-21).

When complete, release pressure. Cables $3/32$-inch or smaller are ready at this point, but larger cables need at least two more compressions. Reposition the tool to the end nearest the thimble, and repeat the process. Finally, compress the sleeve at the opposite end.

Undo the cable clamp and cut the hose free of the cable. Trim away the excess cable on the free end, leaving at least $1/8$ inch sticking out of the sleeve.

To develop full strength, the sleeve must be compressed to at least a certain thickness. Check the dimensions of the long axis with either a Nicopress gauge, caliper, or a micrometer. The maximum thickness after swaging should be:

$1/16$ inch (0.1908 inch), $3/32$ inch (0.2674 inch),
$1/8$ inch (0.3532 inch), $5/32$ inch (0.3965 inch).

That completes the Nicopress process. But the cable has to actually get to something and connect to it. Let's look at cable-related hardware.

Fig. 5-20. With the conventional Nicopress swager, it's often easier to clamp one handle. Note the shackle inserted through the thimble.

Fig. 5-21. Tighten both bolts of the mini-swager equally.

Tangs

A tang is a protrusion of metal intended for attachment of cable. In its simplest form, a tang is a short strap of aluminum with two holes drilled in it. One hole is used to bolt the tang to the structure requiring support, and the cable eye is formed through the other.

They're typically more complex, if only to incorporate a bend between the two holes. A piece of angle or T-section can be used, although the pull imparted by the wire won't be lined up with the bolt holding the angle to the structure. This applies torque to the fitting and requires two bolts to hold it in place.

Tangs are easy to make, typically out of .080 2024T3 aluminum or .050 4130 steel. Because there is a lot riding on these fittings, make sure to follow all the metal-working rules, especially edge margin and bend radius.

In some cases, the AN42 through AN46 eye bolt can act as a ready-made tang. The thread sections are similar to regular AN bolts because they use the same dash-numbers to indicate length. The diameter is equal to the last digit of the AN number plus one (an AN42-10 eyebolt is the same as $2 + 1 = 3$, or an AN3 ($3/16$-inch) bolt).

Cable Shackles

Tangs are great, but forming the cable eye through the tang's hole is a bother, especially if the cable ever has to be replaced. The AN115 cable shackle fixes this problem. It passes through the eye (although far easier prior to forming the eye) and allows the cable to be attached to the structure with a bolt or clevis pin (Fig. 5-22). When using a bolt, don't use a self-locking nut. Use a castle nut and cotter pin, instead.

The dash number of the cable shackle indicates its rated strength in hundreds of pounds. An AN115-21 shackle, for instance, is rated at 2,100 pounds. This is the smallest shackle that can be used with 3/32-inch or 1/8-inch cable. The next smallest size is AN115-8, and its 800-pound rating is sufficient only for 1/16 inch.

One problem with shackles is the lack of throat depth. The side of the shackle might touch the edge of a control horn before the surface reaches full throw. If this is the case, replace the shackle with two flat straps of .050-inch steel or aluminum with holes drilled in either end, as shown in Fig. 5-23. Aluminum for low-load application only, please. Sandwich the horn and cable eye between the plates and bolt them through the hole and eye.

Because the horn is probably thinner than the eye, add washers on either side to keep the straps parallel. The plates shouldn't be allowed to clamp down on the thimble or cable. Use a longer-than-necessary bolt. Or slide a piece of tubing over the bolt to keep the plates apart.

Remember, when using bolts on moving structures like control horns, don't use self-locking nuts. Normal motion can loosen them. Use drilled bolts, castle nuts, and cotter pins. A clevis pin is a lighter and cheaper alternative.

Fig. 5-22. An AN115 cable shackle used to connect a rudder cable. Note the cotter pin in the bottom, holding the clevis pin in place.

Turnbuckles

Because cables apply tension, there must be a way to preload the cable to the proper amount. If the cable hangs slack, it can't do its job. The usual way is using *turnbuckles*.

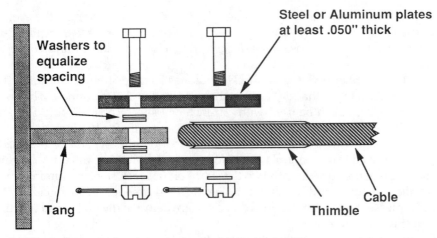

Fig. 5-23. In some cases, flat plates can be used in lieu of shackles.

They're not *entirely* necessary. It's possible to make the cables exactly the right size by building them in place. Run the cable between tangs, tighten it as much as possible, and then clamp in place and compress the Nicopress sleeves. The cable would probably maintain the original tension, but is it the correct amount? What happens if the structure changes slightly? What happens if it's just a teensy bit off? How will you reattach the cable if it must be removed? These are all problems a turnbuckle will solve.

A turnbuckle is a brass barrel with removable fittings screwed into either end. The ends are threaded in opposite directions; when the barrel is rotated, both fittings move inward or outward simultaneously, hence increasing or decreasing the total length of the assembly. Up to three inches of adjustability are available.

Turnbuckle specifications cover complete assemblies and individual components (Fig. 5-24):

AN155 Barrel. Tube-shaped piece of brass threaded at either end. The left-hand-threaded end is indicated by a ring scored around the brass.

AN161 Fork. An end designed to be bolted or clevised directly to a tang, bellcrank, eyebolt, and the like.

AN165 Pin Eye. Designed to fit between the tines of a fork and secured with a bolt or clevis pin.

AN170 Cable Eye. Similar to the pin eye, the hole through this end unit is curved to match a cable and thimble.

The ends are steel, and come in left-hand and right-hand threaded versions. A barrel is combined with two ends to form standard assemblies:

AN130. Barrel, cable eye, and fork: this is probably the most commonly used assembly, and typically is used between the cable and a control bellcrank or tang.

AN135. Barrel, cable eye, and pin eye: this is the most common assembly, used to attach cable to structure.

AN140. Barrel and two cable eyes.

AN150. Barrel and two forks.

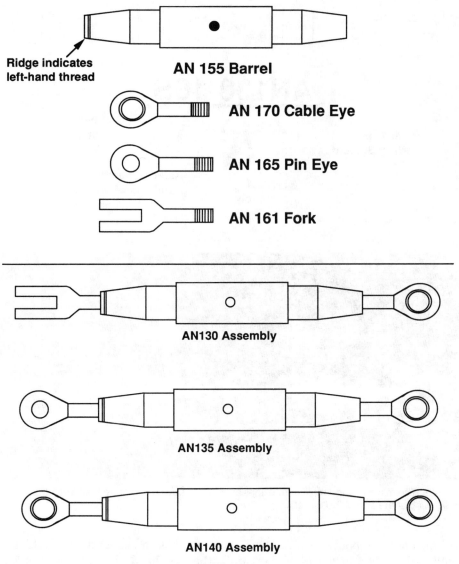

Fig. 5-24. Turnbuckle components and assemblies.

Additional characteristics are given by a dash number. A typical specification is AN135-16S. The 16 indicates the strength in hundreds of pounds, and the S refers to the length category (Fig. 5-25).

The farther the ends are screwed into the barrel, the stronger the turnbuckle. The strength ratings require that no more than three threads are exposed at either end of the barrel. If past that point, the cable must be replaced with a longer one (Fig. 5-26).

Turnbuckles come in two length categories: short and long, indicated by an S or L suffix (AN140-16L, AN135-22S). Short models are about 4^1/$_2$ inches in length and

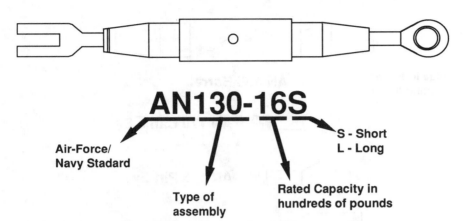

AN130-16S

Air-Force/
Navy Stadard

Type of
assembly

Rated Capacity in
hundreds of pounds

S - Short
L - Long

Fig. 5-25. Turnbuckle specifications.

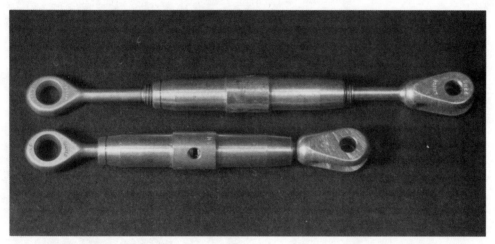

Fig. 5-26. The minimum and maximum extension of a turnbuckle. No more than three threads can be showing at the end of the barrel.

the long ones are about 8 inches. The long version has an inch-and-a-half additional length adjustment, or *take-up*. For example, the AN130-16S has a take-up of 1.125 inches and the AN130-16L is 2.875 inches.

The additional take-up of the long models is a *lot* of adjustment. But they can save you from making a whole new cable assembly if one ever comes out slightly too short. Cut the cable away from the short turnbuckle and install a long one instead. Make sure it has the same strength rating as the short one being replaced.

The barrel and the ends are both longer for the L models. The individual components use a similar dash number scheme to indicate strength and length. So a typical barrel might be an AN155-32S. Turnbuckle end specifications also insert an R or L before the short or long identifier to indicate whether the end is left- or right-hand threaded. Thus, an AN161-16RS turnbuckle end is a right-hand threaded short fork, good for 1,600 pounds.

Remember three main things when using turnbuckles. First, enough screw threads of the cable end must be contained within the barrel to allow the unit to meet fully rated strength. One cannot expect a single thread to withstand a 2,000-pound load. Therefore, make sure no more than three threads are exposed at either end of the barrel.

Second, do not lubricate the threads. Turnbuckle rotation is necessary for installation, not while the aircraft is in operation. Lubrication would only make it easier for the turnbuckle to rotate on its own.

On this note, the third item is prevention of inadvertent slackening. Because tension tries to force turnbuckles to loosen, they must be safety-wired to prevent rotation. The single-wrap method is fastest. Start with a new piece of safety wire about three times the length of the turnbuckle. Never reuse an old piece. It's tempting to just loosen the ends, adjust tension, and rewrap with the same wire. Don't because the changes reduce strength.

Push half the wire through the barrel's center hole. Bend each end in opposite directions (Fig. 5-27), and poke through the cable eye or lay it the across the bottom of a fork. Wrap the ends of the wire around the shanks of the turnbuckle ends at least four times and cut off the excess (Fig. 5-28).

The single-wrap method can be used on turnbuckles connected to $1/8$-inch or smaller cables. Brass safety wire is acceptable on $1/16$-inch and $3/32$-inch cables, but stainless steel is required for $1/8$ inch. Whichever type is used, it must be at least 0.040 inch in diameter. Cables $5/32$ inch and larger can be single-wrapped, but only with .057-inch stainless steel. This thicker wire is harder to bend.

Double-wrapping is similar, but uses two pieces of wire with the ends wrapped in opposite directions around the shanks. Don't wrap one wire directly atop another; instead, move down the shank a bit. Double-wrapping lets you use .040-inch stainless steel on cables $5/32$ inch or larger.

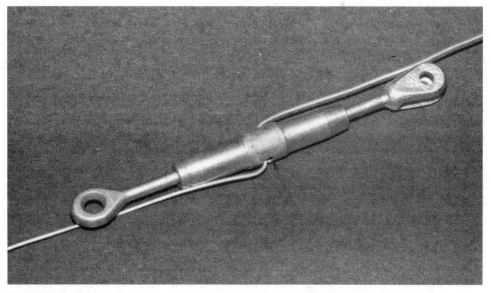

Fig. 5-27. The initial step in safety-wiring a turnbuckle.

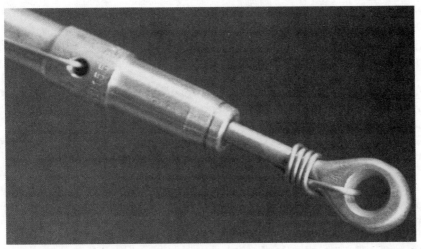

Fig. 5-28. The safety wire must pass through the end of the fitting and wrap around the shank at least four times.

Double-wrapping is actually the FAA's preferred method in all cases, but single-wrapping is acceptable if the above rules are followed. To avoid safety wire entirely, buy the more modern MS212XX-series turnbuckles. These use a quickly-installed safety clip, instead. But like all things fast and neat, they cost more than the ordinary item.

Fairleads and Pulleys

Bracing wires have it easy because they go in a straight line. But control cables must change directions. It would be nice if an aileron cable could go straight from the stick to the aileron without changing directions. But it isn't likely. And you can't just run the cable around a corner because the sawing motion would cut through, unless the friction frayed and snapped the cable first.

This is where *fairleads* and *pulleys* come in. They control and route aircraft cable.

Fairleads are simple objects with no moving parts. They prevent cable sag or change a cable run's direction slightly. They're sacrificial in nature; they wear away before the cable does. They still have to be strong and fairly hard, after all, if they wear through too fast or without warning, the cable could go slack or bind.

The traditional material is *phenolic*, made by a fiberglass-like process that substitutes other substances for glass cloth. It's commonly called *micarta*, a name trademarked by Westinghouse. It cuts easily and doesn't splinter. Other materials used for fairleads include nylon and delrin (an advanced plastic). Anything softer than the steel cable could be used; selected wooden homebuilts just drill small holes through the structure.

Whatever material is installed, friction and wear are a part of its operation. Cable damage isn't eliminated, just slowed. Normally, fairleads shouldn't change the cable's direction more than 5 degrees; 15 degrees is the maximum, but you'll pay the price in excessive wear and stiff controls.

How do you measure the angle? Sometimes you can eyeball it with a protractor, but often the area isn't that accessible. Take a piece of thin, stiff wire, the stuff used for model airplane pushrods, and the like. Bend it to 5 degrees with a pair of pliers. Run one leg of the resulting V shape alongside the cable, placing the bend point inside the fairlead. Tape the leg to the cable.

Now, move the other end of the wire to match the cable on its side of the fairlead. If the V flattens to match the cable, the fairlead installation is acceptable. If it must bend sharper, the cable angle exceeds 5 degrees.

If it failed the 5-degree test, rebend the wire to 15 degrees and try again. If the cable bend is still greater, replan the fairlead location. If it's between 5 and 15 degrees, the installation is acceptable, but expect greater wear and control friction.

If such a sharp angle change is contemplated, make sure the area is easily accessible for preflight inspections. Also, design the fairleads to allow removal of the complete cable. Don't just drill a 3/16-inch hole in a piece of delrin and run an 1/8-inch cable through it. The cable must be replaced someday, and such a fairlead would require cutting the cable.

Why does it make a difference?

In the first place, it's easier to make up a new cable the same length as an existing one. Cutting the cable will make it a little harder to reconstruct the actual length. Secondly, the cable run will not be as easy to get to when the aircraft is complete. Rather than diving headfirst under the instrument panel to Nicopress a rudder-cable sleeve, it's simpler to build the assembly outside and install a completed cable. The fairlead design should either include a slot running to an outer edge, or a wide enough hole to allow passage of a turnbuckle end.

If the angle is too much for a fairlead, a pulley will be necessary. They're generally made of phenolic with a ball-bearing center. The usual specification is by diameter and *bore* (the bolt hole in the center). Flight control pulleys are typically AN210 or MS20220 models. Pulleys with a diameter of 2 inches or less shouldn't be used to change the cable's direction more than 15 degrees. This limitation is for primary flight controls; this angle can be exceeded for secondary applications.

The biggest problem in pulley use is intolerance to misalignment. The pulley's edges should parallel the cable within 2 degrees. Beyond that, excess cable or pulley wear might occur. This can be a problem if the pulley is located close to the part moved by the cable because the angle might change throughout the control range. Fairleads have an advantage in this regard because they are unconcerned with alignment. All they care about is total angle.

One danger with pulleys is the possibility of the cable slipping out. Always include a short strip of bent aluminum or steel, bent in a U shape, slipped over the end of the pulley and held in place by the pulley's pivot bolt.

Putting It All Together

Let's run through a sample cable installation. One end of a 1 × 19 1/8-inch stainless steel cable is going to attach to a tang, and the other end to an eyebolt. An AN115-21 shackle will be used at the eyebolt. Because this is a straight bracing application, no pulleys or fairleads are necessary.

One end of the turnbuckle must be attached to the cable, while the other will be bolted to the tang. This calls for an AN13 assembly, with a cable eye at one end and a fork at the other. The 2,100-pound strength of the cable requires an AN130-22. Because the cable is only being used for bracing, we won't need the extra take-up of the long-series turnbuckle. Hence, we'll use an AN130-22S.

Take your reel of cable and install an eye (using an AN100-4 thimble) at the end. Slip the AN115-21 shackle over the thimble before forming the eye. Because this is $1/8$-inch cable, each sleeve requires three compressions, the first in the middle, the second on the end nearest the eye, and the last on the other end. Because this is stainless steel cable, use a 28-3-M zinc-plated Nicopress sleeve.

Measure the distance to be spanned by the assembly. Cut off that length of cable beyond the newly-installed eye. Because the turnbuckle requires approximately 4 inches of the distance, this leaves enough extra cable. Set the turnbuckle to a little longer than the midpoint of its range.

Slide a piece of shrink-wrap tubing over the end of the cable, followed by a short hunk of rubber hose and a plated Nicopress sleeve. Clip the points off another thimble, and force its ends apart slightly to get it through the turnbuckle's cable eye. Run the cable through the eye and over the thimble, and secure it using the hardware-store cable clamp as discussed earlier in this chapter.

Using a clevis pin or a short AN3 bolt, temporarily install the shackle in the eyebolt and the fork end of the turnbuckle in the tang. Pull on the free end of the cable to tighten it as far as possible, then snug down the cable clamp. We can't develop very much tension yet, that's why the turnbuckle is set a little longer than the midpoint. It gives us a little extra to pick up the slack.

If the cable is used in conjunction with another (an X bracing, or elevator cables, for instance), don't compress the sleeve yet. Similarly install the other cable, too. Adjust the cable lengths so they have equal adjustment ranges.

When satisfied, make the three compressions on each Nicopress sleeve.

Once the ends of the cable are attached, turn the barrel to tighten the cable. How tight? If the plans call for a particular value, you'll need a tension gauge, or tensiometer. Prices range from \$25 to more than \$1,000. The values are given based upon the ambient temperature because the wires change length at a different rate than the airframe. For instance, the plans might specify a setting of 25–35 pounds at temperatures between 40° and 60°F, and 20–30 pounds above 60 degrees.

If a recommended setting isn't available, tighten the wires until they give a satisfying twang. Watch the structure bending if the cable is tightened too far.

Then safety-wire the turnbuckles, and the installation is completed. To summarize:

1. Plain Nicopress sleeves on copper; plated sleeves on stainless steel.
2. Cut the points off the tips of the thimble before use.
3. One compression on $3/32$-inch cable and smaller; larger cables need at least three compressions.
4. Ensure that the free end of the cable sticks out of the end of the sleeve at least $1/8$" after compression.
5. Don't lubricate turnbuckle threads.

6. When in use, no more than three threads can show beyond the turnbuckle barrel.
7. Safety wire all turnbuckles.
8. Install cable keepers on all pulleys.

This chapter has presented some of the basic rules necessary when working with common aircraft hardware. It's time to look at specific construction materials. The next four chapters cover composite, metal monocoque, steel and aluminum tube structures, and wood construction. The wood chapter also discusses fabric covering.

Specifically, each chapter covers the materials, the fasteners, the tools, basic procedures, prevention of degradation (corrosion, rot, UV), and correction of typical errors.

If you've already decided on a construction method, at least skim the other chapters. As mentioned several times, no kitplane uses only one method.

6

Composite Construction

WHEN SOMEONE SAYS, "AIRCRAFT KIT," our first mental image is usually of a speedy composite homebuilt.

For good reason. Composite kits are the closest thing to the plastic scale-model aircraft we've built since childhood (Fig. 6-1)—add glue to the edges of the fuselage halves then clamp them together. Instead of a two-ounce tube of styrene cement, the kitplane builder needs gallons of epoxy.

While composite kitplanes build fast, reduced construction time is just a by-product of the main goal: velocity. While the manufacturer might market the plane as a fun flyer, they won't give up one knot in cruise: no open cockpits and tight, low-drag cowl. Their ads might mention short-field capability, but always follow with ". . . and can still cruise at over . . . ."

Slippery, low-drag shapes are easy to design with composites (Fig. 6-2). Designers turn to them whenever their goal is the fastest airplane on minimal horsepower. A Van's RV-4 gets good performance with a Lycoming O-360. But on the same engine, the fixed-gear Glasair IITD cruises 15 percent faster.

But you pay for the extra performance, literally. The Glasair kit costs twice as much. Once the engine and other non-included items are factored in, the Glasair runs about thirty percent more. But for the extra bucks, you'll be flying a lot sooner.

Some people believe composite kitplanes are faster because the materials are lighter than aluminum. They aren't, for instance, the Glasair II has an empty weight of about 950 pounds, and the RV-4 is about 50 pounds lighter.

Composites *can* produce incredibly light, strong parts. But not light, strong, and *inexpensive*. Cheap fiberglass must be replaced with costly high-tech materials like Kevlar and graphite fibers. Few kitplanes use them, except in critical high-strength applications. Neico has a Kevlar/Nomex option for the Lancair. The option increases the price by 20 percent.

The Glasair vs. RV-4 weight comparison isn't unusual. The composite Pulsar weights 430 pounds. With the same engine and number of seats, the Murphy Renegade Spirit tips the scales at 390. Even the wood-and-fabric Fisher Classic weighs 30 pounds less than the composite design.

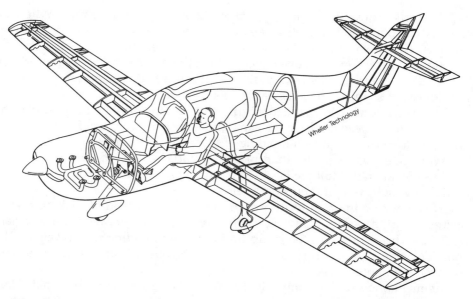

Fig. 6-1. Designwise, the Wheeler Express is little different from a plastic model kit. The main fuse-lage consists of an upper and a lower half, with a bulkhead or two added for rigidity.

Fig. 6-2. Whether fixed gear or retractable, the Glasair series is designed for speed.

The Pulsar is the fastest of the bunch—no argument there. But a composite airplane isn't necessarily the lightest one. Like every other construction mode, a lot depends upon the skill of the designer and the parts manufacturing process.

A kit part begins as a *plug*, which is a full-sized replica made from wood, metal, or other convenient material. A *form* is molded around the plug. When the form is cut open and the plug removed, a negative impression of the part remains. This mold can then be used to duplicate the plug.

Surely you've seen detective movies where the murderer leaves a footprint in the mud. By pouring plaster of paris into the print, the inspector can exactly reproduce the shape of the miscreant's foot.

In kitplane terminology, the foot is the plug and the print in the ground is the form or mold. Instead of plaster of paris, the inspector could use fiberglass cloth and resin to make a high-tech molding of the evildoer's bunions.

While fiberglass is lighter than plaster-of-paris, it's still too heavy for structural use on aircraft. Multiple layers of fiberglass are needed to get the strength sufficiently high. Weight then skyrockets.

This is where the composite materials come in. Instead of making a solid fiberglass part, a piece of lightweight stiff foam is inserted between layers of glass (Fig. 6-3). This core makes the layup thicker without adding much weight. The complementary characteristics of the foam and the fiberglass/resin layup result in stronger and lighter parts.

As mentioned earlier, composite kit production starts with the mold. First, it's covered with a *release agent*, which is a material the resin won't stick to. Otherwise, the newly-made part would end up permanently bonded to the mold.

From this point on, parts can be made in either the *wet layup* or *prepreg* methods. Wet layup is the most common. A layer of fiberglass cloth is laid over the

Fig. 6-3. A Glasair fuselage mold. The foam has just been set in place; further fiberglass layups will be added.

mold and soaked with resin. Another layer is added, and the process is repeated for a given number of laminations. The foam is then set in place, and further laminations added.

Then the layup is covered with a plastic bag and vacuum is applied. The vacuum pulls excess resin through the cloth and away from the layup. When the resin's cured, the part is removed from the mold.

Lancair uses the prepreg method. Rather than making sequential fiberglass layups, they use sheets of fiberglass preimpregnated with the optimal amount of resin. Premature curing is prevented by freezing the sheets until needed. The prepreg sheets are not as limp as dry fiberglass, but are still easily formed. They are laid in the mold, the core material of Divinycell/Nomex honeycomb is added, and the specified number of additional prepreg sheets complete the composite. The assembly is vacuum bagged like the wet layup method, but is cured at high temperature inside large ovens.

Which is better? Only a composite expert without an axe to grind could tell you, and I'm not one of them. Both sides claim advantages. Listen to the manufacturers and make your own decisions. Either method seems to make acceptable aircraft.

This chapter covers the basics of working with composite kitplanes. Details of fiberglass work applicable to every other type of construction area also included:

- Preparing composite kitplane parts.
- Joining composite components.
- Preparing fiberglass components for finishing.

Most composite airplanes still require selected riveting and aluminum work, which is covered in chapter 7.

ADDITIONAL TOOLS

In addition to the tools specified in chapter 4, you'll need a variety of low-cost tools. Prime among them is a *gram scale* reading 0–500 grams and a *hot melt glue gun*. The glue gun can be picked up at any hardware store. The scale might be a little harder to locate, but an ordinary diet scale with metric weights works if you can find one.

Builders using epoxies instead of vinylesters might consider buying an epoxy pump instead of the scale. The pump will dispense the correct ratio of resin and hardener with no bother. Make sure you buy a pump with the same ratio as the epoxy used on your kitplane. The pumps are a bit expensive, around $150.

If you're building a kit that uses vinylester resin instead of epoxy, you'll need one nasty little item: a hypodermic syringe, without needle, of 1–5 cc capacity.

This might require a little delicacy. Walking into a drugstore and asking to browse through the syringes is likely to gain a few raised eyebrows and a possible visit from the local gendarmerie. And *then* you can try explain away the gram scale . . .

All kidding aside, you shouldn't have much trouble buying the one or two syringes you'll need. A bigger problem might be the syringe's capacity. A 5 cc model is pretty large, and ordinary drugstores might not sell them. One cc models are more common (used for insulin) and 3 cc models are about as large as most drugstores go. Check at farm or veterinarian supply stores for the larger models. Whichever you buy, it must be graduated at one-tenth increments.

Ordinary scissors can be used to cut the fiberglass cloth, but spend a few bucks more for industrial model scissors. They'll cut better and last longer.

One thing composite construction does require is more disposable supplies. Find a local source for unwaxed paper cups and tongue depressors. Pick up a bunch of 1-inch disposable paint brushes. Buy an assortment of sandpaper; buy several packs because you'll need them.

MATERIALS, FASTENERS, AND SAFETY

When working with composites, you enter a high-tech world of aircraft homebuilding.

Materials

Cloth. The basic material you'll be working with is fiberglass cloth. Literally, the cloth is formed from fibers spun from molten glass. It's then rather loosely woven to allow stretching and shrinking to fit compound curves without cutting or tearing. In Britain, it's called glassfibre, which is probably a more correct term.

There are three basic weaves of fiberglass cloth: *mat*, *unidirectional*, and *bidirectional*.

Mat isn't really a weave at all. It's just random fiberglass fibers pressed together. Mixed with resin, it's used as a filler. Mat is definitely nonstructural.

Unidirectional, or *uni* for short, has most of the fibers in one direction. It's like a lot of long, strong ropes laid side-by-side. They're kept together by *fill yarns* spaced an inch or so apart. Uni is very, very strong in the direction the fibers run. Properly laid up, it's about 25 percent stronger than an equivalent cross section of 4130 steel. However, it has no crosswise strength, as the fill yarns are only meant to hold the cloth together during handling. Uni is used in places like spar caps, where the designer knows the load will be applied in only one direction.

Bidirectional cloth (called *bid*) has the same number of fibers going in two directions oriented 90 degrees to each other (Fig. 6-4). It isn't as strong as uni, but can take the strain in both directions; it rarely is as strong in both; one direction is usually weaker. The weaker direction is called the *fill*, the stronger is the *warp*. Unidirectional cloth, for instance, has a very high warp strength and a limited fill.

The strength and pliability of *bid* is determined mostly by the weave, count, and weight. The weave determines the stability and pliability of the cloth. For example, the *plain* weave is an ordinary crosshatch pattern. It's not especially pliable, but the pattern minimizes the tendency for the individual yarns to slip out of place under load.

Unidirectional

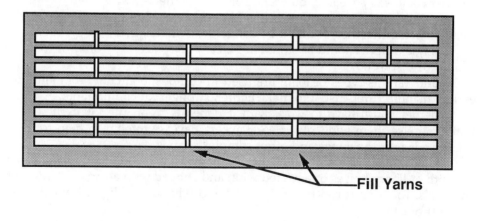

Fill Yarns

Bidirectional

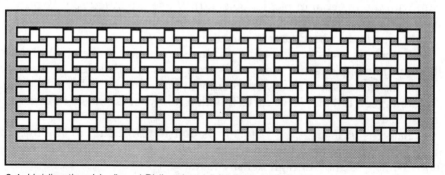

Fig. 6-4. Unidirectional (uni) and Bidirectional (bid) fiberglass cloth.

The count is the number of yarns per inch in each direction. A typical cloth might have 24 × 22 count. The weight is given by the number of ounces per square yard. Coupled with the count, it's an indication of the size of the individual yarns. Like with steel cable, the thicker the individual strands, the stronger and less pliable the material is.

Bid cloth is also referred to as *boat/tooling* or *industrial* cloth. Boat/tooling cloth is generally of plain weave; industrial cloth's weave is more pliable. Otherwise, pay attention to the weight. Lightweight boat or tooling cloth, for instance, comes in various counts and weights.

While this is useful background information, your fiberglass should come with the kit. A problem arises if you run out. There is no AN standard fiberglass;

at least, none that is marketed to homebuilders. As such, it is ordered by the manufacturer's part number. One weaver could sell a cloth by P/N 8810, and another could sell a weaker, different cloth under the same number.

In other words, if you must buy additional cloth, *buy it from the same fabric manufacturer*. Otherwise, it might be dangerously inadequate.

Fiberglass cloth should be stored in a clean, dry place. Any dirty or wet cloth should be discarded.

Foam. Various types of foam are used in composite aircraft. As mentioned earlier, manufacturers use it to make lighter, stronger kit components.

Whether you'll have to work with it depends upon the kit itself. Glasair builders make foam wing ribs. Sea Hawkers have styrofoam to help form the trailing edges of the wings and other complex shapes.

The advantage of foam is its light weight and easy workability. Most can be cut with ordinary knives or a heated wire, and shaped with a variety of inexpensive tools. Foam parts can be easily made and will be very light, but can't take any sort of a load.

The usual practice is to make a component from foam, then cover it with fiberglass. Sometimes the foam is removed after the fiberglass cures (like fuel tanks) but otherwise it's left in place as part of a builder-constructed fiberglass sandwich.

Styrofoam is the original homebuilder's foam. Light blue in color, it has a gritty, rather rough surface due to its large cell structure. Don't confuse true styrofoam with the material in picnic coolers; they use expanded *polystyrene* with a smooth surface. A hot-wire cheese cutter works quite dandy with styrofoam. The technique is discussed later in the chapter.

Styrofoam dissolves in petroleum products and reacts to the vinylester resins used with some kitplanes. In these cases, designers usually select *urethane* foam. It weighs the same as styrofoam, but is smoother due to its smaller cell structure. It's colored green or tan. Never carve it with a hot wire; it emits a hazardous gas when heated.

A relative is *urethane polyester* foam. Its density can be up to 10 times more than styrofoam or urethane foams, which allows it to withstand higher compression loads. *PVC* is another high density foam.

If you run out of foam, pick up an identical replacement. Don't go merely by sight.

Never store foam where it can be exposed to direct sunlight. It deteriorates rapidly.

Other Materials. *Peel ply* is a dacron tape that won't adhere when used in a layup. It's used to prepare a laminate surface for later glassing or to reduce finish sanding.

Microspheres were very big 10 years ago, but have been replaced by *glass bubbles* or *Q-cell*. Both are used the same way, to thicken epoxy or catalyzed vinylester to fill low spots or bond awkward shapes together. *Cabrosil* is similar.

Flox is flocked cotton fiber mixed with epoxy or catalyzed vinylester to thicken the material, similar to glass bubbles. The flocked cotton acts as a binder; flox produces a strong and durable surface when cured.

Use of these materials is covered later in the chapter.

Fasteners

Of course, the primary fastener will be the *epoxy* or *vinylester resin* used to impregnate the fiberglass cloth and bond aircraft components together.

Epoxy. Epoxies are two-part systems, consisting of resin and the hardener. In theory, they're very similar to the epoxy glues you buy at the hardware store. But aircraft epoxies are optimized for fiberglass layups and large-area structural bonding.

Mixing ratios run between 2-to-1 and 4-to-1 resin to hardener, depending upon brand. Each product has different characteristics, such as *viscosity* (thickness) and *curing time*. Curing times vary from a few minutes to several hours. This is the time until the mixture hardens; actual curing for full strength might take far longer.

The resin-to-hardener ratio for epoxies is fixed; variance of more than 5 percent or so might cause problems. Accuracy is important.

Vinylesters. Vinylesters are multipart systems, including resin, *promoter*, *accelerator*, and *catalyst*. To maximize shelf life, the resin is shipped in the unpromoted state; the user is expected to promote the vinylester resin a gallon at a time as needed. A typical ratio is 5 cubic centimeters (ccs) of promoter for each (1) gallon of resin.

To use the vinylester resin, a small amount of *MEKP catalyst* is added. Remember, epoxies are optimized for a given mixing ratio of hardener to resin. Ten percent too much hardener can greatly affect the strength of the bond.

Not so with vinylesters. The catalyst doesn't add to the bond strength, it merely activates the resin. Catalyst/resin ratios can be matched to the environment and the task on hand. By adjusting the ratio, *gellation times* (time until the mixture sets to the point it can't be worked) can be cut in half or more.

Figure 6-5 illustrates the ratio/temperature/time relationships for the vinylester resin used by Glasair. Use only the ratios approved by the kit maker; too little or too much catalyst might result in an understrength bond.

When necessary, accelerator can compensate for lower-than-optimum working temperatures. Above 80°F it isn't needed. Three cubic centimeters of accelerator per gallon of resin usually suffice for temperatures between 65°F and 80°F. Check the specific instructions included with your chemicals.

Store the materials in the proverbial cool, dry place. Vinylesters are often shipped in slightly permeable plastic buckets. Not only does this reduce the shelf life, it makes your shop stink of styrene. If you won't be using the material for awhile, transfer the resin to airtight metal or glass containers.

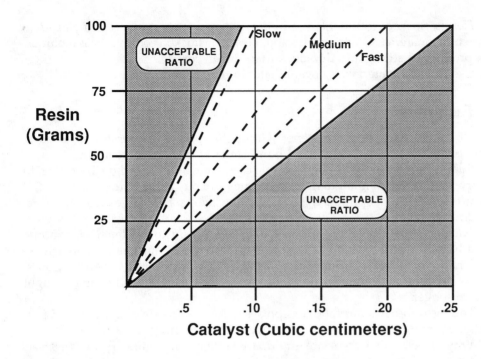

Fig. 6-5. While epoxy has a fixed ratio of components, vinylester proportions can be selected (within certain bounds) for various cure rates.

Catalyst Ratio	Shop Temperature			
	50°	60°	70°	80°
Slow	65	55	30	15
Medium	45	30	20	10
Fast	35	20	15	<10

Approximate Working Time in Minutes

Safety

Composite construction is like mixing a cocktail with the devil's chemistry set. You have to know exactly what you're doing. Composite aircraft are built using a wide variety of chemical substances, some hazardous, some supposedly benign. Proper safety precautions are vital.

Allergic reactions aren't as common as they used to be. But their relative rarity doesn't make them any more fun. Most of the reactions happened during the Long-EZ days. A builder would happily work along for six months or so; then, just entering the shop caused skin to break out in a rash. Some developed breathing problems.

Even those who protected their skin weren't safe, repeated incidental contact would still bring about severe sensitivity. And the effects were permanent.

A drive for safer epoxies arose. Safe-T-Poxy was developed, and eliminated most of the sensitivity problems. A further advance was the use of vinylester systems, for which there are no recorded cases of allergic reaction.

So, technically, if you're using Safe-T-Poxy or vinylesters, you can work with bare hands, right?

The safe levels of exposure to dangerous materials are always being revised. I don't think any limit was ever revised *upward*. We always seem to discover that the dangers are greater than previously believed.

What's considered "safe" now might be defined as hazardous as long-term trends become apparent. So don't apply epoxies or other materials with bare hands. Wear rubber gloves or approved skin barrier cream. Provide adequate ventilation.

Be advised that directly mixing vinylester promoter and catalyst produces a lot of heat and probably an explosion or fire. Keep them well separated. A small reaction can also occur when catalyst is added to vinylester resin. Wear safety goggles because splatters can cause serious eye injury.

Some filler material, such as cotton fiber and Q-cell, are very fine powders that can mess up your lungs if inhaled. Wear a respirator or a dust mask. Sanding the cured material stirs up the same problem, as well as adding fine particles of fiberglass and resin to the air. Wear a mask while sanding, too.

One unexpected danger comes from disposing of excess mixed epoxies or catalyzed vinylester resins. Because they produce considerable heat, don't throw them into the trash until they've cooled.

A good way to dispose of partially-cured resin is to pour it onto plastic sheets. The plastic should be on bare ground, concrete, or other nonflammable material. When the mixture cools, fold up the plastic and discard.

COMPOSITE KIT PARTS

One way that composite kitplanes differ is the inclusion of a large number of identifiable components. A tube-and-fabric kit might deliver a welded fuselage, but everything else is a jumble of generic parts. An aluminum airplane kit will have some metal sheets bent to vague shapes.

But you can look at a composite kit and see the swoopy shape of the fuselage (Fig. 6-6). The wings are obvious, even with the cutouts for the ailerons and flaps.

The primary operation for building the structure is gluing kit parts together and reinforcing the joints with fiberglass cloth. Examination of kit parts would be worthwhile.

Fig. 6-6. Composite aircraft are almost always supplied as complete kits, and come with a high number of identifiable parts.

Coatings and Joggles

The inside and the outside of the parts might differ in color. Cured fiberglass parts are a golden tan. The inside of the pieces of your kit are probably this color.

Outside, there are three possibilities: the same tan as the inside, a dark grey, or a shiny white.

As mentioned in the beginning of the chapter, composite kit parts are laid up in a mold. The curing process generates heat. Any substance applied to the mold before the first fiberglass layup gets bonded to the outside of the finished part.

Want a hard-shell exterior finish? Apply a special paint to the mold, make the part, and it comes out gleaming and protected.

This is how a *gel-coat* finish is applied. In the fiberglass boat industry, it cuts down on the total production time and manpower. The hull comes out of the mold and doesn't have to be painted. Bond the top deck in place, and trim, and the boat is ready for delivery.

Kit aircraft manufacturers previously delivered the exterior surfaces gel-coated. A few still do. But there's a significant difference between aircraft and boats. A one-piece boat hull is easy to do, and the interface between the hull and the deck can easily be hidden.

But an aircraft can't be made in one piece. Gel-coated components must be bonded together, and layers of reinforcing cloth added. It's no real problem, but many of the joints and reinforcements are in plain sight.

However, the surface and edges of the reinforcing fiberglass must be smoothed and blended in. The color won't match the gel-coat, so it'll have to be painted. Matching the gel-coat's exact color and sheen is impossible, hence the whole airplane has to be painted.

And if the whole airplane is going to be painted anyway, why add the gel-coat at all? It just adds weight. Glasair components originally were delivered gel-coated. The company saved 35 pounds per airframe when they stopped using the process.

Instead, many companies apply a grey primer using the same procedure. The primer often contains an ultraviolet block to slow deterioration, or the parts are delivered in the natural honey-tan of the resin and glass.

The manufacturer adds other useful features during the molding process. The most important feature is the *joggle,* which is complementary indentations applied to the two edges that will be bonded together (Fig. 6-7).

Consider trying to glue two playing cards end-to-end. It's a weak joint at best, as the thin edges present little surface for the glue to work with. The edges could be overlapped slightly. This produces a strong bond, but one card is higher. The joint is obvious and hard to hide, and on an aircraft would have to be carefully positioned to minimize the aerodynamic effect.

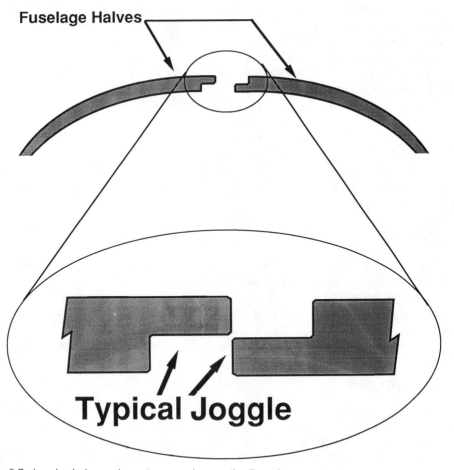

Fig. 6-7. Joggles help produce strong and correctly-aligned components.

However, if the edge of one card were bent down in a Z shape, the pieces could be overlapped and still leave a smooth seam. This is a *joggle*. The tiny seam remaining can be filled and smoothed to completely hide the interface, or layers of fiberglass can be added for additional strength. The joggle also aids in the proper alignment of the two parts. Selected kits use double joggles for additional strength and precision.

Preparation Techniques

The composite parts require different degrees of preparation prior to glassing or bonding, depending upon the manufacturer's process and their location on the aircraft.

At the beginning of the chapter, I mentioned the release agent that prevents the part from bonding to the mold. Sometimes, traces of this chemical remain attached to the exterior of the parts. These traces must be removed. Usually, all that's necessary is to scrub the area with water and let it dry.

The manufacturer's mold produces nice smooth parts. But such a surface isn't the best for glassing because you want a rough surface for the resins to grab. Therefore, any time you glass the exterior of the aircraft, the surface must be first roughed up with sandpaper. If the manufacturer primes the exterior, this coating must be removed as well. Otherwise, the resins will just bond to the paint, which flakes away from the structure at the first chance it gets.

After sanding, wipe the part down with acetone. Be careful if there are bare foam parts nearby, because the acetone will dissolve them.

Little preparation is normally required for the inside of molded kit parts. Follow the kitplane manufacturer's directions.

A last note on preparation: when sanding, roughen the surface, then *stop*. Don't sand into the underlying fiberglass. Any damage to the cloth weakens the structure.

Cutting

The kit parts can be cut with normal tools, such as drills, hole saws, and saber saws. Figure 6-8 shows a typical cutaway area.

Excessive pressure causes heat to build, and heat is the enemy of composites. Let the tool cut at its own pace. If possible, back up the part so you aren't pushing on the part itself.

Fine-shape the parts with the ordinary tools: files, sandpaper, and the like.

Bonding

Bonding kit parts together is very simple. The joggles are coated with the appropriate bonding agent (epoxy or catalyzed vinylester resin) and the parts are pressed together. Enough bonding agent must be used so that the glue extrudes along the entire length of the seam. If it doesn't, there might be some starved areas. Wipe the excess away before it cures.

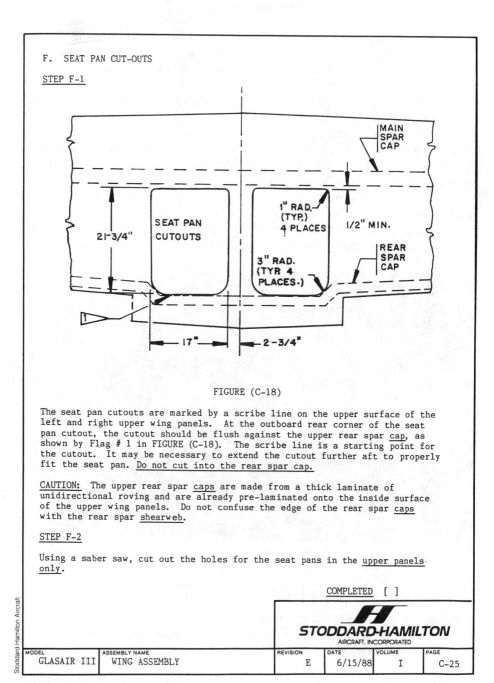

F. SEAT PAN CUT-OUTS

STEP F-1

FIGURE (C-18)

The seat pan cutouts are marked by a scribe line on the upper surface of the left and right upper wing panels. At the outboard rear corner of the seat pan cutout, the cutout should be flush against the upper rear spar cap, as shown by Flag # 1 in FIGURE (C-18). The scribe line is a starting point for the cutout. It may be necessary to extend the cutout further aft to properly fit the seat pan. Do not cut into the rear spar cap.

CAUTION: The upper rear spar caps are made from a thick laminate of unidirectional roving and are already pre-laminated onto the inside surface of the upper wing panels. Do not confuse the edge of the rear spar caps with the rear spar shearweb.

STEP F-2

Using a saber saw, cut out the holes for the seat pans in the upper panels only.

COMPLETED [ ]

STODDARD-HAMILTON
AIRCRAFT, INCORPORATED

MODEL	ASSEMBLY NAME	REVISION	DATE	VOLUME	PAGE
GLASAIR III	WING ASSEMBLY	E	6/15/88	I	C-25

Fig. 6-8. The plans should indicate cutout areas.

The bonding agents can be pretty thin; they have a tendency to flow out of the joints before curing. Often, the resin is mixed with glass bubbles or flox to make a thicker, more viscous material. Later sections discuss the use of these materials.

The pieces must be held rigidly in place until the bonding agent cures. One way is to predrill small holes through the joggle, join the parts, and inset clecos. Hole spacing usually isn't very tight; between six inches to a foot seems typical for large pieces (Fig. 6-9). The holes can be filled once the clecos are removed.

Other methods include wrapping duct or wide masking tape around the part, clamps (atop boards to spread out the force), and weighting down small pieces. No matter the method, take care to equalize the pressure because gaps might form if one area is held together too tightly.

Whether you're done at this point depends upon the required strength. Most joints need reinforcement through additional fiberglass layups.

Fig. 6-9. Clecos temporarily hold this composite fuselage together until the vinylester cures.

LAYUP PREPARATIONS

The basic operation is to cut a piece of fiberglass cloth to the appropriate size, lay it in place, and saturate it with resin. Like many endeavors, the preparations greatly affect the quality of the final product.

The Shop

General rules were given in chapter 4. The need for a warm environment cannot be overemphasized. It's not only because the resins cure slower when cold. They don't flow as well, either. Your goal is to fully saturate the cloth without using excessive resin. The cloth works like a wick in normal temperatures, eagerly absorbing the bonding agent. But cold resin doesn't want to flow and doesn't want to soak into the fiberglass. You end up using more resin, and end up with a heavy aircraft.

Everything associated with the glass work should be warm: resin, cloth, workbench, tools, and the like. Any cold item sucks the heat from the resin. If the workshop must be heated, turn the heater on an hour or so before starting.

Where cold is the primary enemy of fiberglass work, dirt and dust tie for second. They'll contaminate the cloth and resin and deteriorate bond quality. Sanding residue is a primary culprit. Keep the roll of cloth covered until needed, and clean the shop regularly. Those little hand-held vacuums are neat, but beware because they exhaust the air sideways, which can stir up more dust if not carefully handled.

Composite aircraft builders require helpers more often because they are working with large structures and materials that have limited working times. Usual practice has one person mixing the epoxy while two others apply cloth or bond parts. Make sure you have rubber gloves, respirators, and other protective equipment for your assistants, and carefully brief them on the dangers and procedures.

Preparing Epoxy or Vinylester Resins

While you don't actually mix up your epoxy or vinylester until everything else is ready, subsequent information in this chapter will be easier to digest if you understand the process.

To reduce repetition, we'll use the term *bonding agent* or *resin mixture* to refer to the either epoxy resin with hardener added or catalyzed vinylester resin. Similarly, the term *mixing* will also apply to the act of adding the catalyst to vinylester resin. Where differences exist, I'll refer specifically to either epoxy or vinylester.

When dealing with the resin mixture, the primary concern is *pot life*, which is the time until the bonding agent becomes too thick to work with. Many factors affect pot life, which decreases with low humidity or in brisk winds. But temperature is the major variable. The warmer the materials, the faster the material cures.

Both systems have another common characteristic: *exothermic reactions*, they give off heat while curing. The faster they cure, the more heat that is liberated.

The paradox should be obvious. The mixtures cure faster when ambient temperature is high. But curing raises their temperature, which in turn increases the reaction rate, which causes higher temperatures, and so forth. Atomic physicists call this a chain reaction, and with radioactive materials it makes large craters.

Chemicals are somewhat less energetic. The mixture just gets hot, and pot life is severely reduced. This overreaction is called an *exotherm*. High temperatures are a danger to composite aircraft. Too-hot resin can melt foam and weaken the kit parts.

It's not a problem in a fiberglass layup because the bonding agent will be well spread out and the heat will be dissipated. But a mixing cup concentrates the heat and supports the chain reaction.

The solution is to limit the batch size to less than 200 grams (about 7 ounces). The mixture still gets warm, so don't set it on anything that might be damaged by heat, especially completed fiberglass parts.

Sure, it would be nice to catalyze a whole bucket of resin and bond an entire bulkhead in place at one go. But you would quickly end up with a batch of rock-hard goop in a nearly-glowing bucket. Instead, small batches will let you work on the next section even if one part has begun to cure.

Differences exist in the preparation of epoxies and vinylesters.

Epoxy Preparation. The epoxy manufacturer specifies the proper ratio of resin to hardener. It is important to closely approximate the specified ratio. Few if any aircraft epoxies use a 1-to-1 ratio, so you'll need some method to measure out the components.

A common tool is an *epoxy pump*. It includes hoppers for the resin and hardener, and dispenses a given amount of each element with each pump of the handle. Or use a scale to weigh out the proper ratio of materials. A third method uses industrial-sized syringes to draw out and dispense the materials. Buy separate syringes for each component, or you'll be left with epoxied syringes and contaminated materials.

Mix the resin and hardener in an unwaxed paper cup, using a wooden tongue depressor to stir. The cups can be reused as long as the epoxy inside has either completely cured or hasn't become too thick.

Mix for a minute or two. One of the goals of the layup process is to eliminate all air between the laminations. Your first step toward that goal is to keep from adding air into the resin mixture. Don't swirl the stick around like you're cleaning a brush. The more violent the action, the more air ends up in the mixture. The resin and hardener must be thoroughly blended, but keep the air bubbles out. Scrape the sides and bottom of the container with the stick to ensure even mixing.

The curing time depends upon the temperature and the epoxy brand. As mentioned earlier, the mixture will get warm. When it becomes noticeably thicker, mix up a new batch.

Vinylester Preparation. Vinylester mixing is similar to epoxy, with certain differences.

To begin with, make sure the promoter had been added to the resin. Do a gallon at a time, and slap a label on the jug so you know it's been promoted. Add the accelerator at the same time, in the amount appropriate to the temperature.

Because the resin-to-catalyst ratio is so large, a pump makes no sense. Instead, use a scale for the resin and a syringe for the catalyst. A cubic centimeter is about equal to a gram of material, so for 1 percent resin to catalyst ratio add one cc of catalyst for every 100 grams of resin.

A kitplane manufacturer suggested hot-gluing the catalyst bottle and a paper cup to a piece of wood, and keeping the syringe in the cup. That keeps the syringe and bottle together, and the cup prevents the syringe from leaking catalyst wherever it's set.

Mix the resin and catalyst the same way as epoxy; however, don't use a cup with partially-cured resin already in it. It will accelerate the curing of the new mix.

Vinylester resins cure differently from epoxies. While epoxies gradually thicken, vinylesters reach a *gel* point. When gelling starts, it will develop a paste-like consistency for a couple of minutes, then solidify.

At some point during the curing process, epoxies and vinylesters reach the *green cure* state. In this condition, the resin mixture is partially cured and is somewhat rubbery. Loose edges of fiberglass cloth can be easily trimmed away with a razor blade. If you wait too long, the resin mixture takes on a solid cure that takes a lot of work to trim and shape.

Vinylesters reach the green cure state about a half-hour after gellation, while epoxies might take several hours, depending upon temperature. Epoxy and vinylester systems take 24 hours or so to fully cure.

Shaping Foam

The foams used on kit aircraft can be easily cut and shaped using ordinary hand tools. Bandsaws and saber saws work dandy, but the stuff cuts easily with handsaws or even penknives. When making a flat piece such as a rib, glue a pattern to the foam and cut it out.

Final shaping is done with standard abrasive shop tools. Files work fine, as does sandpaper. For the best use, wrap a piece of sandpaper around a chunk of scrap 2 × 4. You can buy a commercial sanding block but the homemade variety works just fine.

A good foam-working tool is the Shurform, made by Stanley. It has a rough, self-cleaning cutting surface, and a pair of handles like a wood plane.

If the piece of foam isn't big enough, glue two pieces together using a thick mix of resin mixture and glass bubbles (covered in detail later in the chapter). However, don't get the mix too close to a cutting line. The cured mixture is difficult to cut, and any attempt usually damages the nearby foam.

Hot Wire Cutting

With styrofoam, and *only* styrofoam, you can use the hot wire method. An electrical current is passed through a fine wire, which heats the wire past the melting point of the styrofoam—like a large cheese cutter, and the hot wire treats the finest styrofoam like so much cheddar.

Hot wire cutting was in its heyday back with the Long-EZ and similar aircraft. The EZ's wings were composed of multiple sections of foam cut out by hot wire separately, bonded together, and fiberglassed. Modern kitplanes don't use this procedure for primary structure, but some require the builder to make trailing edges and other parts from hot-wired foam.

Commercial hot wire systems are available for cutting long shapes that include custom cutting frames and variable transformers and cost about $100 to $200.

But for your typical kitplane, a simple homemade system works well enough. You'll need a spool of stainless steel (.040- or .032-inch) safety wire, a couple pieces of dowel or tubing about one foot long, some nails, a 2 × 4 board about 3 feet long, and a battery charger (Fig. 6-10).

Fig. 6-10. Small parts can be hot-wired using a couple of pieces of dowel or tube, some stainless steel safety wire, a board, and an ordinary battery charger.

Form a giant U shape with the board on the bottom and the tubes/dowels as the uprights. Bore holes into the 2 × 4 to slide the tubes into place. Hammer a nail into the board near the base of each. Drill a small hole through the free end of the tubes. Then string a piece of safety wire from one nail, through the hole in the closest tube, through the hole in the other tube, and back down to the other nail. Tighten the wire by clamping a Vise-Grips to the base of one of the tubes and rotating it.

That's it. Connect the clips from the battery charger to the two nails. Turn the battery charger on. Battery chargers have built-in current limiters, so the meter should be showing a high output without actually pegging the needle. You probably won't see any change in the wire. It'll slacken up a bit with heat, so give the Vise-Grips another turn.

Take a piece of scrap styrofoam and touch it to the wire. You'll hear a hiss as it sinks into the foam. If the styrofoam is visibly melting or the wire cuts a slot larger than its own diameter, the wire is too hot. This leaves pits and melted spots in the finished surfaces. The voltage must be reduced by adding more safety wire between the charger and the tool, or by using a train transformer or other voltage control. Variable resistors are available from the homebuilder's catalogs.

If the wire is slow cutting, the voltage (and hence the wire's temperature) might be too low. Reduce the wire length between the charger and the actual cutting wire. One way would be to clip the charger leads directly to the wire at the end of the tubes. Otherwise, you'll need a stronger power supply.

Fig. 6-11. The cutting wire is guided around templates tacked to both sides of the piece of foam.

When the temperature is right, the hiss of the cutting wire should be faint. When the wire is withdrawn, fine streamers of blue plastic should follow. The surface of the cut areas should be smooth, with no pits, and a lot of little blue plastic streamers.

When cutting a shape, two templates are tacked to the end of a piece of foam. Again, you can slurry two pieces of foam together to get the right length, but don't get the slurry any closer than 1/2 inch from the prospective cut line. Turn on the cutter and sink the wire into the foam until it contacts the templates on both ends (Fig. 6-11). Then slowly move the wire around the outline, taking care that each end of the wire is at the same position on the template at the same time.

Let the wire cut at its own pace, if you pull too hard, the wire will lag excessively and bow *inside* of the foam, distorting the final shape. Cutting is a heck of a lot easier with two people, but coordinate so you're both at the same point on the template.

When done, smooth the foam with files or a sanding block.

Installation

The first step to actually installing parts is determining where they go. The plans might specify a distance from a particular reference point, such as "12.5

inches outboard of wing root." Or there might even be a scored or raised line on the inside of the fuselage or wing to mark the spot.

Wipe the area with acetone and buff it a bit with 80-grit sandpaper. Exterior surfaces will take a bit more buffing to eliminate the smooth surface left by the mold. Make sure all primer and mold release agent has been removed from the bonding areas.

Parts are bonded together with the standard resin mixture. Sometimes, though, the resin is too thin. It can tend to flow out of joints and produce a sub-standard bond.

To counteract this problem, various materials are added to the resin mixture to thicken it up:

Flox. Flocked cotton fiber and epoxy is used to bond parts together and form fillets. It's a white powder about the consistency of lumpy flour. It's mixed in with epoxy at a 2-to-1 fiber-to-epoxy ratio (by volume) to produce a tan putty-like substance. It can be mixed by feel; if the flox doesn't stand up on its own, add cotton fiber; if the mixture is whitish, add epoxy.

Mill Fiber. The vinylester world's equivalent to flox is mill fiber. Instead of flocked cotton, very short fiberglass strands are added to the catalyzed resin. Again, you're looking for a putty-like mixture. Like flox, mill fiber mixtures form strong bonds that are ideal for structural applications.

Micro; Glass Bubbles; Q-cells. Three common names refer to the same basic material, tiny glass bubbles added to the resin mixture for use as a nonstructural glue or filler. The bubbles look like powdered sugar. The term *micro* actually refers to the now-obsolete microspheres, but the term is handy and remains with us. The ratio of bubbles to resin varies with the application. Nicknames describe the three basic ratios.

The first is *slurry* (or *wet mix*): glass bubbles and epoxy/catalyzed resin in about 1-to-1 ratio. It's tan and a bit runny, looking like a cheap chocolate milk-shake. This thin mixture fills the surface pores of styrofoam to seal it prior to applying fiberglass.

A thicker mixture, about the consistency of peanut butter, is called *thick micro* (or *thick mix*). It bonds non-porous parts or acts as a filler (Fig. 6-12). About three times as much balloons as resin mixture is used. The final product looks like pea-nut butter—creamy, please, with no lumps.

Finally, there's *dry micro*, at about a 5-to-1 ratio. It's very thick, and is used to fill the fiberglass weave before painting and to bond blocks of foam together.

All mixture ratios are by volume, not weight. You can learn to recognize the proper consistency and don't have to measure the quantities. The bubbles are always added to mixed epoxy or catalyzed resin, never just to the plain resin. And speaking of nevers, don't allow micro between layers of fiberglass. It weak-ens the bond.

Other materials are used, depending upon the kit. *Cabocil* is a glass powder, similar to Q-cells but much finer. *Chopped strand mat* (usually simply called *mat*) is similar to mill fiber, but still in a cloth-like form.

Kits use different mixtures in different areas. It makes a difference because mill fibers produce a stronger bond. Check the plans carefully.

Fig. 6-12. Thick micro looks and spreads like peanut butter.

Unless you're sealing styrofoam, you'll always use a thick mixture. It's solid enough that you usually don't have to clamp or jig the parts together while the mix cures. Generally, light components don't shift in thick micro because, like peanut butter, it "sticks to the roof" of your airplane. Keep watch, and be prepared to add some temporary external bracing if things don't stay in place.

There are two types of bonding operations. The first involves flat joins, such as two joggled parts or two overlapping flat components. In these cases, there's plenty of bonding surface available; all the builder has to do is apply a bead of thick micro (or whatever mix the manufacturer specifies) along each surface and press the parts together. Hold them together with rubber bands, bungee cords, tape, or whatever, as mentioned earlier. The tried-and-true cleco is often used, as shown in Fig. 6-13.

Some kits call for the "poor man's cleco," the pop rivet. While definitely cheap, they must be removed by drilling. Pop rivets have their places. But if the plans say "use a cleco or a pop rivet," a cleco is usually the better choice. Because the pop rivet does provide a more solid attachment, it's better if the parts are heavy or if the parts will be handled repeatedly.

Make sure the parts are solidly mated and that the joggles lie flat against each other. Wipe off the excess mix that oozes from the seams. If the plans call for it, add the specified number of layups to one or both sides.

The other type of bonding operation attaches two pieces at a sharp angle; one approaching 90 degrees. Examples include wing ribs or fuselage bulkheads. These pieces don't have a premade joggle to hold them in the proper position, and don't produce a nice, flat, easy-to-glass joint.

The method used to join the parts varies with the kit. Most have you apply a layer of thick mix or flox to both parts and press them together. Others temporarily attach the parts with 5-minute epoxy or a hot glue gun, and depend upon the fiberglass layup for strength.

In either case, mix or flox must be used to form a fillet between the parts. A fillet eliminates the sharp inside corner formed by the two components. While

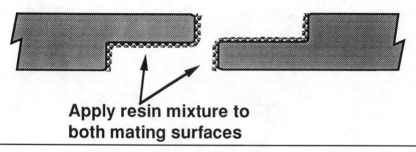

Apply resin mixture to both mating surfaces

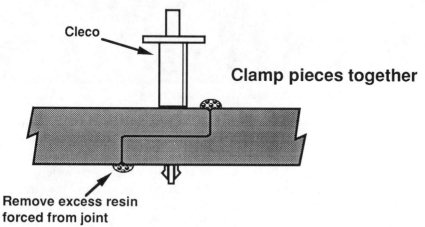

Cleco

Clamp pieces together

Remove excess resin forced from joint

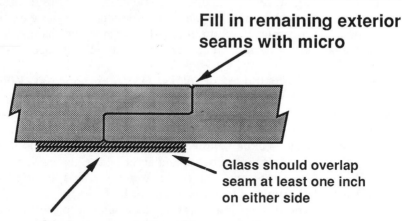

Fill in remaining exterior seams with micro

Glass should overlap seam at least one inch on either side

Reinforce with layers of fiberglass and resin as called for in plans (None, inside only, or top and bottom)

Fig. 6-13. The joggled areas are bonded together with thick micro and held with clecos or other clamps until cured.

fiberglass cloth can be formed to fit almost any shape, it has limitations. It won't tuck into a 90-degree corner. The thick mix or flox is used to make a fillet to smoothly transition between the two surfaces.

If your kit specifies a thick mixture, use a lot of it. When the parts are pressed together, the mix oozes out. Slide the curved end of a mixing stick along the joint to make a concave surface (Fig. 6-14). You'd like to see about a 3/8-inch radius. A fingertip is just about right, but make sure you're wearing gloves or blocking cream. When the fillet is formed, scrape off the excess mix and return it to the mixing cup.

If the kit temporarily attaches the parts using 5-minute epoxy or hot glue, apply thick mix to the joint once the temporary glue has set. You can use the mixing stick, or draw a nice bead with a metal cake-decorating tool (the resins will dissolve a plastic one). Shape the mix with finger or mixing stick into the desired radius. A summary of this process is shown in Fig. 6-15.

Once the fillet has set, it's time to add the fiberglass.

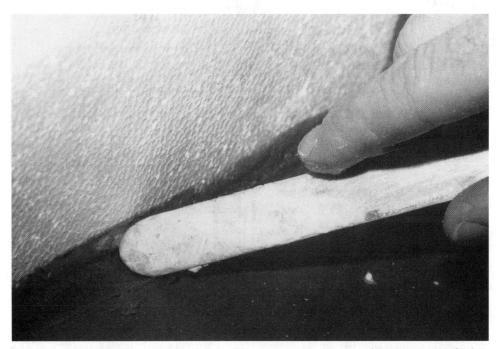

Fig. 6-14. The fiberglass cloth cannot be applied to a sharp corner; a fillet must be added using thick micro. The end of the mixing stick is perfect for making the radius. Excess micro is scraped off and returned to the container.

LAYUP OPERATIONS

When it becomes time to make a layup, you have one overriding goal: saturate the fiberglass cloth with the right amount of resin. Too little, and air bubbles remain in the laminate. Not only do they reduce the strength, they might allow

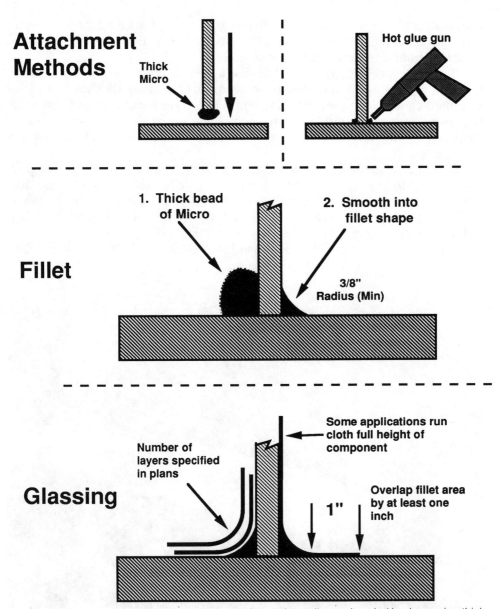

Attachment Methods

Thick Micro

Hot glue gun

Fillet

1. Thick bead of Micro
2. Smooth into fillet shape

3/8" Radius (Min)

Glassing

Number of layers specified in plans

Some applications run cloth full height of component

Overlap fillet area by at least one inch

1"

Fig. 6-15. Attachment methods vary, for instance, Glasair foam ribs are bonded in place using thick mix (as shown on the left), but the prelaminated ribs from another kitplane are first attached with hot glue and a fillet added later.

water to seep between layers of fiberglass cloth. If this water freezes it will cause *delamination* (separation of the layups) and destroy the strength of the bond.

Too much resin adds weight for no gain in strength. Too much resin in a wet layup might actually cause the cloth to float; adding air bubbles as well as moving it out of position.

The following typical layup operation illustrates the step-by-step process.

Foam Preparation

In the rough-and-tumble environment of shop operations, foam parts sometimes get slightly dinged. The cloth tries to assume the shape of the dent and will leave an ugly depression. If it can't fill the dent, an air gap is left as a starting point for delamination. Ordinary dents are easy to fix with a thick mix of micro (Fig. 6-16). Fill the holes, and scrape off the excess. Let it cure, and sand it flush with the rest of the surface.

Styrofoam's open cell structure soaks up a lot of resin, which adds a lot of unneeded weight and might produce a poor fiberglass bond. The surface must first be sealed. This is done with a *slurry*, which is made from resin mixture and glass bubbles in about a 1-to-1 ratio (by volume). Slurry should have a consistency of a cheap milkshake, pouring easily but still thicker than water (Fig. 6-17).

Pour the slurry onto the foam, and spread it around with a squeegee. The slurry is thin enough to fill the cells. Work it around, pausing every few seconds or so to scrape the squeegee across the lip of the mixing cup to remove the excess. Continue until the surface appears sealed.

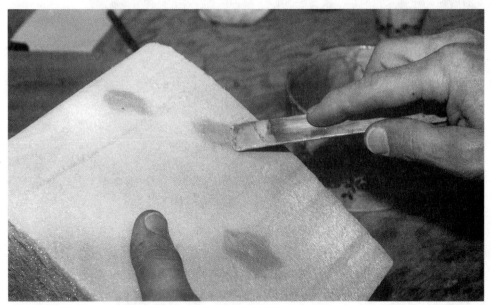

Fig. 6-16. Dents and dings in foam are first filled with dry micro.

Cutting the Cloth

Usual practice is to cut out an oversized piece of cloth and trim to fit. Lines can be marked with a felt-tip pen, but the remaining ink should not be contained within the bond area.

When using bid, the plans should specify the *bias* to which the cloth should be cut. Bias is defined as the angle to the weave. The usual practice is to cut at a

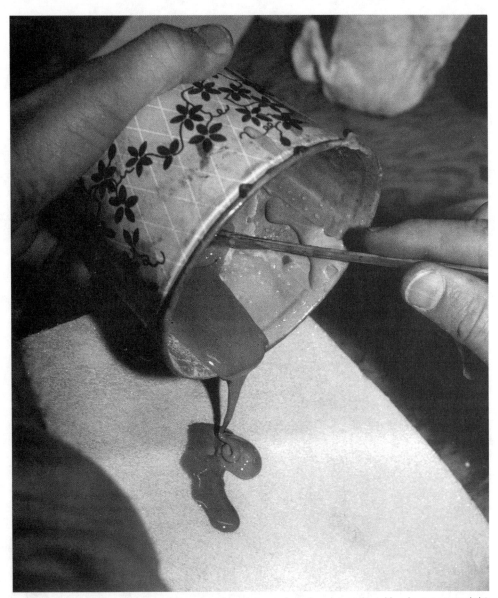

Fig. 6-17. To keep the styrofoam from absorbing too much resin mixture (resulting in excess weight and poor adhesion), its pores are first sealed with slurry.

45-degree bias. That way, the edges of the cut are slantways across a number of threads. A zero-degree bias, on the other hand, makes the cut along the same direction the threads run. The cloth edge unravels a bit easier.

Figure 6-18 shows examples of 45-degree and zero-degree bias cuts. The bias orientation doesn't have to be exact; you don't have to get out a protractor. Eye-balling it is usually good enough. Uni, of course, is always cut directly across the weave.

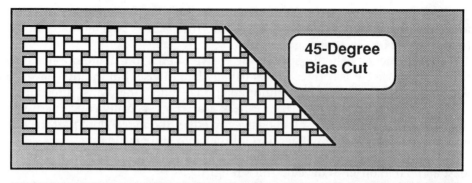

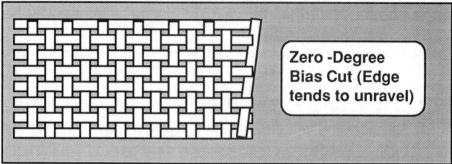

Fig. 6-18. Bias and zero-degree cuts on bid. Note that the yarns tend to unravel on zero-degree cuts.

If the edges of the piece are supposed to be cut away after the glass is laid in place, lay strips of ³/₄-inch masking tape along the borders. Mark the line on the tape, cut, and the remaining tape will hold the edges of the cloth together. Do not use this method if the cloth's edges are to be glassed down because the tape *cannot* be neatly removed.

Two final points: First, don't scrimp. You must totally cover the area specified. Every fraction of an inch contributes to the total strength; if a cloth piece is too short then the part is too weak.

Second, a good, sharp scissors is important because they cut *glass*, not nylon or dacron. A cheapie model dulls quickly, and a dull tool makes ragged edges. Don't fight it. Pay $30 or so for quality scissors and sharpen them frequently.

Shaping and Trimming

Take a look at a square of bid. The cloth has a loose weave; notice the small gaps between threads and the threads that aren't held firmly in place. Pull on the ends, and the piece stretches almost 50 percent in length. The width narrows at the same time. Shove your thumb into the middle. The cloth will assume an even curve without distorting.

That's the magic of fiberglass. The loose weave allows the threads to shift slightly to accommodate compound shapes. As long as no major holes get opened, it'll come out just as strong.

Back to our layup. Hold the cut cloth near the layup area and approximately form it to shape. You're merely making sure the piece is large enough, not getting it perfect before applying the resin.

The instructions should specify the weave orientation, therefore make sure the cloth you've cut is correct. Uni is usually lengthwise, but multiple layers are sometimes laid at slightly different orientations. Follow the plans' directions.

If there is definite excess, trim it back a bit. Be especially concerned where the edge of the cloth will just dangle over an edge. Too much cloth past the edge, and the dangling weight will tend to lift the cloth still on the part.

When glassing two parts together, the cloth should overlap each part by at least an inch in order to develop full strength.

Basic Layup

If you are applying the cloth to a fiberglass kit part, apply a thin layer of resin to the layup area. It fills the rough surface and helps saturate the down side of the cloth. Hidden air bubbles are thus reduced. Don't paint the area with resin if the cloth is being applied to aluminum or another smooth surface. Oversaturation results.

Next, lay the cloth in place (Fig. 6-19). Use your (gloved) fingers, the mixing stick, or anything else cleanable or disposable to push the cloth into all the nooks and crannies. It must lie flat against the surface, or the gap is a starting point for delamination.

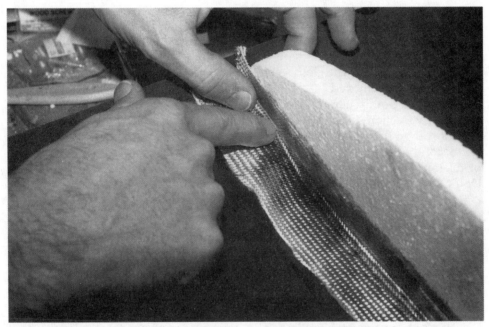

Fig. 6-19. After applying a layer of resin mixture, lay the cloth down. Make sure it makes solid contact on the fillet. The weave should be straight, as waves reduce the strength of the layup. Gloves or barrier cream should be worn.

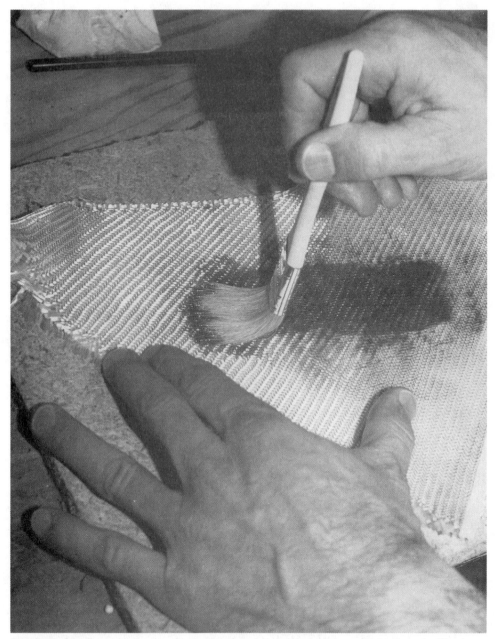

Fig. 6-20. Stippling with the brush ensures the cloth is thoroughly soaked with resin. The force should always be perpendicular to the surface; any slant might shift the cloth.

Stretch and push the cloth as required to eliminate folds and wrinkles. Make sure the weave doesn't, well, "weave." It should run in a straight line across the part. This is vitally important. A wiggle in the weave is like slack in a rope. Get it as straight as humanly possible.

Brush resin atop the cloth, starting in the middle and working out toward the edges, to chase the bubbles away. Take care not to break them into smaller bubbles because that actually makes the situation more complicated.

If possible, use a squeegee to further spread the resin and eliminate bubbles. Don't use too much pressure, that might shift the cloth. If the layup is in a position that can't be reached with a brush, be especially careful not to apply too much resin.

One way to eliminate air bubbles is by *stippling* with the brush. The heel of the brush is pressed down on the cloth to force resin through to replace the air (Fig. 6-20). Note that this counters the usual *painting* procedure, where you don't shove the brush down hard.

Vinylesters have a unique problem. Where epoxies harden gradually, vinylester resin gels suddenly and with little warning. If a part gels before it gets saturated, it must be reworked. Once it has cured, sand away the unsaturated area with 80-grit or coarser sandpaper. Be careful in this process because you don't want to damage the previous layer. Once the unsaturated area is gone, cut a piece of cloth wide enough to cover the area with some overlap and resin it into place.

A well-done layup should look wet, with no greyish-white air bubbles or puddles of resin. The cloth pattern should still be visible. The color is a nice, rich gold. Take care to wipe up any resin dripped on the surface, especially if the drops land in a place where glass will later be applied. Once cured, they're hard to remove without damaging the surface.

If another layer of fiberglass is to be applied, lay it down while the previous layer is still wet. If it has cured, you'll have to sand lightly then apply resin. Take care with the weave orientation because, again, all layers are usually oriented the same way.

If the area is hard to work in, let the area get a little tacky before adding the next layer of fiberglass. Otherwise, your efforts might shift the previous layer, throwing off the layup and adding air bubbles.

Selected multiple layups are awkward to apply to the airplane. There's a sneaky little trick you can use. Cut the layers of glass to size and tack a large piece of plastic wrap or mylar to your benchtop. Make the layups on the plastic, instead of on the airplane. Make sure the edges of each layer of cloth are lined up. When the last layer has been added, resin the target area on the airplane, then apply the multiple layers to the aircraft as one piece.

Figure 6-21 summarizes the glassing process, and Fig. 6-22 shows a completed Glasair rib.

Trimming and Shaping

At some point, the resin-soaked cloth becomes tacky enough to stay together without yet hardening, called the *green cure* stage. At this time, the cloth can be trimmed using a very sharp tool like a single-edged razor blade. When trimmed, press the loose ends of the tacky cloth into the surface.

You can shape the fiberglass once the resin cures, but you won't like it. It's very hard and stiff. Use files and coarse sandpaper.

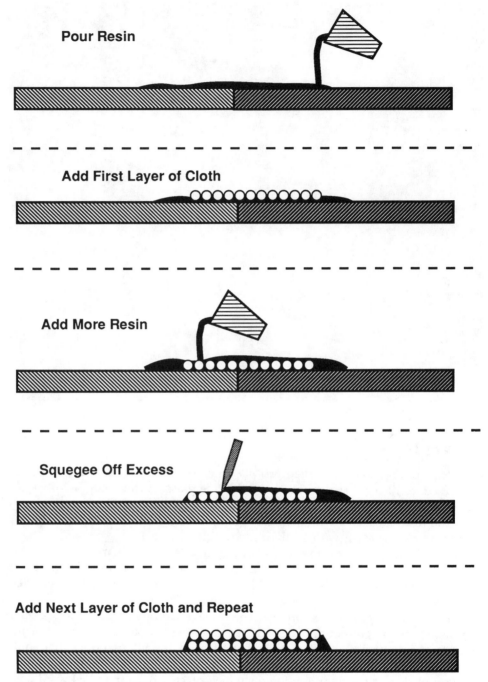

Pour Resin

Add First Layer of Cloth

Add More Resin

Squeegee Off Excess

Add Next Layer of Cloth and Repeat

Fig. 6-21. A summary of the overall process. Peel ply is the last layer.

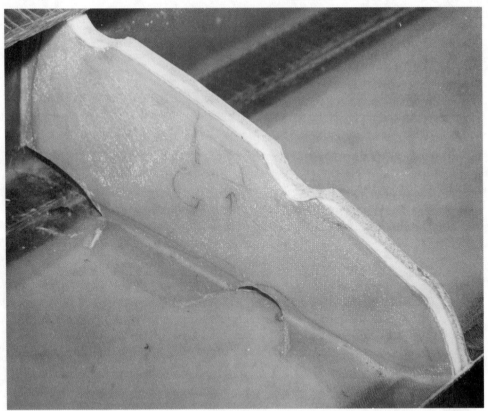

Fig. 6-22. A completed Glasair rib.

Peel Ply

Strips of dacron cloth, called *peel ply*, have several uses in fiberglass work. Peel ply doesn't bond structurally with the fiberglass. It can be applied, wetted out with resin, and left to cure. Even when the cure is complete, the peel ply (and the resin atop and within it) can be removed with a simple tug. The top layer of fiberglass is left with a uniformly rough surface.

Its major use is to ensure a good bond with a layup to be applied later. One problem with layups is that the brush strokes sometimes remain in the cured part. The location has to be thoroughly sanded first to eliminate the imperfections. And as you'll find, sanding cured resin is a tedious task.

An associated problem is comfort. The brush swirls are sharp, called *spikes* in the trade. They might not draw blood, but handling the cured part gets uncomfortable. Apply peel ply to all external layups, and any internal ones that you'll be working around.

Peel ply is also good for holding resin in contact with cloth in awkward positions. For instance, a sharp corner might allow resin to flow downhill, starving the bond area. A piece of peel ply laid on the starved area will keep the resin in place.

Be sure to remove the peel ply after the resin cures. Especially, don't try to glass another part down over peel ply. It will not bond, but might stick just well enough to fool you. Peel ply can be identified by its close, almost invisible weave pattern.

No-Stick Application

Peel ply is great for limited areas, but occasionally you don't want the glass to stick to a surface. Examples include removable fairings or covers. A mold is made from foam or other materials, and fiberglass is applied. When the resin cures, you want to be able to pull the completed part away from the mold with little bother.

The secret is to give the surface of the mold as slick a surface as possible. If the mold could be made from aluminum, you'd be all set. The resin won't stick to it. However, if the mold could be made from aluminum, so could the part.

Because the usual mold material is foam, the surface must be made impervious to the resin. The first step is a heavy coat of slurry. You aren't interested in saving weight because the mold stays on the ground.

The slurry's surface isn't especially slick. Sand it down as well as possible, then add a couple of coats of varnish. Finally, give two coats of ordinary paste wax and shine it up. The paste wax *cannot* contain silicon.

For you belt-and-suspenders types, a quart of mold release sells for under $10. This is the same stuff the manufacturer uses on the kit parts. Just brush or spray it onto the mold after the last coat of wax. After the part has been peeled away, use acetone to remove any release residue.

Other options abound; remember, all you need is a smooth surface. For instance, don't bother with the sealing, simply wrap the mold with wax paper or plastic. Glue it in place or apply cellophane tape to the exterior. Wax paper is dandy for removable parts like fairings. Cover the area with it and form the part. When the resin has cured, remove the part and peel off the wax paper.

FINISHING FIBERGLASS

Just about every kit comes with several fiberglass parts. Usually, they aren't ready to paint and some surface finishing is necessary. An exception is gel-coated parts. Gel-coating is white and shiny, and needs only a light sanding before painting (Fig. 6-23). Some parts might require trimming before installation, as shown in Fig. 6-24.

Most kit parts are made in female molds that produce a reasonably smooth surface. The glass and resin produce the mold's smooth texture. Not so with the shop-made layups because the resin hardens in whatever shapes, patterns, and swirls the builder's brush applied. Peel ply is intended to produce the optimally rough surface for applying additional layers, not a smooth exterior finish. And even with a perfect application of resin mixture, the weave of the cloth will still be showing.

Fig. 6-23. The fiberglass hull of this Avid Amphibian comes gel-coated, which requires only minor sanding prior to painting. The darker areas where the sponsons join the hull indicate where filler has been applied to smooth the seam between sections.

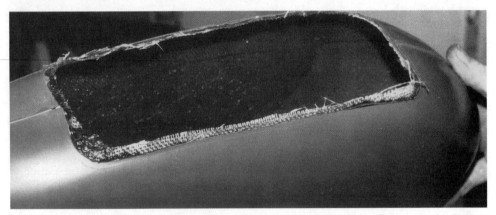

Fig. 6-24. Note the rough, raw edges of fiberglass left on this wheelpant. These can be filed or sanded away, or even cut with a saber saw if the part can be properly supported.

The following section addresses how to get the hand layup to a reasonable state of smoothness, then how to get a glass-smooth finish on either the layups or provided parts.

Smoothing the Shop Layup

If a fiberglass layup is located inside the structure, smoothness doesn't make any difference. Externally is another matter. The excess resin can be sanded down to the surface of the cloth without loss of strength. But the resin generally doesn't fill in level with the weave; the weave pattern remains, and painting makes it even more noticeable.

The solution is to sand, then fill with a dry micro mixture. This is a pretty horrible job over large areas, like older model homebuilts that are totally covered with fiberglass. A KR-2 has about 200 square feet that must be sanded, filled, sanded again, and so on.

Fortunately, modern kitplanes reduce the area that must be filled. An example would be the horizontal stabilizer joint to the fuselage (Fig. 6-25). Fiberglass strips are used on the exterior to strengthen the bond. The weave in the strips must be filled, and the edges faired to make the strip invisible once the area is painted.

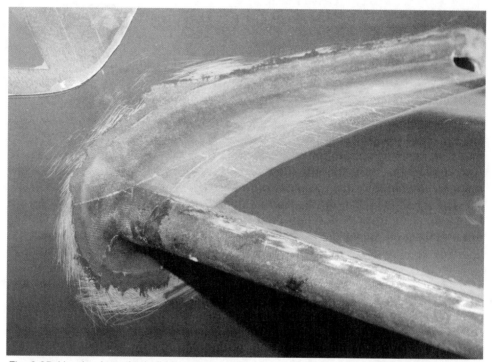

Fig. 6-25. Here's where the builder spends his time when building a composite aircraft. The weave of the fiberglass cloth applied to this horizontal stabilizer must be filled prior to painting.

Begin by sanding the area with coarse sandpaper, 30 to 80 grit. This knocks down the roughness of the cured resin. *Do not* sand into the fiberglass. If you do, the typical repair method is to make another layup, overlapping the damaged area by two inches or so. Check with the kit manufacturer for exact repair methods.

The weave will be filled by covering the surface with dry micro and sanding it flush to the weave. Low areas, or spots that must be built up, should be prepped first by adding a thin coat of resin mixture. Then make a batch of dry micro; five parts or more of glass balloons to one part epoxy or catalyzed resin (mix by volume, not weight). Essentially, just enough bonding agent is added to hold the mass of balloons together. By limiting the amount of resin, dry micro doesn't add much weight.

Use a putty knife to apply. Fill the low spots, then smear on a coat of dry micro. It's not especially easy to work with because it tends to crumble and ball up. Don't be tempted to add more resin because that will just make it harder to sand. When the micro has cured, don your respirator and sand the surface smooth with 150-grit paper. The process is shown in Fig. 6-26.

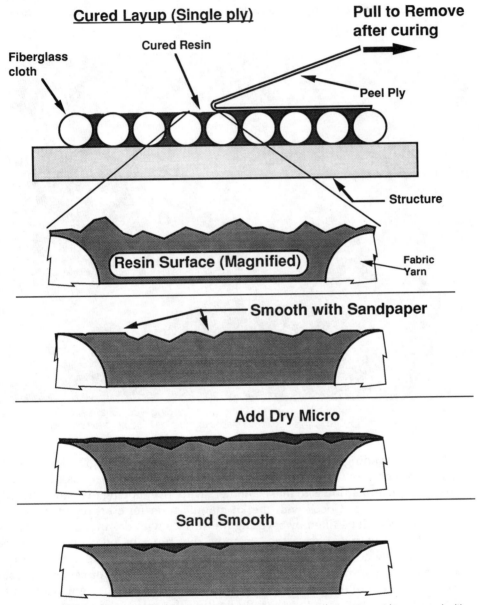

Fig. 6-26. Filling the weave using dry micro. Following this process, the part must be sprayed with a filling primer to fill pinholes too small for the micro.

You won't get it smooth in one application. It will probably take at least two coats.

Most of the micro will end up on the floor and on you. Wear old clothes and a shop coat. Have a hand-held vacuum convenient to slurp up the dust and to clean yourself off before going into the house. A floormat—damp if you can arrange it—reduces tracking the dust into the house. Cover your supply of fiberglass cloth before starting.

As you might gather, it makes sense to wait on your dry-micro activities until they all can be done at once.

Fine Filling

The surface of smoothed shop layups and supplied fiberglass parts still is not perfect. Pinholes can be left by both the mold and the dry micro/sand process.

The traditional way of eliminating pinholes is with the application of Feather Fill or a similar product. It must be mixed carefully according to the directions and applied with a brush or spray gun. Let it cure for two hours or so, then sand with 150-grit. Repeat as necessary, and again, don't forget your respirator.

Feather Fill is one of several products that can be used to fill the pinholes. Ask for a *filling primer*. These aren't aircraft-specific products; in fact, Feather Fill is sold at auto-parts stores. Check with the kitplane manufacturer or fellow builders for alternatives.

However, whichever product you use must be compatible with the primer and topcoat with which the aircraft will be painted.

On first glance, composite construction is the simplest: "All ya gotta' do is glue the parts together." As you've seen in this chapter, it just isn't true. It's hard, dirty, smelly work. But done right, it does produce the slickest airplanes this side of Mach 1. And nothing is *difficult*; it is just a series of logical, easy steps.

Compare it to the alternative. Tube and fabric airplanes fly slowly. Wood airplane kits have far less work already done for you. So it boils down to either slinging a lot of resin, or banging 10,000 rivets.

Composite kitplanes have nothing to apologize for. They're fast to build, and fast in the air. With all the glass airplane kits out there, they must be doing something right.

7

Metal Monocoque Construction

METAL MONOCOQUE COMES CLOSEST to being the "modern" traditional method of building an airplane (Fig. 7-1). Most factory aircraft are built this way: thin metal skins, bent and curved for strength, held together by rivets. Construction usually isn't true monocoque, but a modified form that retains some features (longerons, for instance) from truss-type construction.

This chapter covers those aircraft constructed primarily of aluminum with the metal skin absorbing flight loads. It's amazing the amount of strength a puny piece of sheet aluminum gains from clever design and a rivet or two. Skins and bulkheads are typically .040-inch thickness or less. Certain smaller metal monocoque homebuilts are built from a couple of 4- × -12-foot sheets of aluminum. Cut out and form bulkheads, ribs, and spars; cut out, bend and rivet the skins; and the result is an airframe good for more than six Gs.

The RV series (Fig. 7-2) is currently the most popular metal monocoque kitplane. The examples in this chapter are mostly RV-4s and RV-6s. Zenair has a considerable line of metal kitplanes, as well (Figs. 7-3 and 7-4). All-metal airplanes at opposite ends of the performance scale are the Questair Venture speedster and the Mosler (formerly Monnet) Moni motorglider.

This chapter is mostly concerned with sheet metal techniques:

- Making long, straight cuts.
- Making smooth bends in long sheets.
- Riveting (pop rivets are covered in chapter 8).
- Corrosion-proofing aluminum.

ADDITIONAL TOOLS

In addition to the basic tools specified in chapter 4, more tools are necessary:

- An air compressor of about two horsepower and 20-gallon capacity. You are going to be operating tools with the compressor, so the little $100 tank-

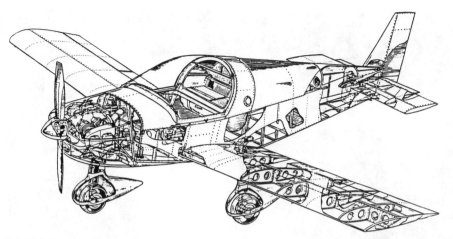

Fig. 7-1. The Zenair series is an example of classic metal monocoque construction. Zenair Ltd.

Fig. 7-2. The RV-4 was one of the first monocoque aircraft kits.

Van's Aircraft Inc.

less models won't do. A good compressor will cost between $250 and $500, or less if bought used.

Two useful compressor accessories are a moisture trap and an oiler. The moisture trap separates water from the air, which is necessary for paint spraying. Pneumatic tools, like rivet guns, need a bit of oil with their air, hence, the addition of an oiler.

Needless to say, the two items are incompatible. If you install an oiler, remove it before painting and use a different air hose because the old one is probably soaked with oil. Or just install a moisture trap, and manually add a drop of oil to the tools before connecting them to the line.

- A pneumatic rivet gun, based on the largest diameter rivet used in the air-craft. A 2X gun is limited to $1/8$-inch rivets, the 3X to $3/16$-inch, and the 4X to $1/14$-inch. A 3X is probably what you'll need; good kit instructions should tell you which to get. Cost: $100–$150.

While the rivet gun supplies a hammering force to drive the rivet, it isn't the same as a pneumatic or air-impact hammer. They look identical,

Fig. 7-3. The Zenair Zodiac was designed for a VW engine, but now uses the Rotax 912 four-stroke.

Fig. 7-4. Most small STOL homebuilts are tube-and-fabric, but the Zenair CH-701 is all-metal. It takes off and lands in little more than 100 feet.

and the hammer costs only $25 or so. But a rivet gun's trigger operates like a variable speed drill. The air hammer's trigger starts the tool at full force, while the rivet gun can be started gradually. You *can* rivet with an air hammer, but it's a tricky operation. A typical kit has 5,000 rivets, and a gun is worth a penny per hole.

• Bucking bars, of various sizes and shapes buy two or three different ones to start. They run around $10–$15.

- Rivet cutter, $20 – $40, plus rivet sets, dimpling dies and blocks as applicable, countersinks, fluting pliers, hand seamers, deburring tools, and a number of other small hand tools costing less than $15 each. And lots and lots of clecos. Find out the sizes of rivets you'll be driving and buy at least 100 clecos per size, about $30 worth. Buy an extra cleco pliers as well.

  One larger-cost option is a hand rivet-squeezer; they're easier to operate than a pneumatic riveter, but can't rivet more than a few inches from a sheet edge. One costs about $100.
- Drift punches in the same sizes as the rivets. These are really nothing more than solid steel rods with one end slightly pointed.

One could go broke buying every tool useful for sheet-metal work. The above should get you started. Others can be picked up as needed.

Several companies offer tool kits for RV-brand aircraft builders. These kits include the rivet gun, sets, bucking bars, and the like. Whether to buy one depends upon what tools you already own and what kitplane you're putting together. Zenair builders, for example, use pulled (pop) rivets [subsequently explained in the next chapter] instead of driven rivets and don't need the rivet guns and bucking bars. But for as many rivets as will be driven, you'd end up buying a pneumatic rivet puller anyway.

Optional tools are suggested in the context of the various operations described.

MATERIALS, FASTENERS, AND SAFETY

Recall that aluminum monocoque construction uses traditional materials and methods.

Materials

The primary materials you'll be working with are sheets of aluminum, either 2024-T3, or less often, 6061-T6. The basic rules, such as edge margin and bend radius, were given in chapter 5.

Few kits supply parts ready to be installed. Most have rough edges that must first be deburred. Details are also given in chapter 5.

Because you are going to use the material for primary aircraft structure, there are a couple of additional precautions. The first is cosmetic. Many pilots like the appearance of polished metal aircraft. If you are planning on leaving the exterior unpainted, take special care not to scratch or ding the skin during construction.

This seems like a trivial precaution, but it isn't. By definition, building a metal airplane means extensive use of metal cutting tools. A moment's inattention while carrying a drill might ruin a skin. The difficulty of the task is indicated by the relatively low number of metal homebuilts left unpainted. If you aren't going to paint, grab some blankets and sheets and cover portions of the aircraft that aren't being worked on.

There are other aspects to handling sheet aluminum. Most references will tell you that a corrosion-resistant coating must be applied to every square inch of aluminum within a closed structure. Look inside a lot of old production planes; I had a 1965 Cessna 150 without internal protection, and it didn't have a speck of corrosion.

The exterior can be left unpainted, as it can be easily inspected and polished. But the inside of the wings, tail, and fuselage can corrode unobserved. If you live in a high-corrosion environment, such as near salt water, you'll probably want to paint the inside.

The techniques for applying corrosion protection are given later in the chapter, but there's one point to remember now: The presence of finger oils on the sheet reduces the primer's adherence. You'll wash and etch the metal prior to application, but the more oils on the sheet, the more chance some might get missed.

Consider wearing thin cotton gloves or rubber surgical gloves when working on unprepared aluminum. Not only does it keep your hands cleaner, but it cuts down on the irritating little scratches and nicks from rough metal edges. Once the primer coating is prepared, you can operate much more freely. All the more reason to apply the primer early.

Fasteners

The fastener of choice for metal aircraft is the rivet. There have been a few experiments with other means, but the lowly little .005-ounce pieces of aluminum are used in everything from Teenie-Twos to 747s.

What else would you use? Small nuts and bolts would be heavy and expensive. Welding aluminum is difficult, and is impossible to undo. Similarly, bonding is permanent and uses dangerous chemicals, besides. Rivets are cheap, light, nontoxic, and easy to install and remove.

A rivet is like a bolt without threads, stuck through both pieces of metal to be joined. The end of the shank is then distorted so the rivet remains in place while pinning the components tightly together.

This distortion can be applied in two ways. Sometimes, the rivet is hollow with a shaft inside leading to a bell-shaped end. A special tool pulls the end into the hollow shank, widening the hollow shank. When the maximum pressure has been applied, the shaft breaks at a predetermined location. This is a *pulled*, or *pop* rivet, and its use is discussed next chapter.

The other type of rivet is the *driven* rivet, and it's the one most commonly used on aircraft. Unlike the pulled rivet, these are solid metal. The end of the shank is distorted by repeatedly ramming it into a hard chunk of steel, called a *bucking bar*. The rivet is softer and flattens instead of the bar. Steel rivets used on buildings and ships are installed red-hot to ensure that they are softer than their bucking bar.

The distortion of the shank end forms what is called the *shop head*. The *manufactured head* is the one the rivet came out of the box with. Two types of manufactured head are generally used: the MS20470 (AN470) with a *universal head* (sort of a flattened oval), and the MS20426 (AN426-100) with the *countersunk head*.

For most applications, the standard MS20470 universal head rivet is used. The MS20426 rivet is the *flush* rivet; when installed properly, no portion sticks into the slipstream to add drag. This requires dimpling or countersinking the skin before installation.

Aluminum rivets come in two alloys, 1100 and 2117-T4. The 1100 alloy is soft and can only be used in non-structural applications. The 1100 alloy rivets have an A suffix on the specification (MS20470A, for example), and the 2117-T4 use an AD suffix. Unless you have a definite call for the softer rivets (they are easier to drive, and work fine for interior trim, for example), always order MS2040AD or MS20426AD rivets.

What if the labels fall off your rivet boxes? The structural rivets will have a small dimple in the middle of the manufactured head. This isn't just for identification, it's an aid to removing the rivet should installation go wrong.

While AN bolts with a variety of head markings can be interchanged, only rivets with *dimples* on the heads should be used for structural purposes. Other markings mean other alloys or other degrees of heat treatment.

Specific rivet sizes are specified by a dash number indicating diameter and length. For an MS20470AD3-2 rivet, for example, the 3 indicates the diameter in $1/32$ of an inch, and the -2 indicates the length in $1/16$. So an MS20470AD4-10 rivet has a standard head, is made of 2117-T4 alloy, and is $1/8$-inch in diameter ($4/32$-inch) and $5/8$-inch ($10/16$-inch) long.

Safety

You can always recognize an old-time riveter's convention because they stand around yelling "WHAT?" and "SPEAK UP!" or "SAY THAT AGAIN!" In other words, riveting is a noisy business. Wear ear plugs as a minimum; hearing protectors are preferred. Wear the plugs for normal work, and put on the earmuffs as well when you pick up your rivet gun.

Don safety glasses before cutting, drilling, or grinding. Metal work produces dust and a lot of sharp edges, so take whatever other precautions you prefer. I wear a shop coat but no gloves, for example. The aluminum dust goes everywhere, so it's nice to have the coat to shrug off before entering the house.

While aluminum work doesn't involve the witches' brew of chemicals that composite construction does, it has its own problems. Most primers are hazardous; the most common, zinc chromate, is suspected of causing lung cancer. You can't even get true zinc chromate in hardware stores anymore, so buy it though aviation suppliers.

METALWORKING

The basic rules for metalworking were given in chapter 5. The following sections concern cutting and forming operations that go beyond making simple fittings as described in chapter 5.

How much you'll have to do depends on how much money you're willing to spend. Complete kits include all major cutting, forming, and bending (Fig. 7-5). But you can save a considerable amount of money by doing these operations yourself. It's the age-old trade: bucks versus building time.

Fig. 7-5. The RV-4 can be built from a complete kit, from subkits, or from plans only. The builder must decide if the time saved is worth the additional money.

Long, Straight Cuts

One advantage of building a metal monocoque airplane is that the metal used is thin; 0.040 inch (¹/25 inch) or less. Aluminum this size is easily cut by a variety of means.

However, what is easy and simple to do on a six-inch pieces gets to be a hassle when the cut is four feet long and when the cut must be a good, sharp, straight line.

Of course, with enough money (or the right friends) the solution to long straight cuts is simple: a *bench shear*. Similar to a bending brake, the bench shear makes large cuts with ease. All it takes is money; these units are expensive. Check with your EAA chapter, find out if someone has one. You probably can do most of your long cutting in a single evening.

With no bench shears, the problem boils down to three factors: finding a believable reference, transferring that reference to the metal, and cutting to that line.

What's straight? Or rather, what do you have around the shop that you can trust as being straight? You might think that the factory-supplied sheet of aluminum could be trusted, but you shouldn't. It can waver ¹/8 inch, or so, over several feet. It might not sound like much, but an uneven edge can really cause a problem if you're trying to butt two edges against each other. But then again, if you can't cut any straighter, you might as well trust the factory edge.

You can buy long straightedges, which are objects with a guaranteed linear edge. Carpenter's squares can be trusted, although they might be too short. With care, you can draw part of the line and slide the square down and extend the line.

But there's one incontrovertable reference: the taut string. When referenced lying flat, it's a fast, easy dead-accurate tool. Note that it cannot be measured from the side because gravity tugs the string into a bowed shape called a *catenary*. Tension can reduce the amount of bow, but never completely eliminate the bow.

So we have our references: taut string, carpenter's squares, straightedges, and the like. Now comes the problem of transferring the reference to the metal.

The archetypical taut string is the carpenter's *chalk line*, which not only provides a straight reference, but leaves a mark on the surface when "twanged." It might work in some cases, but usually the line of chalk is a good 1/8-inch wide.

Instead, get some *1/2 A Control Line* from a local hobby store. This is fine, extremely strong string for controlling small gas-powered U-control models. Stretch it tight across the metal, then spray machinist's ink or paint over the string and metal. Not a heavy coat; just a dusting of color. The string acts like a stencil. When removed, it leaves a fine line of bare aluminum.

For various forms of straightedges, sharp pencils or a scribe work adequately. Apply machinist's ink first so the line is easy to see. If using a scribe, don't make any marks on "good" metal. The Sharpie brand of felt-tip pens is one of several that leave a good line on bare aluminum.

Once the line is marked, you're ready to cut. The use of aviation snips, nibblers, and various power saws are all explained in chapter 5. Larger pieces of

Fig. 7-6. Edges of thin sheets can be smoothed with a deburring tool with a V-shaped notch.

sheet might be difficult to feed through a band saw, and properly supporting the metal for a saber saw might be tough, too.

One tool not mentioned in chapter 5 is *power shears*. These are essentially electric- or air-powered aviation snips. These cost around $150, yet might be worth it if you plan on cutting a lot of metal.

Eventually comes the moment of truth. How close do you cut to the line? The farther from the line, the more additional work to trim it closer. Too close, and the entire piece might be ruined. It depends on how much you trust yourself. For thinner metal, you might prefer to use a hand tool for better control. It's up to you.

Don't forget to file and deburr the edge. Special deburring tools for sheet edges make the job easier (Fig. 7-6).

Cutting Large Holes

You'll have occasional need to cut largish holes in aluminum sheet. Typical uses are lightening holes, for weight reduction, or access panels, for inspections and maintenance.

It's perhaps best to cut round holes, if possible. Square corners, even if rounded, can concentrate stress and give cracks a place to start. In the second (and better) place, there are a number of tools that greatly simplify cutting round holes.

The ordinary hole saw is the first option. Essentially an open cylinder with one edge covered with teeth, sizes are available to two inches or so in $1/8$-inch increments. Larger sizes are sold for specific tasks, like cutting instrument holes. Read the labels carefully when buying hole saws at the hardware store because many are intended only for wood. They'll cut holes in metal, but won't last very long.

One drawback is that a saw must be obtained for every size hole to be drilled. For holes up to $3/4$ inch or so, a *step drill* might be preferred. This is a cone-shaped bit, with steps on the exterior for various sizes of holes. There are a number of other custom hole tools on the market, able to cut or punch holes up to four inches or so. A fly cutter (Fig. 7-7) works well on thin aluminum.

When using a hole cutter, add *cutting fluid* to reduce friction and heat and extend the life of the tool. In a pinch, straight oil will work, but cutting (*tapping*) fluid is cheap and widely available. There are some formulations intended specifically for aluminum. Apply it to the tool/metal interface while cutting, and lift the tool away occasionally and squirt a little on the cut itself.

Access holes aren't always round. Sometimes a rectangular hole gives better access to a particular area.

In this case, remember that you don't *want* nice square corners. The hole must have rounded corners, or it'll be a starting point for cracks. An inch or so radius should be about right.

To make a rectangular access hole, mark a square with dimensions equal to the final size minus the desired corner radius. In other words, for a 5-×3-inch hole with 1-inch radius corners, mark a 4-×2-inch square. Drill a pilot hole at each corner, and follow up with the corner-radius hole saw.

Fig. 7-7. The fly cutter works best in a drill press. This rib is screwed onto a piece of wood to hold it steady while cutting the lightening hole.

Lines drawn tangent to the outer edges of each hole will define the correct final size. A saber saw with a metal-cutting blade is a good tool for removing the metal within the rectangle. Take care when starting each cut because the saw can cross the line quite quickly.

Bending and Forming

Aluminum sheet has many advantages as an aircraft construction material: it's strong, light, and fairly cheap: it's also easily worked, up to a point.

Sheets of aluminum and paper share similar workability characteristics. It's easy to put a simple bend in the material, and the bend adds strength as well.

Take a sheet of paper and draw a clock face on it. Now try to shape it into a bowl. Doesn't work very well, does it? As the "rim" of the bowl rises, the 12 and 6 o'clock positions come closer together. But the distance between the two points *around* the rim (from 12 to 1 to 2, through to 6) stays the same. Something has to give.

A simple bend wouldn't be any problem. The crease would go from 9 to 3. No matter how close the 12 and the 6 were brought to each other, the distance around the rim stays the same.

But we're making a bowl. As the 12 and the 6 are brought closer together, the 1 and the 7 are coming closer, too. So is the 2 and the 8. The 9 and the 3. And so forth.

The net effect is that the diameter of the rim decreases. But where does all the extra paper go? Into ugly folds and wrinkles. You can make a bowl out of paper, but only by cutting pie-shaped wedges from the rim to the center to get rid of the extra material. That's OK for paper, but not for a man-carrying aircraft.

Complex or *compound* shape is the aircraft builder's term for curves beyond a simple linear bend. The stiffer the material, the harder to curve. Fiberglass cloth is limp until soaked in epoxy and allowed to harden. Composite airplane builders don't soak the cloth until it's in the final shape desired. Molten aluminum is limp as well; it can be cast into all sorts of convoluted forms.

Aluminum sheet doesn't like to form complex shapes. (I didn't like school, either, but Mom made me go.) With a little work, you can force aluminum into the shapes needed to build metal aircraft.

We form metal two ways: *stretching* and *shrinking*. Take stretching first. If you take a piece of 1/4-inch 6061-T6 plate and bend it 90 degrees around a 1/2-inch bend radius, the metal on the outside of the bend has to stretch more than 3/8 inch. Where did the extra metal come from?

If you could fit a micrometer in the middle of the bend, you'd find that the metal is now thinner than the original one-quarter inch. Stretching is nothing more than converting thickness into length. Thin sheet doesn't have that much thickness to give up, but at the same time, the difference in radius between the inside and the outside of the bend isn't very much. The minimum bend radius tables are based on the amount of stretching the metal can withstand before tearing.

Stretching is simple enough, then. Now shrinking should be obvious as well: Excess metal is absorbed by a slight increase in the thickness of the material. In other words, if a small bulge in .032-inch aluminum is pounded flat, the metal in the region of the bulge might increase to .033-inch to absorb the extra metal.

Your mind might boggle slightly at the concept of actually forging the metal by hand. But it's nothing more than blacksmiths have been doing for thousands of years. Our high-tech materials are stronger, but don't try to make a horseshoe out of your kitplane's ribs. The hand workability of aluminum has its limits.

Actually, there are simple and effective ways around the shrinking problem. The excess metal can be used up in other ways, such as forming deep wrinkles. These wrinkles add strength as well. Look at a paper cupcake holder. These are circular pieces of paper that can take a bowl shape due to numerous radial creases.

Fear not because you don't have to become a topologist to make metal shapes. Heck, you don't even have to be able to spell topologist.

But how does all this apply to the kitbuilder? The manufacturer says all parts are prebent.

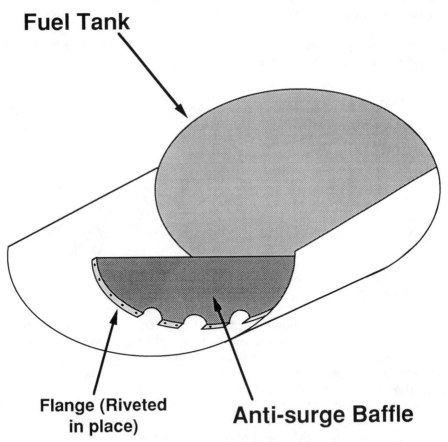

Fuel Tank

Flange (Riveted in place)

Anti-surge Baffle

Fig. 7-8. Fuel tanks require baffles to reduce sloshing of the contents. These are often riveted in place.

Even if you have a "complete" kit, the manufacturer might not completely form all the parts. Rib flanges, for instance, might not have been bent to 90 degrees. The initial crease is there, but the manufacturer expects you to shrink or stretch the material as required to produce the desired angle.

Or maybe you didn't buy the complete kit. It's a lot cheaper to buy the sheet metal and form the parts yourself. Or perhaps you've ruined a kit component, and don't want to go through the delay of ordering a replacement.

Let's step through a metal-forming exercise.

Assume you have to build a fuel tank *antisurge baffle* as shown in Fig. 7-8. The overall shape is that of a half-oval, with a flange on the bottom edge to allow it to be riveted to the inside skin of the fuel tank. A few holes are drilled through it to keep the fuel from surging during rapid maneuvers.

The flange isn't the only possible attachment method. Short pieces of angle could be riveted to the baffle, and thence to the tank (Fig. 7-9). But this would require several additional parts and more rivets. In other words, more weight, more complex, and more expensive—four-letter words to homebuilders.

If the flange were to be added along a straight line, a bending brake would quickly do the job. But the flange is added along a curve. Bending it over means eliminating a bunch of excess metal.

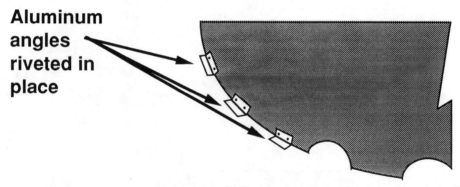

Aluminum angles riveted in place

Fig. 7-9. Riveting angles along the edge of the baffles requires the least metalworking, but adds excessive weight and complexity.

The easiest solution is shown in Fig. 7-10. Just use a hand shears to cut series of notches along the flange, then bend the tabs to the right angle. Or, depending upon the degree of curvature, a series of slots might be sufficient. Either method is acceptable in some cases. But cutting the metal might weaken the baffle. Use this method if allowed in the plans, otherwise do it another way.

The next method is to use up the excess metal by inducing dents, or *flutes* in the flange. Clamp the baffle in the vice between two pieces of wood, with one of the pieces holding the right bend radius for the metal.

Tap the metal lightly with a rubber or plastic hammer to bend the flange portion. Because of the curve of the baffle, you won't be able to work on more than a

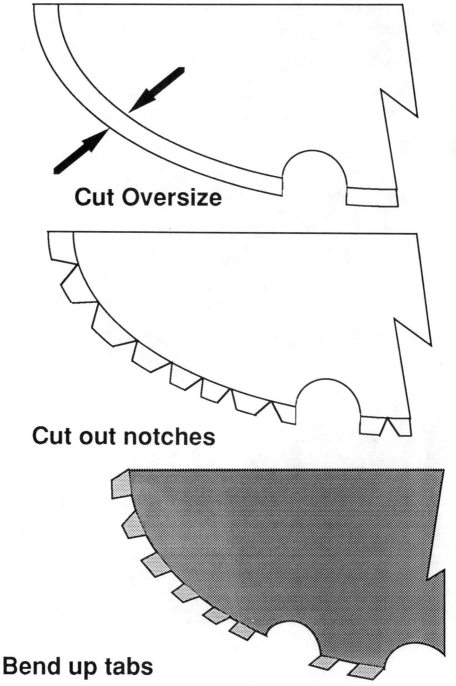

Cut Oversize

Cut out notches

Bend up tabs

Fig. 7-10. Cutting slots or V-notches in the flange area produces tabs that can be bent over and riveted.

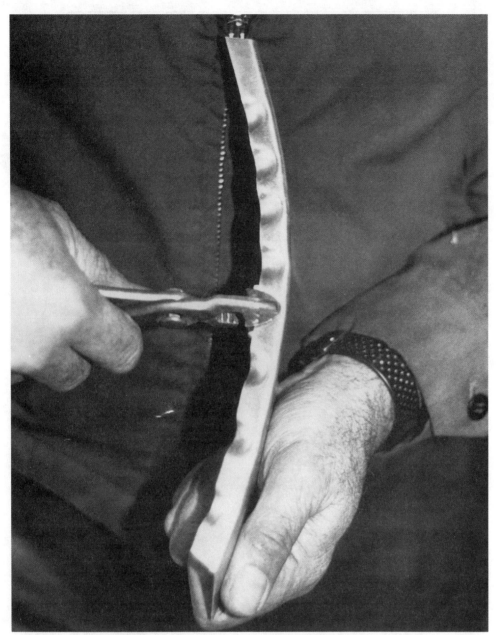

Fig. 7-11. Note how the flutes in the bent portion curve the entire structure. This tendency can also be used to straighten a curved sheet.

small section at a time. Do the actual bending in stages; bend the working section over 20–30 degrees or so, then loosen the vise and slide the baffle so you can work on the next section. Bend it the same amount and repeat. When the whole flange is bent to the same angle, bend one section over a bit farther and repeat the process. Eventually, the whole flange will be bent to 90 degrees.

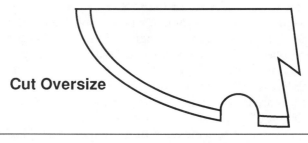

Cut Oversize

Bend up flange as far as possible

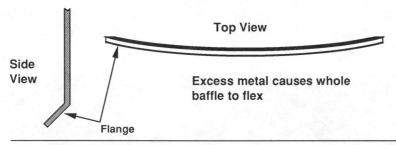

Side
View

Top View

Excess metal causes whole
baffle to flex

Flange

Fluting pliers use up extra metal

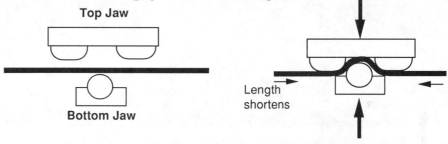

Top Jaw

Bottom Jaw

Length
shortens

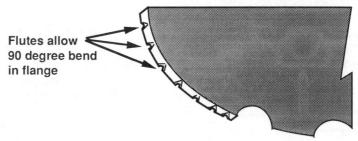

Flutes allow
90 degree bend
in flange

Fig. 7-12. The fluting process.

But the flange and the baffle look awful. Both are wavy and distorted because of the excess metal on the flange.

This is where the *fluting pliers* come in. They look very similar to ordinary pliers except for the jaws. Looking end-on, one jaw has two bumps and the other has one, which fits between the other two when the jaws are closed. The bumps are smooth, and taper towards the tip of the tool (Fig. 7-11). Don't buy fluting pliers anywhere except an aviation supplier. The nonaviation variety are often intended only for heating ducts, where damage to the metal is acceptable.

When the fluting plier's handles are squeezed with the jaws pressed on the edge of a metal sheet, it makes a tapered U-shaped dent. This dent isn't made by stretching the metal. Rather, it's formed from available material.

So the fluting pliers are used to straighten the flange. Crimp carefully on the edge of the flange, and you'll see it start to work straight. A series of small flutes, rather than numerous small flutes, makes a nicer-shaped part. However, don't put a flute anywhere a rivet or a bolt will be installed. If you use a series of small flutes, make sure they don't interfere with later rivet installation.

Eventually, the baffle will straighten out and the flange will be completed (Fig. 7-12). The strength isn't affected. It looks a little ugly, but if it's on the inside of the wing, no one will see it (Fig. 7-13).

Fig. 7-13. Flutes have been added to this rib's flange to straighten it out. Note that the rivet hole positions are marked between the flutes.

The last way uses a combination of shrinking and stretching to end up with a smooth flange without flutes. This uses a *backup block* and *pointed hammer* to shrink and stretch the metal as necessary as the flange is bent. It's a lot of work and not recommended for 2024-T3. It's a rather complex procedure, best learned in person from an experienced metalworker. Check with your local EAA chapter and get some "dual instruction."

RIVETING

At last we get to the nexus of metal aircraft building. You might be able to make pretzels out of $1/2$-inch 2024-T3; punch a round 3.125-inch instrument hole with one blow of your mighty fist; or cut a 10-foot sheet of aluminum into $25/64$-inch strips with only your steady arm and a penknife.

But if you can't drive good rivets, your airplane ain't gonna' hold together. Period.

Riveting isn't that difficult. It requires good judgment and some attention, but with a little practice it's easy.

One point to recall: Usually, the metal must be protected from corrosion before being riveted. Corrosion protection is covered later in the chapter.

Holding In Position

Sometimes, you have the luxury of being able to take the pieces to your workbench to rivet them. Most of the time, you're trying to rivet an awkward-shaped piece of aluminum to another awkward-shaped piece that's already installed on the aircraft.

Obviously, when you start drilling holes, you want the new component to be sitting exactly in its final location.

There are a number of items that help. C-clamps and spring clamps immediately come to mind, and there are several types of spring clamps worked by cleco pliers. Be careful, though, because clamps can apply a surprising amount of pressure. Placed directly on metal, they'll leave marks. Never place metal directly against metal. Stick a largish scrap of wood between a clamp's jaws and the aircraft component. If only moderate pressure will be applied, wrap the clamp jaws with a couple layers of duct tape.

Other options include rope, string, tape, rubber bands, and the like. The important issue is to hold the pieces securely.

But only apply enough force to hold the piece in position. Too much pressure can cause subtle warping of the sheet, not enough to be permanent, but it slightly changes the component's relationships. If riveted in this position, the piece will retain an unsightly warp.

Figuring Out Where the Hole Goes

The first thing to do is determine the hole position. It's nice when this is supplied on the plans, in the form of a template to be taped to the piece. Nice, but rare. Instead, the plans might say "Rivet the bottom edge to the bulkhead." Leaving the number of rivets and their exact location up to you.

And in some cases, you might not want to drill the hole where the template says. It's usually better to drill both pieces at once, to ensure hole alignment.

The first thing to remember is the edge margin: at least three times the diameter of the hole between holes, and twice the diameter to any edge. The equation for how many rivets can be installed is

$$N = 1 + (W - 4 \times D)/(3 \times D)$$

Where W is the width of the riveted area, and D is the diameter of the rivet. Drop any fractional rivets from N; if the answer comes out to be 12.789, drop .789 for the resultant 12.

The actual spacing between rivets is given by the equation

$$S = (W - 4 \times D)/(N - 1)$$

Now, those equations are a *real* pain, aren't they? Relief comes in several forms. First, as long as you observe the edge margin requirements, the exact spacing doesn't really matter. Even spacing looks better, but it's not necessary for strength. Put rivets in a nice row, and slight unevenness won't show.

Second, there's a handy little tool called a *rivet spacer*, which is a bunch of metal strips pivoted together. It looks like a complex pantograph. Once you know how many rivets to install (and most plans tell you that much), lay the spacer against the metal, and it'll directly indicate the hole locations.

But the cheapest way is to go to a sewing shop and buy a strip of white elastic for about $2. Lay out the elastic and mark half-inch intervals with a felt-tip pen. Then clamp one end to the start of the metal, and stretch the elastic until the marks are at proper intervals.

The usual first step is to draw a line upon which the rivets will be installed. You won't use a scribe, of course, but pencil marks don't show up very well unless you've already applied a primer coat to the aluminum. Bare aluminum can be marked with a fine line felt-tip pen and cleaned with MEK. Or you could apply a strip of wide masking tape and mark on that.

Draw the line with a straightedge, and either calculate the spacing or use the rivet spacer to indicate correct locations. Put a small x at the location of the holes to be drilled. Position the automatic centerpunch at the center of the x, and press down.

Sometimes you have to install a sheet atop components that are already drilled for the rivet holes. In this case, use a hole duplicator. Select a duplicator based on the size of the hole already drilled. Slide the end with the stud under the sheet being added, drop the stud into the predrilled hole, then drill the sheet through the bushing.

Drilling the Hole

If the hole is excessively large, the rivet will fit sloppily. It might let the metal shift. Stresses could concentrate and cause premature failure.

The proper procedure is to drill an undersized pilot hole then enlarge it to the final size. Reams can be used for this enlargement, but they work slower and are expensive. They do leave a smooth hole, though.

There must be some small amount of clearance between the hole and the side of the rivet to allow it to pass. This clearance can't exceed .004 inch. So for the pilot drill, we'll use a bit of the same size as the rivet. The final bit will then be the next size larger. This works out as follows:

Rivet	Pilot	Final
3/32	3/32	#40
1/8	1/8	#30
5/32	5/32	#21
3/16	3/16	#11

When drilling the pilot hole, keep the drill bit perpendicular to the metal. Don't use too much pressure; there's no rush. If at all possible, support the metal on the back side with a block of wood. Otherwise, pushing on the drill might place an indentation around the site of the hole.

As each pilot hole is drilled, insert the proper-sized cleco. For the larger sizes, first drill and cleco all holes to 1/8 inch. Then use the respective pilot bit and replace the clecos with the larger size.

When the holes are drilled to the final size, remove the clecos and deburr the holes. If it isn't possible to disassemble the parts for deburring, you could use a *chip chaser*. It's a strip of metal with a hook at one end; slide it between the parts and use the hook to pull away the raised metal. *But they can severely scratch the aluminum. Disassemble if at all possible.*

Countersinking and Dimpling

If the plans call for flush rivets, the hole must be countersunk or dimpled.

For countersinking, metal is cut away so the sides of the hole match the cross-section of the rivet. Quite logically, this is done with a tool called a countersink (Fig. 7-14). It's like a wide-angled cone (actually, it's 100 degrees, to match the rivet) with flukes to do the cutting.

The biggest restriction on drill countersinking is the minimum allowable thickness of metal. It must be at least as thick as the rivet head is deep. This

Fig. 7-14. A countersink in action. Note the tapering hole.

makes sense; otherwise the bearing surface (the portion of the metal actually contacted by the rivet) is drastically reduced. The approximate minimum thickness values are:

Rivet	Metal
3/32	.036
1/8	.042
5/32	.055
3/16	.070

Note that only the metal to be countersunk must have this thickness. Thinner sheets must be on the same side as the shop head.

After final drilling of the hole, use the countersink to remove enough metal to let a rivet sit flush in the hole. It's very important to apply the tool perpendicular to the metal because any angle will produce an oval shape that won't match the rivet head.

The major problem is cutting too deep. A little too far isn't critical; the rivet head is recessed instead of flush. But eventually the tip of the countersink will begin to enlarge the hole's diameter. The rivet will be loose. If this happens, ream out the hole for the next size larger rivet.

Countersink with a drill press whenever possible and use an automatic-stop countersink. These units can be adjusted for maximum depth.

However, they are thrown off by curved surfaces. And most of the time a drill press isn't practical. Instead, use a regular countersink in a hand-held power drill at slow speed. Switch back and forth with the drill and a rivet until the proper depth is reached. With practice, you'll be able to eyeball the hole and won't have to check as often. When the hole approaches the proper size, finish it by hand using a countersink mounted in the end of a wooden dowel.

Dimpling is faster and easier than countersinking. Rather than carving metal away to match the rivet head, dimpling forms a calibrated dent in both sheets of metal. Equipment consists of a punch and a die (Fig. 7-15). The punch includes a projection of metal shaped like the flush rivet, and the die has a corresponding indentation. (The angles are actually a bit sharper, to allow for springback.)

Don't final-drill holes to be dimpled. Dimpling stretches the metal, and actually increases the size of the hole. If you final-drill before dimpling, the rivet will be loose.

Instead, pilot drill the hole to the size specified, then place the die on the back side of the hole. The die must be solidly supported. Insert the punch and use the rivet gun to dimple the hole. Hand-squeezers and pop riveters also can dimple, with the proper dies. Don't use too much force, as it tends to ripple the skins. Follow the instructions for the particular system you buy.

Then final-drill the hole to the final diameter.

Dimpled rivet holes are stronger, but don't look as good as countersunk holes. The dimpling effect isn't confined to the hole itself, and tends to add a wide, shallow depression to the surrounding metal. Use whichever process the

Fig. 7-15. The dimpling punch and die. Note how the end of the die is shaped like a flush rivet. The die has an indentation to match.

kitplane manufacturer recommends, or try each process on a piece of scrap and pick the one you prefer.

In some cases, a combination of methods is used. For instance, when riveting skins to heavier structure, the structure will have to be countersunk, while the skin must be dimpled.

Making the Shop Head

Now comes the moment of truth. The shop head can be formed by either a pneumatic rivet gun or a rivet squeezer (Fig. 7-16). Squeezers make the job easy, but they can be used only near the edge of a sheet and are subject to other limitations. Up to 3 inch throats are available in hand and pneumatic models.

Actually, there's nothing magical about driving rivets. All you need is a continuous or momentary (shock) force and a hard surface to form the shop head. A carpenter's hammer can supply the force. But for building airplanes, pneumatic rivet guns are the tool of choice.

Fig. 7-16. The rivet squeezer in action. They work best when the work can be brought to the tabletop and the depth of the throat limits which rivets they can drive.

The rivet gun is like an electric drill; it supplies the motion, but interchangeable units supply the action required. Drills need bits; rivet guns need *rivet sets*.

The set has two functions: transmit the hammering action from the gun to the rivet and prevent damage to the surrounding skin. Standard and flush rivet sets operate in different ways. The sets for standard rivets are slim rods, with an indented cup in the end that fits the manufactured head of the rivet. When properly used, this cup doesn't touch the metal skin. Because head dimensions change with rivet diameter, a separate set is required for each rivet diameter.

Flush rivet sets approach the problem differently. The end of the set is large; at least an inch in diameter. It's flat with rounded edges and has a polished, smooth surface. The flat set doesn't gouge the metal if slightly off-center. Some flush-rivet sets cover the edge with rubber to reduce the possibility of skin damage.

A variety of sets is shown in Fig. 7-17.

The last necessary item is the *bucking bar*. All it really has to be is a piece of steel of the appropriate hardness and weight. If the bar will be hand-held, it should weigh a minimum of 25 to 30 times the rivet diameter. Hence 3/32-inch rivets need a bucking bar at least 2.5 pounds, 1/8 inch calls for 3.5 pounds, etc.

If the bucking bar is too light, the edge of the sheet might distort. Bending its

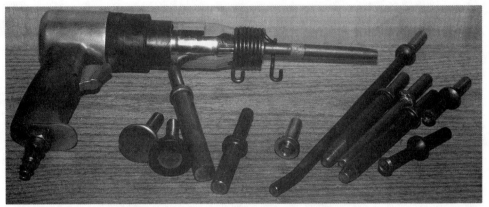
Fig. 7-17. A rivet gun and a variety of sets. The flush sets are on the left. Note the long set with the angle on the end; it's designed to reach difficult places.

edge down before riveting will stop the warping, but it's not easy to do on long sheets. Select a heavy bucking bar instead.

All the bucking bar needs is a flat surface that can be held perpendicular to the rivet. One thing you'll notice about commercial bucking bars is weird shapes (Fig. 7-18).

The weird shapes accommodate a lot of rivets that are often in places with poor accessibility. The flange of a rib, for instance, might be only a half-inch wide. The rib probably curves, so only the corner of a wide bar could be used. The rest of the bar just gets in the way.

Instead, a manufacturer might make a tapering bar, to make it more suited to those tight corners. The other end is usually designed for another kind of tight spot; one, for instance, where the bar can only approach from an oblique angle. Metal airplanes have a lot of nooks and crannies that the builder must snake a bucking bar into, hence metal-airplane builders tend to collect varieties of bucking bars.

Which leads to another point: Plans specify the order in which the rivets should be driven. Follow it exactly. Otherwise, you might wind up with a rivet in a corner that you can't reach. It's not the end of the world because the rivets could be removed and you could start over, or the right kind of pop-rivet could be substituted.

Similarly, lightening holes are often not *just* lightening holes. Sometimes they provide sole access to hard-to-reach spots. Don't arbitrarily decide not to make a lightening hole when the plans call for it. If it's listed as optional, fine. Otherwise you might be hurting later.

Now you might also see why certain ribs' flanges face left and others face right. The flanges usually face away from the first rib to be riveted; therefore make sure you use the correct rib on the appropriate side. If all flanges face one direction, the two wings are probably built with different procedures. The plans should make the flange orientation clear.

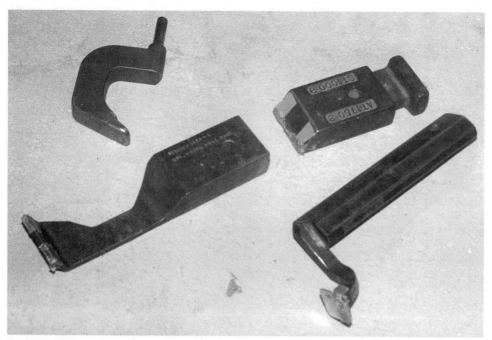

Fig. 7-18. Bucking bars. Again, the odd shapes let the builder ease them into tight spots.

The plans should also specify the rivets, or at least the general type (regular or countersunk) and diameter. If the plans just say, "Use $^3/_{32}$ inch AN470 rivets," it's up to you to determine length.

The rivets should be at least one-and-a-half times the diameter of the rivet, *plus* the thickness of the metal to be joined. So if you're joining a .025-inch skin to a .040-inch rib with a $^3/_{32}$-inch rivet, the rivet would have to be about .025 + .040 + 1.5 × $^3/_{32}$ = .205 inch, or a bit less than a quarter-inch-long. Because the dash-numbers for rivet diameters are in $^1/_{32}$-inch increments, and rivet lengths are in $^1/_{16}$-inch increments, you'd need AN470AD3-4 (MS20470AD3-4) rivets.

If you don't have the correct size rivets, a rivet cutter can shorten longer ones. Some builders buy only a few different sizes and shorten them as required, rather than maintain a stock of all lengths.

If the plans call for it, or if you desire, dip the rivet into primer before placing it in the hole. However, the paint acts as lubricant. The set and bucking bar tend to slide off the rivet.

With all this preparation, the actual driving of the rivet is anti-climatic. Remove the cleco and insert the rivet. If using flush rivets, hold them in place with a strip of masking tape directly over the heads. The two pieces of metal to be joined must lie solidly against each other. If one sheet flexes away when the cleco is removed, use clamps or other means to rejoin them. Otherwise, the rivet material will flow between the pieces resulting in a weak and unsightly join.

Insert the correct set into the rivet gun, then check the gun's power level setting. Too high, and there's a danger of punching the rivet right through the skin

Fig. 7-19. Basic riveting procedure. The bucking bar is clamped to the tabletop, although they can be hand-held instead. Note how the set and gun are directly in line with the rivet.

if the bucking bar slips. Too little, and the rivet will *work harden* (become more brittle) and crack.

When properly adjusted, the gun should completely drive a rivet in two or three seconds. Practice with various rivet diameters, and mark the gun's dial at the correct settings.

Place the set on the manufactured head perpendicular to the metal's surface. For standard rivets, the set should make contact with the top of the head only. It shouldn't touch near the edges.

The riveter must apply the force directly along the length of the rivet. Any angle will tend to push the rivet sideways. An excessive angle might cause the set to contact the metal skin and place a little half-moon-shaped smile alongside the rivet.

Place the bucking bar perpendicular to the other end of the rivet. Hold it by hand if necessary, but it's easier if it can be clamped to a solid surface (Fig. 7-19). Squeeze the rivet gun's trigger.

The gun works by impact. You don't have to hold the gun and bucking bars with a death grip. Only push hard enough to keep the tools in place. If hand-held, the bucking bar vibrates in time with the gun; let it. Inertia does the work, by the time the bar is moving away, it's already done its job. Your hand pressure just moves it back into position in time for the next blow.

Some builders can buck their own flush-headed rivets, which is difficult because the set tends to slide unless held in place with the other hand. Rubber-edged sets can reduce the problem. Or if the riveting can be done at the work-bench, clamp the bucking bar down and use both hands on skin side. Otherwise, count on needing some help.

As mentioned, the rivet should be set in two to three seconds. The change in sound is a good clue. Release the trigger, *then* remove the bucking bar. If you take away the bar before the gun stops, it'll bang the heck out of the skin and may even drive the whole rivet right through it.

It's in, but is it right?

Checking and Removal

The shop head should be at least one-half the rivet diameter high, and one-and-a-half times the diameter in width, as shown on Fig. 7-20. You can buy or make rivet gauges to quickly make this determination. If the shop head is still high but not wide enough, place the gun and bar back into position and hit it a few more times.

The shop head should be uniformly flat. Neither the shop head nor the manufactured head should have any cracks or untoward deformation. Both heads should be vertically aligned with each other; if one is displaced sideways, either the set or the bucking bar was held at an angle.

Check flush rivets by placing a straightedge along the metal across the manufactured head. If it doesn't lie flat, the hole probably needs additional counter-sinking.

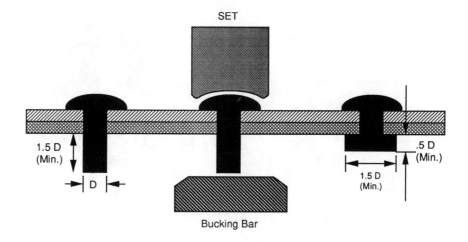

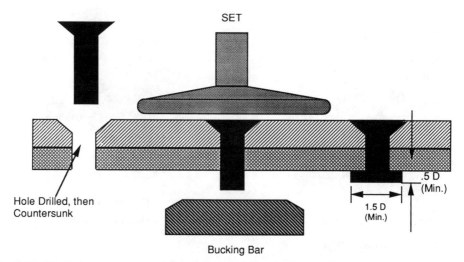

Fig. 7-20. Riveting procedures and dimensions for both regular and flush rivets.

The joined metal shouldn't show any distortion. If the metal is bulged toward the shop head, the bar was too light. If it puckers around the rivet, you probably drove the rivet too far.

If a rivet is bad, it will have to be drilled out and replaced. Structural rivets have a dimple on the head for guiding the drill. Use a bit slightly smaller than the rivet itself. Place it in the dimple and drill until the head breaks free. Only friction is holding the rest.

With a good-fitting drift punch, the remainder will be knocked out. However, the metal around the hole must be supported or residual friction might bend the skin before the rivet pops clear. Take a piece of wood with a hole drilled

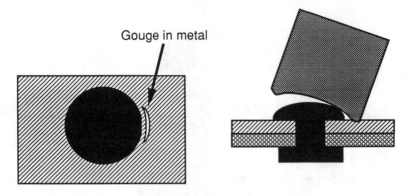

Incorrect Set or Set not perpendicular to sheet

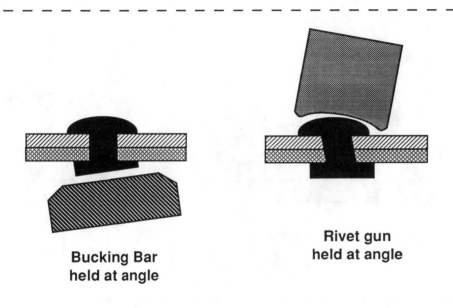

Bucking Bar held at angle

Rivet gun held at angle

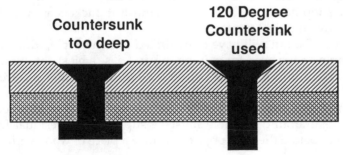

Countersunk too deep

120 Degree Countersink used

Fig. 7-21. Riveting errors and how to spot them.

in it slightly larger than the shop head of the rivet. Place the wood flush to the back of the metal with the shop head inside the hole. Then knock the rivet through with a hammer and punch.

Note that this procedure takes at least three hands, unless you can place the offending structure on a bench or other solid surface.

Touch up the hole a little bit with a file or deburring tool, and you're ready to drive another rivet.

Sometimes the rivet doesn't pop away cleanly, and distorts the hole. It's allowable procedure to replace the rivet with the next higher size. Drill the pilot hole, ream, and drive the bigger rivet.

Typical errors are presented in Fig. 7-21.

If you're building a metal airplane, you're going to drive thousands of rivets. The above information makes it sound like you're going to spend five minutes per rivet. You won't. Like many other tasks involved with building a kitplane, the preparations take more time than the actual operation. Ninety percent of your time will be spent preparing to rivet: priming, drilling, countersinking.

In fact, most of that 90 percent will be spent making sure the airplane won't corrode away. Let's look at the protection process.

PRIMING

Everybody loves to see pictures of unpainted kitplanes turned into grand champions by a few gallons of Imron. Painting is glamorous. It's the crowning moment of the kitbuilder's project because the airplane is painted just before it starts flying.

But if you're building an aluminum airplane, you'd be better off spending more attention on priming than painting. It's a dirty job that gets little publicity. While *finish painting* is often done in a paint booth (either permanent or temporary), priming takes place in the dirty, gritty environment of the workshop.

The need to prime surprises some people. Aluminum doesn't crumble away into reddish powder. Why prime?

Aluminum Doesn't Rust, or Does It?

We tend to think that the major advantage of aluminum over steel is that aluminum doesn't rust. Well, that's true because when exposed to moisture and air, aluminum doesn't form ferric oxide. But aluminum does *corrode*, with the same visual and structural impacts. Fancy trim colors be damned because a properly applied prime coat is going to make your airplane last.

Proof is available no farther than your local flight line. In the late '70s, one major lightplane manufacturer changed their corrosion-proofing procedure, going to a simpler, cheaper, faster method. The procedure they used was not approved by the paint manufacturer.

A few months after delivery, certain airplanes' paint started cracking. Owners soon discovered the problem wasn't in the paint—the aluminum under the paint was being eaten alive by *filiform corrosion*. The owners stripped the paint off the airplanes, primed them properly, and repainted them.

The problem returned. The corrosion was too deeply set; aftermarket fixes couldn't reach everywhere.

Within a few short years, a significant number of these aircraft were unairworthy. All from cutting corners; saving a few bucks on priming.

Your kitplane uses the same 2024 alloy. Without proper care, it can happen to you, too.

How Much Is Enough?

The degree of corrosion protection varies from builder to builder. Most apply a primer to the outside of the aircraft prior to painting. They have to, as paint won't adhere properly without it. But internal corrosion protection is another matter.

One designer has built about six metal airplanes of his own design. He didn't apply internal priming on any of them. He's had no problems; in fact, one of his airplanes hangs in a major museum. Alclad is designed to stop corrosion. In his case, it had done its job.

Certain kit manufacturers feel the same way. Others demand a full corrosion-protection regimen. There's nothing *different* about their airplanes, it's just the personal opinion of the designer. Some builders assemble the structure, then spray it with primer just prior to skinning. Others prime each and every piece before installation. Some even use separate kinds of primer on different parts (Fig. 7-22).

Fig. 7-22. This builder used different primers on the bulkheads and skins.

If the aircraft is made from 6061 alloy (which is highly corrosion-resistant), you might easily get by without priming. But otherwise it depends on a broad set of factors. Climate is the major player. An Arizona kitplane can get away with procedures that would make a Florida kitplane crumble to dust within a few years. Builder care is another factor. The fewer scratches made during construction, the more Alclad the builder leaves in place, the fewer places major corrosion can start.

Most of us won't be that careful. And today's mobile society might end up moving you and your kitplane into warm, salty, corrosion-rife tropical breezes. Priming all the pieces before assembly has a few other advantages. The primer acts as a shield against scratches and minor damage. Most primers can be written on using a pencil, which makes marking rivet patterns far easier. Epoxy primers are impervious to most shop accidents.

There are a few caveats where the prime/no prime decision is made for you. First, steel rusts quickly and always requires priming. Steel priming is covered in chapter 8 under steel-tube fuselage preparation.

Second, if two different metals are to be joined, both must be primed. Dissimilar metals joined and exposed to salt air, exhaust, or even dew can build up an electrical potential. The resulting current flow accelerates corrosion. Aluminum airplanes incorporate a lot of steel fittings, from landing gear parts to control bellcranks. Ensure each piece is well-primed.

Third, do not prime any surface to which something will be bonded, or the inside of fuel tanks. Neither the bonding agents (epoxies, etc) nor the substance used to seal the fuel tanks will bond to the primer. This has a definite impact in the tanks, as the sealant may come loose and clog the fuel lines.

Otherwise, you can decide for yourself, based on the kitplane manufacturer's recommendation and discussions with other builders. The following pages describe for full protection. It's your decision how much to implement.

The Basics

Every item made from 2024 or 7075 alloy should be *cleaned*, *etched*, and primed before assembly. Steel parts need similar preparation, but don't have to be etched. Aluminum alloy 6061 is more corrosion resistant than other alloys; follow the kit manufacturer's recommendation. And don't take it into your head to substitute 6061 for 2024; 6061 is only $2/3$ as strong.

Cleaning means the removal of surface grime and corrosion. The corrosion is the toughest. Don't attack it with steel wool, emery cloth, or a wire brush. It breaks the corrosion free, but imbeds little pieces of it into the metal. Try a mildly abrasive household cleanser. When finished, wipe it down with MEK or acetone to get rid of fingerprints. Don't handle the metal with bare hands because the whole purpose is to clean off fingerprints and similar impurities.

Paint doesn't like to stick to smooth, shiny surfaces. So before applying the primer coat, we etch the surface using a weak acid solution. Apply the etchant, let it sit for awhile, then rinse it off. The metal takes on a rather dull, hazy appearance.

There are several quite satisfactory self-etching primers on the market that require only minor cleaning prior to application. Check with your fellow home-builders and the kit manufacturer for recommendations.

The primer is then sprayed on. Zinc chromate is the traditional primer, but several types of epoxy primers are gaining favor.

The above is just a general set of guidelines; whichever primer you chose, follow the manufacturer's instructions. Read the instructions and cautions carefully before you decide. For instance, one primer maker declares that his product doesn't require etching before application. But if you read carefully, you see the maker requires *another* process before priming, and *that* product requires etching.

The instructions are all easily followed, in theory. But the kitbuilder's workshop is a pretty nontheoretical place. Let's see how to apply primers outside the laboratory.

The Real World

Cleaning and etching the metal are pretty easy to handle. Remember, one of the purposes is to eliminate finger oils from the metal. Gloves are a good start.

You'd rather not lay the parts flat on the floor during the painting process. Coat-hanger racks have been mentioned, but even small plywood easels would be adequate. The nice thing about racks and easels is they let you check if the paint is dry without touching the part itself. Touch the rack instead.

The etchant is generally a phosphoric acid solution; be sure to dilute it according to the instructions. Small parts can be dropped into a bucket of etchant. Brush it onto larger pieces, working the acid into the metal with a Scotchlite pad. It stinks a bit. After five minutes, rinse the part under running water. My ecological conscience bothers me a bit about rinsing; it's impractical to catch the runoff. One could wipe off the etchant, but you're still left with the disposal of the paper towels or rags.

With the metal cleaned and etched, you're ready for priming. It's great if your workshop is big enough to set up a permanent paint booth. It's not too likely, either.

There are two major problems to overcome. Anyone who's done any sort of spraying knows the particles go everywhere. At the absolute minimum, a backdrop to catch the excess paint is necessary. For low-volume painting with spray cans, a large cardboard box makes a cheap and easy spray booth. But a compressor-type sprayer can kick out a lot of paint at high speed. You can "curtain off" your painting area using plastic sheets.

Painting in a basement means a houseful of paint odors. Even an attached garage seems to let the odors seep into the living areas. Some folks do their priming outside. It's not a bad idea, although spraying is restricted to windless days.

The second problem is the hazards associated with the paint itself. Wear a respirator specifically designed for painting, and ensure an adequate flow of fresh air. Note, a *flow* of fresh air, not just an open window. An open window allows the paint to linger; ventilation clears the air faster. Some primer manufacturers even specify a system to supply fresh air to the painter.

Fig. 7-23. With all the talk about composites, only metal monocoque construction is used for the primary structure of modern airliners. Properly designed kitplanes like this Zenair Zodiac are strong and enduring.

One common idea is to use a fan to move the air through the painting area, or build a paint booth using an old high-power fan. Perhaps, but the paint spray is explosive as well. At the right concentration, the fan's motor can ignite the suspended particles. Check supply houses for explosion-proof fans. If you rig up an exhaust system, run the exit pipe high enough so the paint isn't deposited back on the house.

If any gas appliances that use an open flame (such as a furnace or water heater) are located in the workshop area, keep the painting area as far away as possible. Turn off the appliances until the air clears.

The actual primer application is no different from any other kind of painting. Primers are available in spray cans, but it isn't as cost effective as using a spray gun and a compressor. The nice thing about the cans is the minimal cleanup afterwards. But a spray gun can be set to kick out a tight stream of paint to minimize overspray.

Avoid thick coats because they don't stick quite as well and take a lot longer to dry thoroughly than a couple of thin coats. Concentrate on the edges because they don't include any cladding. Also, corrosion tends to start between two sheets, so ensure good coverage on areas to be riveted together.

Let the pieces dry. If it's a calm sunny day, let the sun bake them. The coating always seems harder when I do this. While they're dry to the touch in just a few minutes, it takes 12 hours or so to reach full hardness.

Metal monocoque airplanes have their drawbacks for the kitplane builder. The process was originally intended for factories, which can use expensive tooling to reduce manhours and costs. A press with custom-made dies or a numerically-controlled milling machine can crank out complex aluminum shapes a lot easier than a homebuilder's vise.

But the kitplane designers know this, and make allowances. Traditional plans-built designs like the Thorp T-18 Tiger use simple curves throughout. Planes like the RV series supply the curved components as part of the kit. Metal monocoque is the standard for commercial airliners and military fighters. And if it's good enough for Mach Two, it's good enough for your 100-mph pride-and-joy (Fig. 7-23).

The skills aren't that hard to learn. Just pick up a few scrap pieces of aluminum and get some practice before starting on the kit. One kitplane manufacturer suggested that a builder fabricate a handy tool carrier using practically all materials and techniques necessary to construct the airplane.

8

Steel and Aluminum Tube Construction

TUBE-AND-FABRIC IS ONE TRADITIONAL METHOD of homebuilt construction (Fig. 8-1). Steel tubing is cut into appropriate lengths, bent as necessary, and welded to form the fuselage structure. Any kit should be prewelded. But if you'd care to try it, there are still plenty of plans-built designs on the market.

Early homebuilts were mostly steel tube or wood. The drive to reduce building time begat more modern construction materials and methods. The '60s brought the metal revolution with the Midget Mustang and the Thorp T-18, and the '70s saw the composite stampede led by the VariEze, KR, and Glasair. Tube-and-fabric designs seemed destined to fade into history.

But the '80s saw a resurgence. Two friends in Idaho developed the Avid Flyer (Fig. 8-2), a light STOL design. The kit supplied (and still does supply) a completely prewelded steel-tube fuselage and everything else necessary to complete the aircraft. This inexpensive kit led tube-and-fabric's comeback.

Their production figures are larger than most realize. About 40 percent of the kitplanes on the market are tube-and-fabric designs—versus 25 percent for composite, 20 percent wood, and 15 percent metal monocoque. Glasair has sold the most kits (more than 1,000), but two different steel-tube designs have each sold almost as many. Tube-and-fabric kitplanes include the Avid Flyer line, the Kitfox, the RANS line, the Protech PT-2 (Fig. 8-3), Challenger, Murphy Renegade, Light Miniature Aircraft series, the Sonerai, CGS Hawk, and other ARV designs.

Popularity is fed by two factors. The first is cost. Steel-tube manufacturers don't need expensive composite molds or hydraulic presses. Because the initial investment is less, the kit sells for less.

The second factor is the ease of construction. A tube-type kitplane is the easiest to build.

That's a strong statement. Right now, the composite, aluminum, and wooden airplane buffs are rising to refute it. What about the Pulsar? The Teenie Two? The Fly Baby?

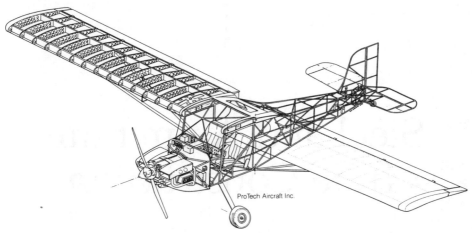

ProTech Aircraft Inc.

Fig. 8-1. Today's tube-and-fabric kitplanes are little different in design than the homebuilts of 40 years ago.

Avid Aviation

Fig. 8-2. In the early 1980's, the Avid Flyer marked the beginning of the modern tube-and-fabric kitplane.

Steel-tube airplane kits supply a structurally complete fuselage. No other type of construction does. Composite fuselages assemble fast, but you have to bond the fuselage halves together. Metal or wood structures are built from hundreds of separate parts. The kit might include prebent/precut bulkheads, and precurved skins. But the builder must still rivet or bond them together.

Fig. 8-3. The ProTech PT-2 is unusual among tube-and-fabric kits in that it uses the Revmaster VW conversion instead of the usual Rotax.

But the builder pulls a complete fuselage out of the delivery crate (Fig. 8-4). It's welded by the factory, just like production airplanes like the Citabria and Super Cub. A lot of work remains. The fuselage must be protected against corrosion. The wings must be built. But most of the work left for the builder is of the simple "bolt-in-place" variety.

Several tube-type kitplanes don't use welded steel tubing. The Murphy Renegade uses aluminum tube, joined by pop rivets and special fittings. The CIRCA Nieuport replicas (Bebe single-seater and N-12 two-seater) use aluminum-sheet gussets instead of the special fittings.

In this chapter, you'll learn some of the basic skills involved with steel and aluminum-tube airplanes:

- Checking and preparing welded steel-tube fuselages.
- Pop riveting.
- Cutting, drilling, and bending aluminum tubes.

Tube-type kitplanes require techniques from most contruction disciplines. You'll probably be making some sheet metal structures; the details of this process are in chapter 7. Composite work is in chapter 6, and wood preparation and fabric coverings are discussed in the next chapter.

Wing construction varies between types. Many use aluminum-tube spars (Fig. 8-5), others build the wing from aluminum C-section and aluminum or wood ribs. A few have metal-covered wings like all-aluminum airplanes. Between this and the other construction chapters, you'll be prepared regardless of kit design.

ADDITIONAL TOOLS

The primary addition to the basic tool list in chapter 4 is a *pop-rivet gun*. Prices vary from a few dollars at the local hardware store to $300 for a pneumatic gun.

Fig. 8-4. Steel-tube kitplanes are delivered with the fuselage structurally complete.

Avid Aviation

Fig. 8-5. The long tubes in the center of the picture are the wing spars. The wooden ribs shown install between the fore and aft spars.

Which model you buy depends on the amount of use. Murphy Renegades and Circa Nieuports build most of the airplane with pop rivets, while welded steel-tube kitplanes can probably get by with a cheaper model.

Selecting a pop riveter is easier the second time around. I bought a $15 unit from the local hardware store, and it has worked well, but its ergonometrics are poor; the handles are too far apart to be comfortably worked by one hand. I'll know what to look for on the next one.

If there's a lot of riveting, consider springing for the pneumatic tool. Sonerais, for instance, pop rivet the wing skins. There's only so much squeezing that your hand(s) can take.

Hand or pneumatic, make sure the riveter can handle the variety of rivet sizes to be used. Most supply a selection of nosepieces that adapt the same gun. If the plans call for Cherry rivets, a special gun is needed.

Most other tools are trivial and cheap. You'll need a wire brush for your drill. A tubing cutter. Extra clecos. If there's a lot of sheet metal work, check the list of additional tools in chapter 7.

MATERIALS, FASTENERS, AND SAFETY

Steel or aluminum tube airplanes are straightforward and traditional.

Materials

Tubing is the material you'll be working with, either steel or aluminum. Characteristics of each alloy are discussed in chapter 5. Recall that 4130 steel is the most common. It's strong and weldable. It also rusts freely, and must be protected.

Most aluminum-tube aircraft use 6061-T6 alloy. This isn't as strong as 2024-T3, but is significantly more corrosion resistant. Wide-diameter 6061-T6 tubing is used for wing spars on several kitplanes.

Tubing is sold by diameter and wall thickness. The inside diameter can be determined by subtracting twice the wall thickness from the diameter. The alloy, diameter, and wall thickness are printed along the tubing, and repeat every foot or so.

Fasteners

The primary fastener in steel-tube aircraft is the *weld*, which you probably won't be doing. A steel-tube *kitplane* should come with a prewelded fuselage. I have no objection to those who want to learn to weld, or build up their own fuselage. But everything on a kit should be prewelded.

But you should know *something* about welding. You'd like to be able to check the welding on the fuselage. And if you ever have an accident and have to get the fuselage repaired, it's nice to be able to intelligently discuss the problem with the welder.

The welding operation consists of placing the two components to be joined in close proximity, then heating the junction into a molten state. The metal of the two pieces then flows together. When the joint cools, the two pieces of metal have been turned into one continuous unit. A filler rod supplies additional metal to the weld area to add thickness and ductility.

There are two major welding methods. The first and most common is *gas* welding, which produces steel-melting heat by burning acetylene with oxygen under pressure. The second method, *heliarc* welding, isn't as common in the home workshop. It heats the metal with an electric arc, while excluding oxygen (which causes embrittlement) by supplying a forced flow of argon gas.

Gas welding is easier for the occasional worker. Heliarc is more difficult, but gives professional welders far better control. An oxyacetylene flame allows the area to be heated and normalized after the joint is welded, while heliarc reduces oxygen embrittlement.

Which is better? When properly done, either a gas-welded or heliarced joint will stand up to the load. The tubing itself will probably fail before either type of weld. For a kit buyer, the skill of the manufacturer's welders is more important than the process itself.

While you won't be doing any welding on your kitplane, you'll probably install a lot of pop rivets. These inexpensive fasteners, also called pulled or blind rivets, are ubiquitous on most tube-type airplanes.

The advantage of pop rivets over driven rivets (see chapter 7) is that they can be installed one handed in areas where a bucking bar can't reach. Conventional rivets form a shop head by ramming the rivet repeatedly into a steel bucking bar. The rivets can't be installed into a tube because there's usually no way to get a bucking bar inside. Hence, pop rivets instead.

Their big disadvantage is their strength, or lack of it. Most aren't as strong as driven rivets. They're a bit unsightly, too. The shop head is pretty ugly, and the stem leaves a hole in the manufactured head.

Stronger monel and stainless steel rivets are available. *Cherry* rivets are the most commonly known. A $3/16$-inch stainless steel structural Cherry rivet is rated at 1,650 pounds in shear. However, the harder shank means more pull resistance. They take more effort to install.

Other than pop rivets, the most common fastening hardware is standard AN nuts, washers, and bolts. To recap the rules of their use:

- If possible, install the bolt with the head up or forward.
- The grip length should be at least as long as the combined widths of the material to be bolted.
- No more than three washers per bolt.
- Use self-locking nuts or castle nuts with cotter pins for all applications.

Chapter 5 details the "nuts and bolts" of AN hardware.

Safety

Other than suffering a few gouges and nicks from sharp edges, tube airplanes are fairly benign. If any welding is done in the shop (for instance, you hire a welder to redo a bad joint) isolate flammables from the area and keep a fire extinguisher handy. It's a hot flame, a backfire can spit out bits of molten steel. The cheaper extinguishers don't turn off once started; spend a little extra for the multiple-use kind. Carbon dioxide works well.

If you're going to watch or help, wear old clothes, gloves, and safety goggles. Cotton shirts and denim jeans might be best. Polyester and other man-made fabrics can melt or ignite. No leisure suits in the shop, please . . .

The other safety hazard involves the health hazards of the anticorrosion primers applied to the tubing. Follow the instructions, wear a respirator, and ensure adequate ventilation.

WELDED STRUCTURES

As mentioned at the beginning of the chapter, most tube-type aircraft use welded steel fuselages. The fuselage is structurally complete; all you have to do is attach fairings, controls, and other items that don't affect load-carrying ability.

But you can't just ignore the fuselage. You have to verify that it's been done properly, and follow a few basic rules during construction.

Basic Construction

Steel tube fuselages begin as straight lengths of tubing. Pieces of the proper diameter and wall thickness are cut to length, bent as necessary, and installed on a jig that holds the tubing in the desired shape. Figure 8-6 shows an example.

The main tubes that run fore and aft are called *longerons*. Pieces that run vertically between them are called *uprights*, and *laterals* run horizontally between longerons on opposite sides. *Diagonals* are mounted aslant between uprights, laterals, or longerons.

The tubes are welded together wherever they meet. Each tube must be shaped to make contact over the maximum amount of area. For example, on a simple T-joint, the end of the vertical bar must have a semicircular notch to fit snugly over the other tube. If there's a wide gap, the steel can't intermingle when the two pieces are heated. The filler rod adds some additional metal, but there's a limit.

The usual practice is to lay one side of the fuselage flat on the work table, tack-weld all the joints, then build the other side. A tack-weld melts the material in just a small area, to temporarily hold the components for shop handling. One could do the finish weld on the table, but all-around access is usually impossible without lifting the pieces.

After the other side of the fuselage is tack-welded, the two sides are jigged upright and laterals and diagonals are tack-welded between them. Once the entire fuselage is tack-welded, the builder does a finish weld on each join.

In addition to welding up the structure, the builder adds tabs and plates in areas where items will later be bolted. The tubing used on the lighter homebuilts is pretty small, typically around 1/2- to 3/4-inch in diameter. Drilling a 1/4-inch hole through a tube for an AN4 bolt weakens the tube drastically. Instead, the builder welds a small plate into place, bracing it with additional tubing or sheet as required.

Fig. 8-6. Wood blocks nailed to the tabletop hold the steel tubing in position for tack welding.

Buying Tack-Welded Fuselages

Sometimes the manufacturer offers a tack-welded fuselage as part of a lower-cost kit. This generally isn't an option for complete kits; rather, it's available as part of certain materials kits. Wag-Aero sells both fully-welded and tack-welded fuselages for their ragwing Piper replicas, for example.

Cost is the main reason for ordering a tack-welded fuselage instead of a fully-welded one. Typically, Wag-Aero charges several thousand dollars less for tack-welded Sport Trainer and Sportsman 2+2 fuselages. Depending upon the complexity, a certified aircraft welder might want between $1,500 and $2,500 to finish-weld the fuselage. A fellow EAAer might do the job for half that.

The main point is to find someone experienced in welding aircraft tubing. Your best bet would be an FAA-certified repair station (Fig. 8-7). Even if the station (as a matter of policy) doesn't do welding for experimental aircraft, the welders might take after-hours projects on their own. After all, you don't need the certification signoff; just someone who knows aircraft tubing.

Don't just page through the yellow pages under welding. Commercial welders normally work with far thicker materials than the .045-inch walls of 4130 tubing. You want someone with demonstrated aircraft expertise.

Or, you could spend $100 for a community college welding course and learn how to do it yourself. The most time-consuming process is cutting and fitting all the tubes. If you buy a tack-welded fuselage, that's already done. You might make mistakes, but all you'd have to do is cut away the ruined tubing and weld in a replacement.

Fig. 8-7. For a professional job on a tack-welded fuselage, or repairs on a prewelded fuselage, contact an FAA-certified welding shop: FAA Repair Station plus the certificate number.

Welded-Fuselage Checkover

Whether you buy a tack-welded fuselage or the finished product that comes in a complete kit, check the fuselage carefully before proceeding.

The first step is an overall examination for dented, bent or crimped tubing. This type of damage shouldn't occur during normal shipment. If the exterior of the crate is broken, note the fact on the form when you sign for delivery. Immediately inspect for internal damage. The shipper is liable in this case and should pay the repair costs.

If the crate didn't show any sign of abuse, the damage was probably caused during assembly. Take pictures and contact the kit maker.

Repair methods depend upon the type and location of damage. Usually, a two-piece sleeve is welded over the damaged area. You'd prefer an experienced welder to do these repairs. Don't say, "Well, I'll fix it myself." Have it professionally repaired; an experienced welder will not only fix it properly, but might spot further damage. In severe cases, it might have to be shipped back to the factory and rejigged. Work out the details with the kit manufacturer.

After checking the tubes, examine the welds. I've seen some kits delivered where the manufacturer missed finish-welding some joints. Here's what good

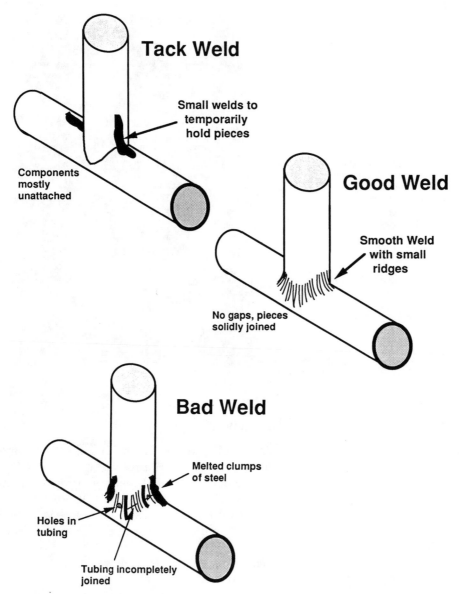

Tack Weld

Small welds to
temporarily
hold pieces

Components
mostly
unattached

Good Weld

Smooth Weld
with small
ridges

No gaps, pieces
solidly joined

Bad Weld

Melted clumps
of steel

Holes in
tubing

Tubing incompletely
joined

Fig. 8-8. General characteristics of welds.

finish welds should look like:

- The weld should go all the way around the joint in a continuous bead. If there are just a couple of melted spots holding the tube in place, it's a tack-weld that didn't get completed.
- The melted area should be smooth. Gentle, continuous ripples are normal and the sign of a good weld. Watch for sharp edges and generally rough

and dull appearances. No matter how ugly a weld might look, don't file it. The metal is necessary for strength.

- The weld should taper smoothly into the surrounding metal.
- There should be no cracks, in either the tubes or the weld itself.
- No holes in the welds or tubes.

Figure 8-8 presents some of the characteristics of good and bad welds.

If you suspect a problem, call your EAA technical counselor or someone else knowledgeable. If they confirm your suspicions, contact the manufacturer.

After the inspection, ream out all the drilled holes to the proper size and deburr (Fig. 8-9). Most will be for AN3 ($3/16$ inch) or AN4 ($1/4$ inch) bolts, but check the plans.

Avid Aviation

Fig. 8-9. Kit manufacturers leave the deburring of holes to the builder. Note the burrs around the holes in the plate.

Internal Corrosion-Proofing

Steel-tube fuselages are strong and light. Properly protected, they'll last a long time. But 4130 steel tubing is prone to rust and must be protected. The outside is easy enough. But what about the inside?

Properly-done finish welds make a complete airtight seal and stop exterior moisture from entering. The air inside retains whatever moisture it contained when the structure was welded. There might be enough to support corrosion. In any case, bolt holes and cracks can allow moist outer air to enter and accelerate the process.

Steel-tube structures are internally corrosion-proofed by a coating of *line oil*. The old-timers used hot linseed oil, but line oil combines corrosion-proofing with an ability to seal tiny holes. The oil is introduced into the lower longerons then sloshed about by rolling the fuselage from side to side. Pour out the oil and close the holes with self-tapping machine screws and a sealant, such as Permatex.

Should you line-oil? It's a weird situation. Most of the old-time homebuilders I talked to were quite adamant about the necessity, yet the kit manufacturers generally don't make it a required operation. Most list it as optional.

I feel the lower longerons need the protection, especially on amphibians and floatplanes (Fig. 8-10). These tubes are most likely to be rusted-out on old production aircraft, so the same is probably true for kitplanes. It's a longevity issue that probably won't make much difference over the five years you'll probably own the airplane, especially if it's hangared.

In any case, don't leave open holes into the structure. Atmospheric moisture is bad enough without allowing rain to get inside. Seal any bolts or screws that penetrate the tube wall.

Fig. 8-10. Note the spray pattern behind the floats; any holes or gaps in the lower fuselage fabric mean the lower longerons will be drenched. Proper corrosion protection is vital, especially if the kitplane will be operated from salt water.

External Corrosion-Proofing

Applying a corrosion-resistant coating to the outside of the fuselage is one of the first things you should do. Alloy 4130 starts rusting almost immediately, and every day you delay priming just adds to the rust and scale that must be removed. The kit might have been shipped with an oil coating on the tubing to retard the rusting process, but many kits aren't shipped with a coating. In any case, the oil is easily disturbed during handling.

The degree of cleaning and the tools required depend upon how rusty the fuselage has become. Sandpaper or steel wool might suffice if it's fairly clean. Or it might take a steel brush on a power drill.

The best way to clean the fuselage is to sandblast, which presents its own problems, primarily sand everywhere. Usual practice is to pick a nice day and do it in the driveway or back yard.

Sandblasting setups for air compressors can be bought for $50 or so. If you don't have access to a compressor or yours isn't big enough, rent an all-in-one unit from the local U-Rent-It store. Or check with auto body shops, then take the fuselage in ready to blast, and they'll charge $150 or so. Paint it as soon as you get it home, or have the shop do that as well.

If you sandblast it yourself, set the unit for the narrowest stream possible. Because the fuselage is made of a bunch of thin tubes, most of the sand will be wasted anyway. Buy extra; you can always return it if unopened, and it eliminates the hassle of having to stop and go out for more.

Whatever the method, the intent is to remove all contamination. The nice things about steel compared to aluminum are that it doesn't have to be etched prior to priming, and doesn't have an Alclad surface to baby. You can bear down on the sandpaper and the sand blaster. The rougher the surface after cleaning, the better the paint will stick.

Once the rust is knocked off, put on gloves and clean the tubing with MEK or acetone. This gets rid of the residual grease, dust, and preservative oil. Until the primer has been applied, don't touch the fuselage with your bare hands.

Prime as soon as possible after sanding, otherwise, the tubing will just rust up again. Application is similar to priming aluminum, as given in the previous chapter. Note also the safety precautions given in chapter 7; aircraft primers are rather nasty stuff, and you'll want to protect yourself properly. Make sure the primer is compatible with the planned fabric-covering process.

Because the tubing isn't an external item, appearance of the primer coat isn't a factor. It can be applied with a brush. If you decide to spray, set the gun to the narrowest setting. If possible, rig up a rack that allows you to rotate the fuselage to be able to hit all the nooks and crannies.

Apply a light coat, let it dry, and apply another. Then grab a flashlight or trouble light and go hunting for thin spots. Make sure the bottom of the fuselage is well covered, and especially the area around the tailwheel. These are the areas where water will collect.

A typical primed steel-tube fuselage is shown in Fig. 8-11.

A lot of aluminum parts will be attached to the fuselage; these must also be protected against corrosion. Details on preparing and priming aluminum are in the preceding chapter.

Whenever different materials must be joined together, there is a chance of accelerated corrosion. Where steel and aluminum are to be attached, each must be thoroughly primed to prevent dissimilar metals corrosion. That's one of the reasons for the cadmium plating applied to aircraft hardware.

Fig. 8-11. A primed RANS S-10 fuselage with wheels and engine attached.

And remember that wood retains a significant amount of moisture. When a wood component is joined to a metal one, varnish the wood, in addition to priming the metal.

Steel-Tube Summary

The nice thing about steel-tube kitplanes is once the fuselage is primed, you can forget about it. There are a few things to remember, like only drilling holes specifically called for by the plans. But the fuselage just becomes a structure to hang other parts on. And many of those parts bolt to prewelded attachment points.

Building the wings is another problem. There aren't too many steel-tube-winged airplanes out there (the Dyke Delta is one). Most tube-and-fabric kitplanes' wings use either aluminum or wood construction. Working with wood is covered in the next chapter. The following sections discuss aluminum structures.

ALUMINUM-TUBE STRUCTURES

Probably the biggest thing the ultralight fad did for homebuilt airplanes is popularize and prove aluminum-tube structures. Instead of a wooden spar, the ultralighters used a tube of light aluminum alloy. Instead of a complicated longeron-and-bulkhead fuselage structure, they just ran a thicker aluminum tube from nose-to-tail.

Until the ultralight movement came along, the empty weight for the smallest conventionally-built single-seat homebuilts was about 400 pounds. Ultralights cut this figure in half, and more. When the first version of FAR Part 103 was due to come out, the rumored maximum weight was about 150 pounds. Many vehicles in production already met that limit. The actual limit turned out to be 254 pounds. A blizzard of comfort-type items like fairings and closed cockpits then appeared in the sales catalogs.

The ultralight boom has subsided, but its effect on homebuilding lives on. The RANS series, the Avid Flyer, the Kitfox, and many others use an aluminum tube as a combination main spar and leading edge. Many bend ribs from 1/2-inch aluminum tubes. Those with conventional wood or metal ribs include them in the kit nearly ready-to-install.

For the homebuilder, aluminum tubing has several advantages over more traditional methods. It's light. It's easily cut with simple hand tools. It can be bent into graceful curves with little effort. It's more resistant to corrosion.

But its drawbacks also hit home. Aluminum isn't as strong or hard as steel, so some of the weight advantage is lost by the need for a beefier structure. It's tricky to weld, too. Steel's color depends upon its temperature, which makes it easy to heat it to the proper condition. Set a torch to aluminum, and it looks the same right to the point it collapses into a molten puddle.

Because welding is difficult, aluminum tube structures are held together by other means, such as bolts and pop rivets. These labor-intensive processes are usually left to the builder. The following sections describe cutting, shaping, drilling, and riveting aluminum tubing.

Cutting

A horizontal bandsaw is the best way to cut aluminum tube. The clamp ensures straightness, and the cut requires only a moderate amount of cleaning up.

But they are expensive. Hand-held tubing cutters (Fig. 8-12) cost around $10–$20, and consist of a movable cutting wheel positioned opposite a set of rollers. Get the kind for tubing 1/8 inch to 1 1/8 inches in diameter.

Place a pencil mark on the tube at the desired cut line. Turn the cutter's knob to retract the wheel until the tube slides between it and the rollers. Position the wheel just outside of the cut line, and tighten the knob. Hold the tube stationary, and rotate the tool around it. Watch the tracking when you start because sometimes the tube doesn't seat properly and the tool starts cutting a spiral.

As the wheel cuts a groove, the tool's resistance to rotation lessens. Tighten the knob every couple of turns to maintain the same level. Eventually the wheel breaks through and the tubing separates. This can happen suddenly; be careful not to drop the tool.

The operation leaves a pinched-in area at the end of the tube; the pinched-in area must be removed. Clamp the tube down and run a coarse file across the end a few times to get rid of some of the excess metal. A hook-type deburring tool (Fig. 8-13) takes care of the remaining thin lip with little fuss. Polish with emery cloth to eliminate any scratches or file marks. As discussed in chapter 5, these can be starting points for cracks. A 1-inch belt sander is a useful bench tool for smoothing aluminum.

Fig. 8-12. The tubing cutter in action.

The tubing cutter won't work in all cases. There must be enough tube past the cut line to allow the rollers to make full contact. The rollers on either side are of unequal length, and you can sometimes get closer to the end by flipping the tool the other way to put the shorter roller on the tube's end. If the amount to be removed is too small for the cutter, a file is about the only way.

Most tubing cutters can't open wide enough for tubes much greater than an inch. You'll occasionally have to resort to a hacksaw. The problem is making a straight cut because hacksaws sometimes seem to have a mind of their own.

The first step is to apply a cut line all the way around the circumference. A single pencil mark isn't enough because you need a definite reference to keep the cut perpendicular to the end of the tube.

Start with a piece of paper shaped like a grocery receipt. The length should be about four times the diameter of the tube, and the edges should be perfectly straight. Wrap it around the tube. When it starts to overlap, place the additional paper directly atop that previously applied (Fig. 8-14). Tape the end down.

You've just made a paper tube with an inside diameter equal to the outside diameter of the aluminum one. By carefully overlapping the paper, the end of the paper tube is square and true.

Fig. 8-13. The tubing cutter leaves a pinched-in burr around the circumference of the tube. File it down a bit first, then scrape it away with the deburring tool as shown.

Slide the tube to the cut line. The paper tube should cover the end of the good metal, leaving the cut-off end bare. Slather machinists' ink, spray paint, or a magic marker over the area where the paper tube meets the aluminum one.

When dry, remove the paper tube. The paper acted as a stencil, keeping the "good" tubing clean while marking the excess. If you cut away all the tubing marked with paint, the tube end will be true.

Clamp the tube, leaving the end accessible to the hacksaw. Protect the tube from the vise jaws. Actually, vises aren't the best for holding round objects. A Black & Decker benchtop Workmate has triangular plastic jaws that are great for holding tubes.

Start the hacksaw on the painted area, about 1/8 inch from the paint line. You aren't trying to cut off all the end, you're just trimming it down to lessen the amount of filing.

Cut slowly and carefully. If the saw comes too near the line, apply a bit of twist to take it away. Keep close watch on the back of the tube, too, because it's very easy for the saw to take a bit of slant and cut the other side of the tube differently.

When the end is removed, use files to trim away the remaining painted metal.

Where to Drill Holes

You can't drill a hole anywhere you want. A hole will weaken the tube; how much depends on where you put it and how well it's aligned.

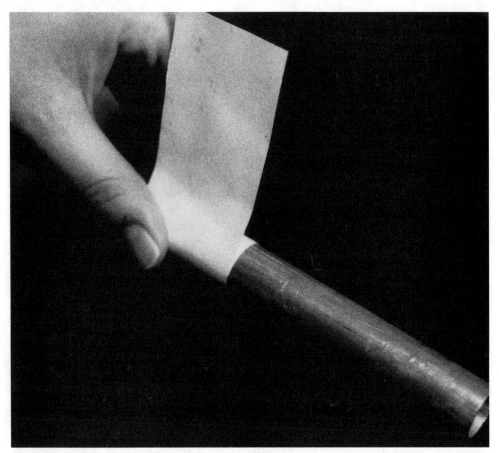

Fig. 8-14. A strip of paper with a straight edge can be used to mark the end of a tube for cutting. Lay the edge of the paper directly over the previous wrap, then paint the interface area. When the paper is removed, the sharp line remaining can be used as a reference when using the hacksaw.

To help describe the right way to drill holes, we'll reference points around the circumference of the tube similarly to a compass. Looking at the tube end-on, it's a circle. North is straight up, or 0 degrees; east is directly to the right, or 90 degrees; south (180 degrees) is straight down; west (270) is directly to the left.

Imagine a long tube supported by two sawhorses. Hang a bucket from the midpoint of the tube, and it flexes a bit. This flexing has similar effects as when aluminum sheet is bent, as described in chapter 5. The south (bottom) edge of the tube has to stretch slightly, while the north edge compresses to match the curve. Add weight to the bucket, and amount of stretching and compressing increases. Too much weight, and either the south edge stretches too far and breaks open, or the north edge crumbles.

Looking at a cross-section, there must be two points where the force changes from stretching to compression. The points are approximately halfway between north and south; in other words, east and west. At these positions, the force is neutral, the metal isn't under stress.

If a hole is drilled in the top or bottom, the surrounding area is under stress and the hole will cause the metal to fail early. But if the hole is drilled horizontally from east to west, no problems result. Because the area isn't under stress, drilling doesn't affect the tube's load-carrying ability.

As mentioned, many smaller kitplanes use aluminum tubes for spars. While the above description is drastically simplified, the lesson is unchanged: Do not drill holes through the top or bottom of the spar. They should go fore and aft, though the neutral portion. This is illustrated in Fig. 8-15.

Some instructions call for drilling holes at various positions around the circumference. Anything closer than, say, 45 degrees to north or south might cause problems. But you'll find the designers compensate in various ways. They'll nest another tube inside the drilled area, for example. Or the portion might not be under much load, like the tip of the spar.

Both the Kitfox and the Avid series install a couple of pop rivets on the bottom of the spar underneath the strut fitting (Fig. 8-16). The strut fittings themselves provide considerable reinforcement along the spar to compensate.

The manufacturer can specify hole drilling in these seemingly forbidden locations because they have the facilities to analyze and test to ensure the strength isn't affected. You don't. So keep the holes out of the top and bottom

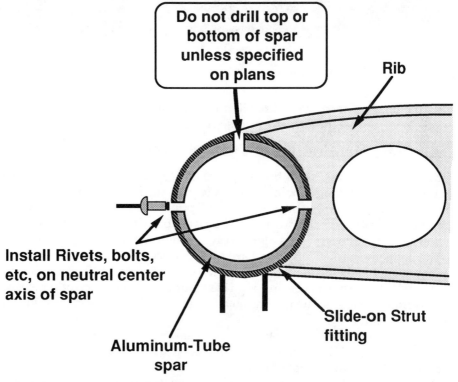

Fig. 8-15. In most cases, the builder should never drill holes on the top or bottom of the wing spar. The same holds true for other tubes, as well.

Fig. 8-16. An exception to the rule. The thin sheet reinforces the fabric around the strut attachment, and is pop riveted to the bottom of the spar. But the large wing strut attachment plates riveted and epoxied to the tube provides more than enough compensation.

unless the plans specifically call for them. If the situation feels funny, call the manufacturer. You might have misread something.

Accurate Positioning of Drilled Holes

Knowing where holes can be safely drilled is one thing, accurately placing them is another. Drilling a hole in one side of a steel or aluminum tube isn't much of a problem. Dimple the spot with the punch, drill a pilot hole, and ream to the final diameter. That'll work for pop rivets or self-tapping screws, or any other fastener that doesn't pass all the way through the tube.

The problem arises when holes have to be correctly oriented with each other. Take the spars for the RANS S-10 Sakota, as shown in Fig. 8-17. Aluminum rib fittings and aileron nut plates are pop riveted at the 9 o'clock position at given intervals from the root to the tip. Install a few at 10 o'clock instead and the wing aerodynamics suffer.

Even aircraft with different rib systems come against a similar problem. A fitting might be riveted to the root, and a fiberglass wingtip to the other end. Get the holes off line, and the tip will have an odd twist.

Drilling the first hole is easy enough. But how do you position the rest of the holes at the same clock position? The labeling printed on the tube can't necessar-

Fig. 8-17. The rivet holding the rib and the nut plate for the aileron must be exactly aligned, or weird things will happen to the wing aerodynamics.

ily be trusted. The tube might have rotated slightly while the label was being applied, or the tube might have already been painted.

And what happens if the plans call for bolting through the spar? The bolt must pass directly through the center, perpendicular to the tube's long axis. The second hole must be drilled directly across from the first at the same distance from the tube end.

The farther the second hole is off, the weaker the attachment. If one hole is at three o'clock and the other is at eight, the tube will warp as the bolt is tightened. If the bolt slants one way or the other along the length of the tube, other localized stresses can be induced. In either case, the bolt head, nut, and washers won't rest flat on the surface.

We need to establish a pair of reference lines the length of the tube exactly opposite each other. It's easy enough for a pair of short lines because all you have to do is wrap the tube in carbon paper and close a vise gently on it. If the vise's jaws are parallel, their first points of contact are (by definition) exactly opposite to each other. But vises aren't really precision pieces of equipment, and this procedure wouldn't work for tubing longer than a few inches.

Instead, place the tube on a flat, smooth, surface and hold it rigidly in place. One good way is to pin the tube between pieces of wood nailed to the surface.

Holding it by hand won't do because with any little wiggle, the reference will be skewed.

Place a combination square on the table with the ruler vertical, like you were measuring the height of something. Slide the square until the ruler touches the side of the tube. Take a look at the ruler line equal to half the tube's diameter. You'll find that line is the point in contact with the tube.

The actual dimension doesn't make any difference. If the base of the square is flat on the table, the ruler *must* make contact at the point 90 degrees around the circumference from the lowest and highest points.

Because we are only seeking two reference points exactly opposite to each other, the relationship to the 6- or 12-o'clock positions are moot. As long as the tabletop is flat, a square held to either side will make contact at two points exactly opposite to each other, as Fig. 8-18 demonstrates. The tube doesn't even have to be level. The only reason we attach it tightly to the tabletop is to make sure it doesn't roll between markings.

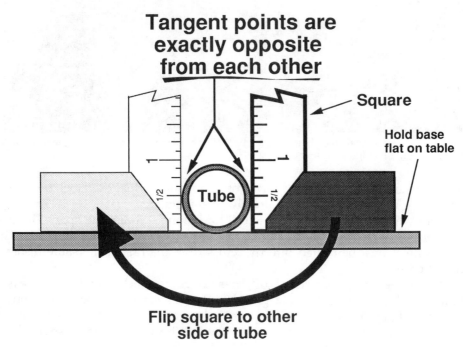

Fig. 8-18. No matter the diameter of the tube, a square held flat on the worktable on either side will touch the tube at directly opposite points. The measurement itself is immaterial because it is the points of contact we're interested in.

But just marking single points of contact on either side doesn't do us much good. We need to define reference lines on both sides along the entire length.

Figure 8-19 shows the solution. Bring the square into position, then slide it along the length of the tube while remaining in contact. As long as the base of the square stays flat on the table, it will apply a long, straight scratch. This

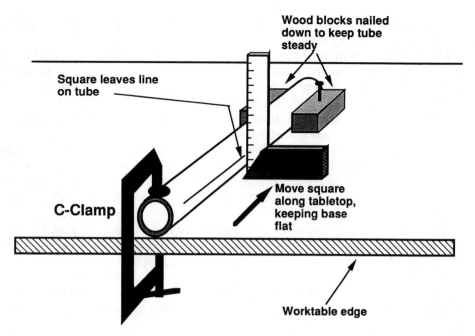

Wood blocks nailed down to keep tube steady

Square leaves line on tube

Move square along tabletop, keeping base flat

C-Clamp

Worktable edge

Fig. 8-19. Opposite centerlines can be drawn by dragging the square along the tube on either side, keeping the base flat. Keeping the tube from moving or rotating during this process is vitally important.

scratch and the one we immediately apply on the other side become our reference lines.

But because you don't want to actually scratch the tube, paint the sides with machinist's ink first. Let it dry, then slide the square with enough pressure to scrape the paint but not enough to excessively gouge the underlying aluminum. Then slide it along the other side as well.

It isn't as damaging as it sounds. If the tube is 6061 alloy, it doesn't have an Alclad surface to disrupt. The scrape will be along the neutral, unloaded axis of the tube. And if you don't use too much pressure the scratches should buff right out with emery cloth.

There's nothing magical about machinist's ink in this case. Just about anything will work: old spray paint, shoe polish, whatever. I use a wide-tipped marking pen. Just as long as it can be cleaned off before priming.

The tube and whatever's holding it in place shouldn't be moved until both sides of tube are marked. Holder placement must be planned so the areas to be marked aren't obstructed.

With the two reference lines in place, any number of holes can be drilled exactly along the same axis. However, we still have a problem with bolts. The holes through both sides must be directly across from each other along the length of the tube, as well.

I know what you're thinking. Why not mark on one side, then clamp the thing in the drill press and drill both sides at once?

Go ahead. Everybody's got to try that once. Use a cheap piece of scrap tubing for the first attempt, though.

The problems are many. First, the tube has to be set up so that the drill point hits the exact center. Any inaccuracy, and the bit tries to crawl away, chewing the surface as it goes. You'll have to position the tube so the point where you want the hole is the highest point, which is the point closest to the drill press.

Second, the tube has to be dead level when it's being drilled. Any sort of slant puts the opposite hole somewhere else.

Third, drill bits aren't stiff. They flex. The slightest off-center pressure between the first hole and the opposite inside wall will put the hole in some other location. If the drill's a little off center, the curve of the tube will change the aim. When it hits the opposite wall, the bit will walk around until it digs in at a random location.

A few of you might be able to do this perfectly every time. I hit about 50 percent correctly. Even if you can do it 90 percent of the time, every tenth hole will ruin the part. Do yourself a favor and drill the holes separately.

There are a couple of ways to find the opposite point on the other side of the tube. The best way is by measurement. If the bolt is supposed to be installed 30 inches from the end, measure the distance on both reference lines and mark the points.

An alternate method uses a drilled pattern. Take a 2-inch piece of tubing with an inside diameter equal to the outside diameter of the tube to be drilled. Working very slowly and carefully, mark reference lines on the tube and drill two small reference holes directly across from each other. This short bit of tubing then becomes a pattern for marking bolt holes.

Slide it over the tube to be drilled. Position one reference hole at the point to be drilled, then insert the punch in the hole on the other side of the pattern and mark the location. You could even drill pilot holes through the pattern, but this tends to enlarge the reference holes.

With the positions marked, it's time to drill.

Drilling

Once the hole is accurately marked, drilling is a snap. Dimple the metal with the automatic centerpunch, then drill a small pilot hole. Use the drill press, or hold a hand drill perpendicular to the surface while drilling. Then drill and ream to the desired size.

But before you reach for the larger drill bits, take a look at the pilot hole. Is it where it's supposed to be? If not, some degree of correction can still be made. When using the larger bit, drill at a slant. Point the drill in the direction the hole needs to be moved, and apply a moderate amount of pressure in that direction. Think of it as pushing the hole in the right direction; if the hole has to go to the right, the drill should be slanted to the left. Experience will teach you how much slant and force.

Large holes present a problem. The big bits have a tendency to grab and make wavy edges. A countersink makes smoother holes. For 1/2 inch, for instance, drill to 3/16 inch then follow with a 1/2-inch countersink.

Dress the edges of drilled holes with a larger-diameter countersink to deburr and chamfer the edge. A hook-type deburring tool removes the burr on the inside of the tube. Or if the hole is near the end of the tube, slide a round or half-round file inside to knock down the inner burr. Polish with emery cloth.

It's especially important to eliminate the interior burr when another tube is going to slide inside the drilled one. A burr on the inside might stop the smaller tube. Even if you can force it past the hole, the sharp metal will scratch its exterior.

Other Shaping

Sometimes the end of the tube must be shaped to fit other parts of the structure, rather than cut off square.

As shown in Fig. 8-20, where one tube joins another, its end must match the other's curvature. For example, the end of a 1/2-inch tube forming a T with a larger-diameter tube should be saddle-shaped to make contact with the maximum surface on the larger tube, to spread out the compression loads. If not, the tubes would only make contact at two small points. Either tube might then crumble under heavy loads.

This process is called *fishmouthing*. Smaller or thin-walled (.035-inch) tubing is best fishmouthed by a round or half-round file. Start a notch with a rat-tail file, then enlarge it until it starts matching the tube to join. Then notch the other side and file it down until both sides are the same. Deepen each side sequentially, never letting either get too far ahead of the other.

Bench tools can help, especially when the intersection is at other than 90 degrees (Fig. 8-21). Fuselage diagonals are good examples. These can be done by hand, but it's a slow process. Instead, chuck up a reamer or a grinding stone of the same diameter of the longeron in a drill press. Mount the tube in a drill table clamp at the same angle (relative to the ream or stone) at which it is supposed to join the longeron. Start the motor, and slowly lower the ream to the tube.

If you have a bench press clamp with a table that cranks in and out, mount the longeron-sized ream in the chuck, and use the table to bring the diagonal into contact. This method can also be used with a metal lathe.

Whether fishmouthing by hand or power, deburr and smooth with emery cloth when the metal has been properly trimmed.

More complex-shaped notches can be made by a variety of methods. Both the Avid Flyer and the Kitfox need oddly-shaped slots in the root ends of the spars. Use a template to mark the area to be removed. For larger areas like these, a Dremel Moto-Tool is probably best. Cut close to the line, and take out the rest with files.

Hold the Moto-Tool firmly—if the cutter or wheel catches an edge, it'll jerk the tool, chewing up the surface. If possible, position the tool so the tendency is to pull away from the spar instead of into it. As always, deburr when finished.

Fasteners

Two fastener types predominate aluminum-tube construction: pop rivets and bolts.

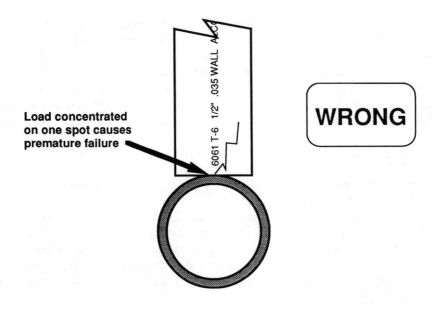

WRONG

Load concentrated
on one spot causes
premature failure

6061 T-6 1/2" .035 WALL ALCO

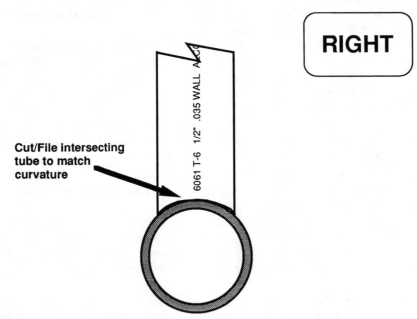

RIGHT

Cut/File intersecting
tube to match
curvature

6061 T-6 1/2" .035 WALL ALCO

Fig. 8-20. Intersecting tubes must be fishmouthed or loads will be concentrated.

Driven rivets form the shop head by repeatedly ramming the end into a bucking bar. Pop rivets do it differently.

As shown in Fig. 8-22, a pop rivet has a hole running its length. A stem passes from the shop end through the center of the rivet, and sticks out above

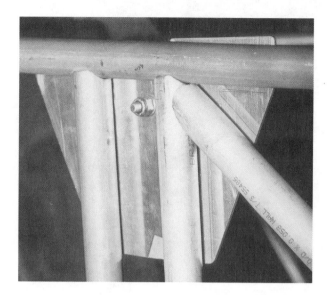

Fig. 8-21. The intersection of the two verticals with the longeron are easy to do, but the diagonal presents another problem.

the manufactured head. At the shop end, the stem is wider to form a *mandrel* wider than the center hole. This mandrel is wider than the center hole, and sits flush against the shop end.

The rivet tool grabs the free end of the stem and pulls while holding the rest of the rivet stationary. The force tries to drag the mandrel toward the manufactured head. Because it's wider than the center hole, the only way the mandrel can move is to crush the softer metal of the rivet shank.

When the shank is completely compressed and can't collapse any more, the stem of the rivet breaks away.

Where a flat plate must be pop-riveted to a tube, the line of rivets should run straight down the centerline (Fig. 8-23). Any offset weakens the connection.

Countersunk pop rivets are used in airplanes such as the Sonerai. However, be advised that the head angle is different from that of standard rivets. Countersunk pop rivets require a 120-degree tool. Countersinks are available in this angle, as are sheet-metal dimplers designed to work with the pop rivet tool. The countersinking/dimpling procedure is identical to that of driven rivets, which was included in the last chapter.

There are MS/NAS standard pop rivets, but they don't seem to be widely marketed. So it becomes even more important not to diverge from the plan's specifications. The standard pop rivet is the Cherry rivet, sold by most homebuilder's supply outlets. These are available in different styles, sizes, and materials. Use only approved substitutes because rivets of the wrong material might be only half as strong as required.

One advantage of certain pop rivets is an ability to expand and match misdrilled holes. Don't depend on this too much, as the stronger rivets specified for aircraft generally don't expand as much.

Preparation is similar to that of driven rivets, given in the previous chapter. The hole is pilot drilled then reamed to the final size by using the correct bit. Like

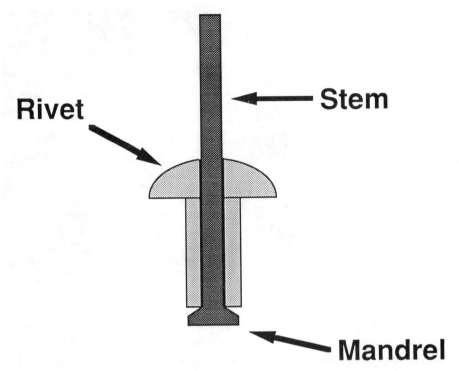

Fig. 8-22. The pop, or pulled, rivet.

Fig. 8-23. Where a plate must be riveted to a tube, the holes should be drilled on the tube's centerline. (This photo shows at least three violations of the practices described in chapter 5. Can you spot them?)

driven rivets, pop rivet sizes are exact, so this final drill must be slightly larger than the rivet diameter. The pilot and final drill sizes (incremental inches) for popular rivets are:

Rivet	Pilot	Final
$3/32$	$3/32$	#40
$1/8$	$1/8$	#30
$5/32$	$5/32$	#21
$3/16$	$3/16$	#11

Debur the holes, and cleco the components together in the usual fashion. When all the holes are ready, remove a cleco and insert a pop rivet. If the plans specify, dip the rivet in zinc chromate first. Some types of rivets don't have smooth shanks and resist sliding all the way into the hole. A piece of wooden dowel with a stem-sized hole drilled in one end can be used to apply a little force.

The rivet's manufactured head should rest solidly on the surface. The components being riveted must be held tightly together during the process because the rivet might try to expand into any gap. This results in wrinkled aluminum and a weak joint.

Make sure the correct head or nosepiece is installed on the rivet gun and slide it down the stem until the tip is solidly against the rivet head. Some types of Cherry rivets require a special tool.

Squeeze the handles of the tool together (Fig. 8-24). The jaws inside the riveter will clench the stem, then start drawing it back. The mandrel crushes the shell of the rivet. When it has moved as far as it can, the stem suddenly breaks away. Figure 8-25 summarizes the process.

Unless the material is especially thick or the rivet is short, the tool's handles will meet before the rivet pops. Release the handles, slide the nosepiece down the stem until it rests on the head again, and repeat.

When the stem breaks away, lift the tool away from the metal, release the handle, and twitch the tool in a safe direction. The stem should release from the riveter, and you're ready to go again.

Pop rivets are designed to work only in shear, so never install one in a position where the load tries to pull it directly out of its hole. Also, the rivets used on the Zenair and CIRCA Nieuports have rather soft heads; don't lift the top sheet away from the material once the first rivet is in place. Lifting the top sheet away bends the head.

If a pop rivet must be removed, drill out the center with the pilot drill. The driven-rivet method of using a smaller drill to remove the head and punch out the remainder doesn't work, due to the shank's expansion tendency.

Installation of bolts is perfectly straightforward, and follows the same basic rules explained in chapter 5. However, there's another factor to consider when bolting aluminum tubes.

Screw threads are force-multipliers; they convert a small rotational motion into high compression. The old hand-operated printing presses are a prime

Fig. 8-24. The rivet puller in action.

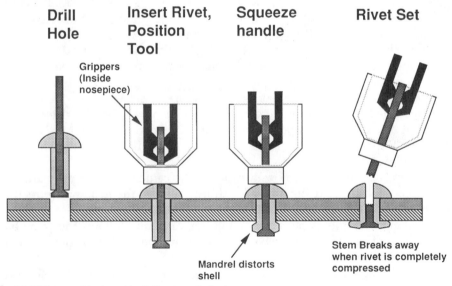

Drill Hole   **Insert Rivet, Position Tool**   **Squeeze handle**   **Rivet Set**

Grippers (Inside nosepiece)

Mandrel distorts shell

Stem Breaks away when rivet is completely compressed

Fig. 8-25. The mechanics of installing a pop rivet.

example of this effect. A few foot-pounds via the handle generated hundreds of pounds of force at the platen.

Bolts work the same way. But most commonly-used sizes of aluminum tubing aren't strong enough to withstand it. As the bolt tightens, the tubing pinches in. You could completely squash a .049- or .058-inch aluminum tube using normal hand tools. Of course, an aircraft bolt would bottom out on its threads first. But the tube would still distort and recommended torque levels couldn't be reached.

The solution is to increase the tube's wall thickness in the area of the bolt. This can be done two ways. Some kits apply reinforcement plates to the exterior. This is especially common on kitplanes with aluminum-tube spars. Installation of these vary with the type of aircraft.

The other method is to install reinforcement inside the drilled area. To be effective, the outside diameter of the insert must be nearly equal to the inside diameter of the tubing.

The easiest way is to use a short piece of the next-size-smaller tubing. This works out fairly well; one-inch diameter, .058-inch wall has an I.D. of .884-inch, which is just a tad larger than the O.D. of 7/8 inch (.875) tubing. If the 7/8-inch tubing also has a .058-inch wall, the total wall thickness has been doubled. Twice .058 inch is .116 inch, just a tad less than 1/8 inch. Therefore, for tubes with .058-inch wall thickness, select an insert with 1/8-inch shorter diameter.

Other reinforcement methods are possible. A solid aluminum plug will be strong, albeit heavy. An excellent compromise is a wooden plug. You want an old hard wood like ash, maple, or oak. If the exact diameter can't be found, use a piece of the next highest available size. They can be turned in a wood lathe, or even a drill press (using a file). If a lot of wood must be removed, take a knife or saw and carve away some of the excess first.

The absolute minimum length of the insert is four times the diameter of the hole to be drilled. This allows the appropriate end margin in both directions. I like to make them a bit longer because their resistance to compression also depends on length. Mine are usually about twice the minimum; about 1 1/2 inches long for AN3-sized bolt holes.

On the tubing to be reinforced, mark the position of the bolt holes. Drill a pilot hole on both sides. Deburr as well as possible. Drill a single, same-sized hole midway on the insert. Slide the insert into the tube until its hole matches one of the pilot holes. Cleco the pieces together, then pilot-drill the other side of the insert through the other pilot hole in the tubing. Drill and ream the holes to final size, then disassemble (if possible) and deburr.

This method works when the insert can be rotated to match its pilot hole with the tube's. This isn't possible on square tubing or the ovals used on some ARVs. The insert in these cases is usually carved wood. To install, drill the pilot holes in the outer tubing. Mark the center of the insert using a felt-tip pen. Draw a line all the way across.

Watch through a pilot hole as the insert is slid into place. When the centerline appears, drill the insert's pilot holes. Drill and ream to final size.

Bending Tubing

Bending aluminum tubing is very much like bending a thick aluminum plate. The inside of the bend radius stays pretty much the same, while the metal on the outside stretches to remain continuous.

There are a couple of differences. First, you're manipulating less metal. A 1-inch diameter tube is far easier to bend than a 1-inch-thick piece of aluminum plate. Believe me.

Second, the tube has a lot of ways to react if it doesn't like the bend. As mentioned earlier, a tube under a bending load is compressed on one side and stretched on the other. The center remains neutral. When overbent, or bent incorrectly, the outside of the bend can tear open and/or the inside can crimp.

How the tube reacts depends on the bending procedure. Proper fixtures and practices minimize problems.

To begin, let's understand how a crimp is formed. Take an ordinary drinking straw and slowly fold it in half while watching the inside of the bend. The material under compression has to go someplace, so it folds inward. But because the surface is curved, the material expands sideways as well. When the straw is fully folded, the crimp is one-half the circumference of the straw across, and there *is* no inside diameter at the fold point. And because tubes depend upon their circular cross section for strength, the area of the crimp is very weak.

The goal when bending aluminum tubing is to avoid crimps. One way is to maintain the cross section of the tube throughout the bending process. There are coiled-spring tube shells that slide over the outside, but they usually aren't available for the larger sizes. Or you can fill the tube with sand prior to bending. This works, but long or wide tubes need a lot of sand.

There are custom tools on the market for bending tubing. Before you buy one, make sure it'll handle both the diameter and material of the tubes you wish to bend. Some are intended for soft annealed aluminum brake lines. An adequate tool to bend a single size of large-diameter aluminum tubing costs about $100; those that can handle a wide range of tubing sizes cost four or five times that.

The cheapest solution is procedural: Bend the tube gradually around a radius in such a manner to stretch the outside surface rather than crimp the inside edge. This can be done in the shop with few tools.

There are two types of bends used on kitplanes: *radius bends,* where bends of particular characteristics are required; and *shape bends,* where the final shape is given rather than a set of radii. We'll look at radius bends first.

A benchtop jig is necessary. The kitplane's instructions will specify the radius, and the total angle of bend can be eyeballed. Take a bit of scrap plywood, draw a circle of the proper radius on it, then cut it out with a saber or band saw.

A full circle, not just 90 degrees or whatever. The tube will have a bit of springback, meaning the disk must continue the radius a few degrees more than the angle the plans call for. You'll typically have to draw nearly a full circle on the wood anyway, so you might as well cut the whole thing out; you might need a greater angle at the same radius later. Smooth the edge a bit with sandpaper, and nail the disk to the worktable.

Of course, it would be even better if the edge were grooved like a pulley to match the tube. It's not easy to do, but if you've got a router table it might be worth a try. Another option is to keep your eyes open for cheap pulleys of the proper sizes.

When ready to bend, place the tube against the disk and nail a piece of 2 × 4 against the free end on the opposite side. Position the tube so the point at which the bend is to begin is the point touching the disk.

Draw an imaginary line from the point the tube contacts the disk through the disk center: call this the axis of contact. To make the bend, grab the free end of the tube and pull *parallel* to the axis of contact, as shown in Fig. 8-26. *Never* apply any sort of pressure *towards* the disk. If you do, the metal on the inside tries to compress and starts a crimp.

Bend the tube to the desired angle while watching carefully for crimping. If it looks like the tube is beginning to flatten, relax pressure and slide the tube slightly to move the point of contact. If table space becomes a problem, move the scrap holding the free end and resume until the desired angle is reached.

Heavier tubing requires a lot of pressure. The longer the work end, the more force you apply with less effort. If possible, don't cut the tube to length before bending. If the tube is already to length, slip a smaller diameter tube or a wooden dowel into the work end to give you more leverage.

Shape bends are usually easier. Begin by stapling a piece of brown wrapping paper to the tabletop. Transfer the shape shown in the plans to full-size on the paper.

The shaping process consists of bending the tube, then comparing it to the desired shape. Bends are applied where necessary and as necessary. If a piece gets bent too far, you can bend it back as long as it hasn't started to crimp.

The same rules still apply. Bend all tubes around a radius. Too sharp an edge will start a crimp. But these shapes usually involve gradual curves that don't start problems. Slight bends at many points are less angular than a few big ones.

A very common shape is a rib formed of bent 1/2-inch aluminum tubes. A common bending tool in these cases comes in pairs, your knees. These smaller tubes can be easily shaped by hand.

Where heavier tubes must be shaped, a combination approach works best. Bending jigs can make gradual curves by making closely spaced small bends. For very wide-radius shapes in heavy tubes, a car's full-size spare tire can be used. Prop up the tire and place the tube on top. You and a friend then push downwards, seesawing back and forth over the tire. It sounds ugly, but it works.

Building Structures

Drilling, pop-riveting, and bending are all steps in the construction of aluminum structures. As mentioned earlier, aluminum tube construction is used for

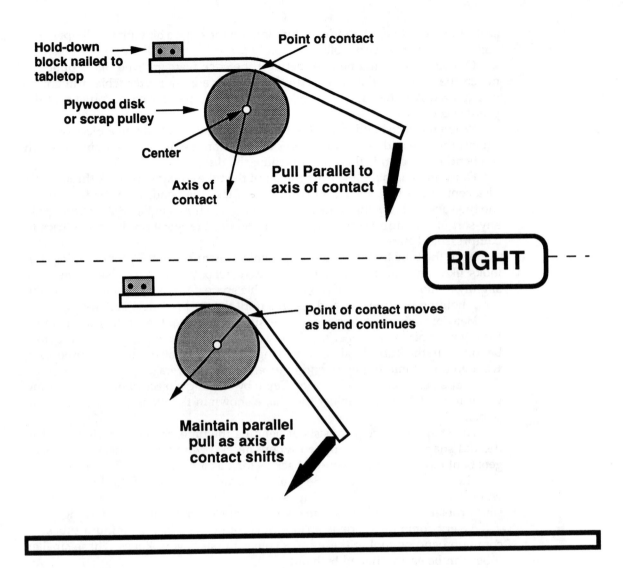

Hold-down block nailed to tabletop

Point of contact

Plywood disk or scrap pulley

Center

Axis of contact

Pull Parallel to axis of contact

RIGHT

Point of contact moves as bend continues

Maintain parallel pull as axis of contact shifts

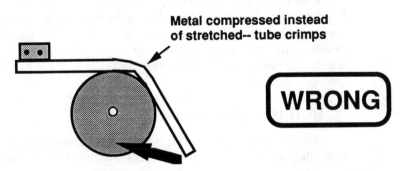

Metal compressed instead of stretched-- tube crimps

WRONG

Fig. 8-26. Smooth tubing bends are made if the pressure is applied parallel to the axis of contact.

the wings of many fun-type kitplanes. The wing of an Avid Amphibian, as seen in Fig. 8-27, is a good example. Several in the ARV class also have the builders assemble truss-type fuselages from aluminum tubes (Fig. 8-28).

The first step in building a full structure is to draw a single long reference line atop of the worktable. This is your basic reference line for measuring positions and angles of all elements. You could use a chalk line, but go over it with something more permanent. Or pull an ordinary string tight, mark positions every foot or so, then connect the dots with a straightedge and pencil.

Fig. 8-27. A typical kitplane aluminum structure is this wing of an Avid Amphibian.

Fig. 8-28. The Murphy Renegade's fuselage is aluminum tubing. The builder assembles the entire structure, using special fittings and pop rivets.

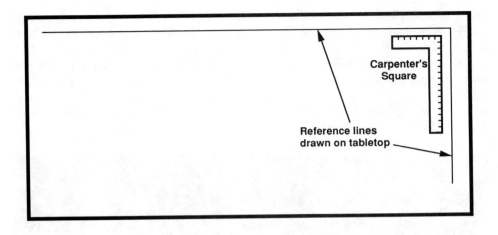

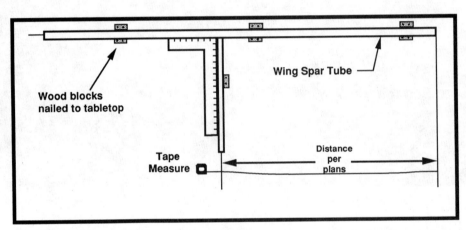

Fig. 8-29. In building a large structure like a wing or fuselage, two perpendicular reference lines are first drawn on the surface of the worktable. Then other components are put in place in reference to the dimensions shown in the plans. Parts are held in place with short pieces of wood nailed in place.

Define one end of the line as being the firewall or wing root (depending upon what's being built) and use the carpenter's square to draw another line perpendicular to it.

The plans will give the distances at which verticals, compression struts, or ribs must be installed. Transfer these dimensions to the table, and draw perpendiculars at the points with the carpenter's square.

Cut the components to size and install in relationship to the patterns drawn on the table. Hold them in place with pieces of 1 × 2s or 2 × 4s (called *jigging blocks*) nailed to the tabletop. Don't pound the nails all the way in; leave a half-inch or so sticking up so they'll be easy to remove.

This process is shown in Fig. 8-29.

Corrosion Protection

If the fuselage is made from 6061 alloy, you might be able to forego the corrosion protection regime required for 2024 aluminum or even 4130 steel. It depends on where the plane will be stored, your proximity to salt water, and how strongly the kitplane manufacturer recommends it.

Priming aluminum is detailed in chapter 7. To recap, the metal must first be cleaned to eliminate grease and finger oils. Then it's etched with a mild phosphoric acid wash to roughen the surface to enhance paint adhesion (engineerese for making the paint stick well) and the primer is sprayed on.

Tube-fuselaged kitplanes are about the fastest to build. They're intended as fun airplanes, so you don't get involved with large amounts of subsystems work. As such, they're generally pretty light as well, which makes them easier to manhandle around the shop.

The second half of the following chapter discusses installing fabric covering.

9

Wood and Fabric

WOODWORKING HASN'T BENEFITTED MUCH from the technological revolution. Wood is passé. Industry spends millions of dollars for developing faster, more precise methods for cutting aluminum, or for new composite construction materials and methods. Mainstream aviation was the catalyst for some of these activities, and the kitplanes industry reaps the benefits.

Wood suffers from other faults, as well. Manufactured materials can be easily checked for flaws—samples of 2024-T3 aluminum should be identical. Any difference is bad. It's easy to design equipment to detect imperfections and reject faulty products.

But wood is a natural material. Two samples from the same spruce tree will show differences, yet each might be perfectly adequate aircraft material. Computers can't tell minor variances from dangerous flaws. It takes a trained human eye. And if you can't detect the bad pieces with a computer, modern manufacturers aren't interested.

And as a natural material, wood is subject to nature's course. Everyone has seen rotted wood fences and other structures. Knute Rockne's death caused wooden spars to be banned for commercial aircraft. Like the Hindenberg disaster, a single well-publicized failure negated many years of proven service.

When talk turns to wooden homebuilts, our mental image is that of an old geezer in bib overalls surrounded by ancient saws, puttering happily along in clouds of sawdust. Somewhere in the corner of the shop, a Fly Baby or a World War I biplane replica is slowly taking shape.

You might think wood has become a dusty footnote in aviation history.

There's one problem: Wood is a perfect material for aircraft. It's strong, it's light, it forms complex shapes with ease, and can be cut and shaped with inexpensive hand tools (Fig. 9-1). A Rand KR-1 (Fig. 9-2) uses foam for shape and fiberglass for surface toughness, but underneath is a wooden structure Glenn Curtiss would recognize. The Sea Hawker's wings are composite shells, but a hunk of shaped balsa adds compression resistance to the leading edge. The Lancair and the RV-4 might be battling for the small-engine speed crown, but the all-wood Sequoia Falco is breathing down their necks.

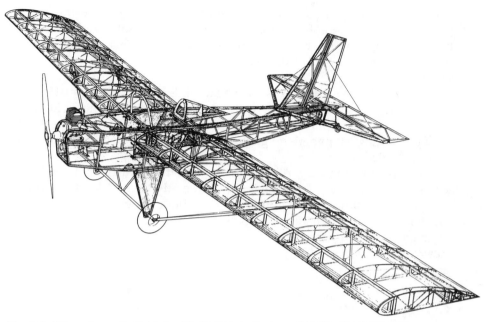

Fig. 9-1. Although an ultralight, the MiniMAX is built of wood in the classic manner. TEAM, Inc.

Fig. 9-2. Beneath the fiberglass and foam of this KR-1 is a conventional wooden structure.

Wood rot hasn't been eliminated, but modern finishes provide far better protection. And who leaves a brand-new homebuilt out in the rain, anyway?

Wood is probably the safest and most satisfying medium to build a kitplane from. Sawdust smells good and won't kill you. Varnish might make you high, but won't give you hives. Buy your tools at the hardware store. Dip a strip in hot water and it loosens up enough to tie a pretzel. It reacts predictably when carved or sanded, and mistakes are easily corrected.

Worried about strength? Pound for pound, wood has double the tensile strength of a aluminum. Tools? You don't have to order exotic framistats from homebuilder's supply outfits because good woodworking tools are sold at most good hardware stores. At decent prices, to boot.

Wood is nice.

But there are a few disadvantages for the kitbuilder. Your pieces must be protected from temperature or humidity extremes. Airplanes like the Avid series come with prerouted and glued ribs, but normally the kitbuilder is expected to do most of the cutting, shaping, and gluing. Thus a typical wooden kitplane will take longer to build.

No matter. It's fun and easy. And you'll have all those dusty Fly Baby-building geezers coming to your shop, giving you free advice, and lending you all sorts of useful tools. If there's one thing the EAA builder network can offer, it's expertise in building wooden aircraft.

This chapter has two parts. The first half presents the techniques of building wooden aircraft:

- Cutting wood parts.
- Shaping by hand and power tools.
- Gluing components.
- Preserving wooden structures.

The second half of the chapter discusses the basics of fabric coverings. Fabric isn't just wooden aircraft operation, but the techniques are similar, no matter the structural material.

The following sections discuss the basics of working with wood.

ADDITIONAL TOOLS

Additions to the basic tool list given in chapter 4 depend upon the kit being built. The smaller kits like the Fishers (Fig. 9-3) and the Loehle 5151 Mustang can get by with the basic list, with an emphasis on wood-cutting blades for the power tools. As always, a bandsaw would be useful. Typical wood-cutting speed is about 2,000 feet per minute; find one that can also run at half that speed for cutting aluminum.

Larger, more complex wooden aircraft might need a table saw. Or you'll need one if you decide to build a simpler aircraft with minimal reliance on precut parts. A table saw is irreplaceable for turning thick lumber into thin capstrips or stringers. Long, straight cuts are its forte. Benchtop table saws can be had for $100 or so, and floor models start at about $50 more. Good-quality tools reduce the frustration factor in kit building, and for a wooden aircraft, the table saw is even more important. Check your recommended tool list; if one is necessary, break the bucks loose and buy a good one.

You'll be doing a lot of sanding, so get a *pad-type power sander*. A *rotary* type works, too, but I feel the pad models are gentler and allow better control. There are other types of sanders you may wish to try.

Fisher Flying Products

Fig. 9-3. Smaller wood kitplanes like this Fisher Koala need only the basic tool list.

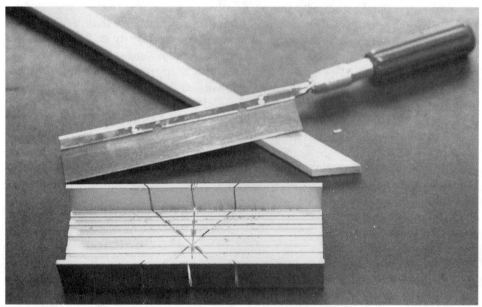

Fig. 9-4. Razor saws make smooth cuts in thin wood.

One thing you'll need is a lot of clamps. One or two types predominate, depending upon the kit. You might be able to get by with just a few, but you'll spend a lot of time waiting for glue to dry so you can remove the clamps and use them on the next components. It's not farfetched to need 25 or more of each variety.

Add the usual assortment of inexpensive hand tools: chisels, saws, files. One nifty little item is a *razor saw*, available at hobby stores. These are small, incredibly sharp saws that work great for cutting small pieces of wood. They sell for approximately $2 each, and a few dollars more gets a slick miniature miter box (Fig. 9-4).

The final item is a shop vacuum. Maybe the little hand-held unit works great for other types of construction, but woodworking generates tons of sawdust. Might as well bite the bullet and get an adequate vacuum from the start. Professional models of some hand tools (like sanders) come with an inlet for hooking up a shop vac. This cuts down on the airborne dust, but really jacks up the price.

MATERIALS, FASTENERS, AND SAFETY

The materials and fasteners are quite straightforward, and the additional safety precautions are minor.

Materials

The premier wood used in homebuilt construction is sitka spruce. The trees come from Alaska and British Columbia, and the timber is carefully sawed, dried, and milled to produce aircraft-quality material.

Make no mistake about it, aircraft spruce doesn't come from orange-crates. Just like aluminum and steel, there are extensive specifications that certified lumber must meet. Type A wood must have a minimum density of 24 pounds per cubic foot, moisture content between 10 and 17 percent, and the grain slope must be less than 1 inch in 15 (Fig. 9-5). Type A wood is approved for all applications. Type B specifications allow slightly less density and a bit more grain slope (1 in 15), and shouldn't be part of the primary load path.

If you botch a part, you can't just run off to the hardware store for another piece of wood. Face it, aircraft quality lumber does not end up at "Wood Is Us" at 49 cents a board.

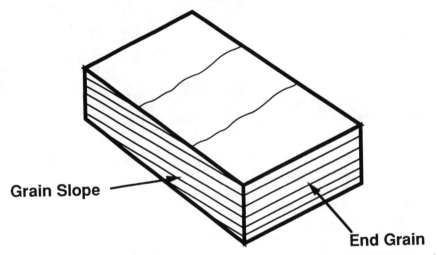

Grain Slope

End Grain

Fig. 9-5. The grain slope must be less than 1 in 15 for wood used in spars and other structural elements.

This isn't to say that spruce is the only wood that can be used. Western hemlock, for example, is 14 percent stronger. AC 41.13 allows direct substitution for spruce. But its quality varies more (good hemlock is more difficult to find), and it is about seven percent heavier. Certified lumber in other than spruce is hard to find.

You can use the hardware-store stuff for trim and interior purposes. For instance, I've bought thin fir molding strips for use as turtledeck stringers. But don't replace anything resembling a structural component with nonapproved wood.

The other wood commonly used in aircraft consists of multiple thin plies of bass, mahogany, birch, fir, or poplar bonded glued together under heat and pressure: plywood, in other words. It comes in various combinations of thickness, plies, and grain orientation. Between three to seven plies are used. The grain refers to the relative orientation of the grain for each ply, usually either 90-degree or 45-degree.

Plywood is used several ways. The most common use is to cover a structure while adding strength. Simple wood airplanes' fuselages are built as a wood truss. Gluing plywood to the exterior makes the truss far stronger. It's often used to cover wings, even in composite kitplanes like the Pulsar.

Proper storage is vital. While metal corrosion is a chemical process, dry rot is caused by a fungus. It is an infection, pure and simple. A piece of contaminated wood will contaminate any nearby lumber. Infected wood shows whitish dots and streaks at first. If left unchecked, the wood turns black and crumbly.

Rot is stopped by keeping lumber moisture content below 20 percent. Keep it well-ventilated, but you don't want the wood to become absolutely dry. A moisture content of about 12 percent is about right. Too little will result in embrittlement and overabsorption of glue.

Store wood in an area with a relative humidity of around 12 to 15 percent. Because the cut ends of lumber dry out faster, slap a quick coat of varnish or paint over the ends of stored wood. Keep it dry, and don't store it near a heat source.

Wood must be stored flat, not leaned against the wall. Any small but steady pressure will warp the pieces over time. Make sure it's well supported; don't just lay an eight-foot piece of plywood across two 2 × 4s. It's an exasperating exercise because you want the wood supported over its entire length, but you must still allow air circulation or the unequal drying will cause it to warp.

Plywood can be stored flat against the wall, but clamp it to prevent drooping and warping. A ceiling rack gets it out of the way but complicates access.

Fasteners

The primary fastener is glue. There are several varieties. Casein glues are powders that are mixed with water. Resin or resorcinal glues are usually two-part synthetics that show superior resistance to moisture, and thus are popular with amphibious aircraft.

There are two main problems with both of these glues: They don't fill gaps very well, and they require pressure to bond properly. The joints must be tight to begin with, and the pieces must be clamped while the glue sets. Usually, also, these glues require the shop temperature to be at least 70 degrees during the multiday curing period.

Again, modern epoxies come to the rescue, in the form of structural adhesives. These consist of two parts, resin and hardener. When mixed, they form a thick glop that fills gaps and bonds tightly under moderate pressure. They cure quickly, and can be used at lower temperatures (although the cure time is affected). The most common example is T-88 (Fig. 9-6).

No matter which type of glue is used, one requirement is that it must form a bond stronger than the wood itself. If the joint fails, it must fail due to a splintering of the wood rather than a break in the glue bond. Make up test pieces out of scrap wood. Bond them together, let cure, then clamp one end in a vise and break them apart. Save the pieces because the FAA inspector might want some sort of proof of the glue's effectiveness. You might hang on to one or two unbroken test joints as well.

Other fasteners for wood aircraft are familiar to most home craftsmen: nails and wood screws. The nails look like what the home craftsman would call *brads*, thin and less than an inch long. They are not used to hold the airplane together. Rather, they hold plywood in place until the glue sets. The plywood can either be a large sheet, like wing covering, or a small gusset used to reinforce a truss joint.

Wood screws are either the AN545 round-head or the AN550 flat-head variety. They aren't used for structural purposes.

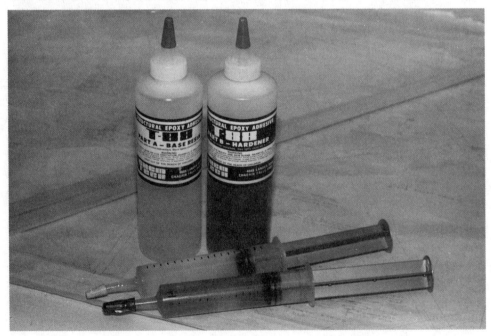

Fig. 9-6. T-88 is a popular structural epoxy. The syringes aid in dispensing the proper ratio.

Safety

The main danger in building a wooden aircraft is from your tools because they all have nasty sharp edges. Even if you're using T-88, you aren't going to be awash in chemicals like composite aircraft builders are. The varnishes and other spirits used are pretty benign, unlike the primers necessary for metal airplanes.

However, the moisture content of wood can cause its own problems. Steel bolts through wood will tend to rust (Fig. 9-7). In most cases, this won't cause a significant problem for quite awhile, but it's a good idea to coat the hole with varnish before inserting the bolt. Or use stainless steel hardware, if your budget can stand it.

The nails mentioned earlier come with antirust coatings. That's a good reason to use genuine aircraft nails than hardware store brads.

Epoxy glues are occasionally sensitive about mixing ratios, so use a scale or a pair of syringes, as shown in Fig. 9-6. Read the label because some don't use a simple 1-to-1 ratio. I've seen a glue that requires a 3-to-2 mixture of resin and hardener. Read the label for safety information, as well. Chapter 6 details the problems of chemical sensitivity.

Fig. 9-7. The steel bolt and the washers will tend to corrode if kept in contact with bare wood. Varnish the structure and the bolt hole before assembly.

Again, don't be tempted to substitute hardware store wood for aircraft materials. And use the size specified. Substituting $1/16$-inch plywood or $1/32$-inch will double the weight. Builders have been known to use $1/8$-inch marine plywood instead of $1/16$-inch aircraft grade. While you might be able to find a break on price, it just isn't worth the extra weight and the reduction in flexibility.

As a last point, keep in mind that any airplane has a lot of metal parts. These have to be protected from corrosion, as described in chapter 7.

CUTTING

Just about everyone has sawed a board or two. But building an airplane isn't like slapping a birdhouse together.

The Grain

When cutting wood, the grain orientation is a key factor. The grain is a remnant of the annular rings, caused by the variation in growth rates through the years. The long wood cells, and the natural cellulose cement that binds them, makes the wood strongest in the direction of the grain.

The plans should specify the grain direction for each piece. Make sure to get it right, as it makes a great difference in strength. If the plans don't give the direction, align the long axis of the part in the same direction as the grain.

The difficulty of the cut is dependent upon its direction relative to the grain. The cement is actually harder than the wood fibers themselves. Cross-grain cuts are therefore hard work for the saw because it must cut through multiple cement lines. While the cut takes longer and the multiple cement lines tend to dull the blade, the saw is easy to guide.

Cutting with the grain is another story. The cement lines act like grooves on a roadway, they complicate lane changes. On the highway, the path of least resistance is along the same direction of the grooves. When met at a shallow angle, they try to force a car straight.

Thus it is with the grain. When a blade hits one of the cement lines, it tries to deflect the saw into a path of lower resistance, along the grain.

If the grain ran dead-straight in the direction you want to cut, this wouldn't be much of a problem. But even the most stringent certification standards allow a 1 inch in 15 inches slope, approximately 4 degrees. If you're cutting a slot 1 inch deep, just following the grain will result in a $1/16$-inch gap, which is the maximum allowed when using high-tech epoxies. This gap would be intolerable with conventional aircraft glues.

Another byproduct of this process is the tendency of the wood fibers to break away from the cement, forming splinters. Splinters can make a shambles of a cut line. Some woods are worse than others, pine, for example, splinters easily. Spruce's popularity is due in no small part to its splinter resistance: *resistance*, mind you; it will still splinter.

While aluminum-airplane proponents have to watch the bend radius and composite-kit adherents watch the thermometer, the wood airplane builder must keep grain orientation foremost in his mind. There aren't any magic cures. You just have to take care to cut and trim to the line, fighting the grain all the way. A sharp tool is less likely to be deflected, so sharpen/replace as needed.

One way to reduce problems is to carefully select the cutting direction. if the grain is going to divert the cutting tool, cut so the deflection will be *away* from the piece, not into it.

Laying Out the Shape

Before a part can be cut out, its outline must be marked. The usual practice is to use a pencil and draw the shape directly on the wood. Remaining pencil marks don't affect the varnish, but as the varnish is transparent, you'll want to erase the marks if the part will still be visible after completion of the aircraft. (I used a piece of hardware-store hemlock for a seat spacer, and neglected to sand off the rubber-stamped price before varnishing. The whole world can see where $1.09 of my money went . . .)

Or draw the outline on a piece of paper, glue it in place, and cut out the outline. This is best suited for complex components that have a lot of notches, holes, and the like. It's a lot easier to sit at the dining room table and draw on paper than to try drawing the piece on the corner of a 4 × 8 sheet of plywood.

Some plans include full-size templates. Photocopy the template and glue the copy to the wood. If you botch the part, it's then easy enough to just grab another copy and start over.

Simple Cuts

The vast majority of cutting on a wooden kitplane consists of cutting lumber to length or shapes out of plywood. This process is little different from scale models. Thin plywood can even be cut with a modeller's knife. Run the blade along the same line a few times, and break the wood at the line.

For heavier pieces, a number of tools can be used: saber saws, drill presses, table saws. Long, straight cuts are best handled by a table saw. Its unencumbered surface makes it one of the few tools that can handle large sheets of plywood. The blade housing on bandsaws often gets in the way.

Alternatives include the circular saw and the saber saw. Table saws have a *rip fence* to guide straight cuts, and it's a good idea to add one when using these hand tools. Commercial models are available, but a long piece of angle aluminum (or wood, if you can find a straight piece) and a pair of clamps work just as well. Measure the distance from the outer side of the saw's shoe to the opposite edge of the blade, and set the fence this distance away from the cut line. Clamp it in place. When cutting, maintain slight pressure against the fence.

Curving cuts are generally left to the band and saber saws. Circular saws can't make tight turns. Really sharp turns might call for manual tools like coping or keyhole saws.

While the cutting process is similar to aluminum, wood has a couple of differences that require changes in technique.

The first is the rapid rate of cutting. Wood cuts easily, that's one reason it's so popular. But a misdirected saw will do far more damage before your reflexes respond. One jiggle and the part is ruined.

A factor in your favor is that wooden airplane building, even wooden kit-plane building, isn't a manufacturing process. It's more a function of making pieces to fit other pieces. Even if you cut exactly to the line, you'll probably do some shaping to get the parts to fit. Just don't cross the cut line.

The second problem is the saw's tendency to splinter the wood at the point that the blade exits the wood. As each tooth exits, friction pulls the surrounding wood with it. The wood lifts away and splinters (Fig. 9-8). Sharp blades reduce the problem. In addition, clamp the wood to a piece of scrap and cut through both at the same time. The pressure from the scrap will hold the work piece's wood fibers in place.

Again, don't cut exactly next to the line. Leave a little extra wood to splinter instead of the piece itself. Any splintering usually doesn't propagate far in spruce. There are also special plywood-cutting blades which have small teeth to reduce the problem.

Speaking of plywood, the glues used between the plies are tough on blades. Be prepared to sharpen or replace them more often.

Fig. 9-8. Wood tends to splinter on the backside of the cut; when using a saber saw as shown here, it'll splinter on both sides. To prevent splintering, clamp the wood to a piece of scrap and cut through both pieces at once. Or cut far enough from the outline so the splintering doesn't reach the part.

Fine Cuts

While the information in the previous chapter is suited for general cutting, there are often times when more exact trimming is necessary. The typical example is cutting a notch to join another piece of wood.

A simple L-shaped notch on a corner is easy enough. Use the same tools and procedures mentioned in the last section.

A square U-shaped notch is another matter. The sides can be cut with any convenient saw: saber, band, razor. But how do you make the last cut across the inside?

It's pretty easy with wide notches. Cut the notch as a rounded-U, running the blade along the marked lines on each side as far as practical. Then there will be enough room to cut out the corners.

If the notch is narrow, your options are many. Cut out the sides first, in any case. This leaves an unsupported tab that is pretty flimsy. You can cut across the base with a razor blade. Or drill some holes near the corners and slip a coping saw into place.

Or cut additional slots between the sides (Fig. 9-9) and clean up the pieces with a *wood chisel*. Wood chisels are wedge-shaped knives that come in a variety of widths. Set the chisel along the cut line, with the flat side toward the part and the sloped edge facing the tab to be removed. Hold the tool vertical, then tap on the end of the handle with your other palm. A good, sharp chisel should cut right through, leaving few splinters (Fig. 9-10). Harder woods or duller tools might need a hammer, and leave the edge a bit ragged.

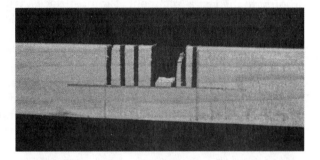

Fig. 9-9. When cutting small notches, run the saw between the edges to make a sort of comb. Then cut away the teeth and sand or file the sides smooth.

Fig. 9-10. A chisel can be used to trim the notch.

Chisels come in a variety of widths, starting at 1/4-inch and working upwards. It's best to use a smaller width than the notch, even if you have a chisel of exactly the right size. This gives you a little extra control.

Most cuts will have to be smoothed afterwards. Chisels work well with larger notches. Be careful with files because sometimes they're *too* effective. Emery boards are nice notch-cleaning accessories.

Holes

Hole cutters for wood are widely available. A drill press is almost a necessity, but use of the tool is straightforward. The main thing to watch out for is overheating, caused by trying to cut too much too fast. It dulls the tools and chars the wood. If you smell something burning, back off on the press and let the pieces cool.

Large-diameter holes, especially in plywood, are the saber saw's forte. Start with a hole saw slightly larger than the width of the saber saw's blade. Cut a hole near the cut line on the portion of the wood to be removed. Then insert the saber saw and cut towards the cut line at a shallow angle.

FINAL SHAPING

At some point, you'll have to add the final shape to the wood pieces. This might amount to the last adjustment to joint surfaces, trimming away a blob of glue, or a variety of other actions. This requires hand tools of various varieties.

Smoothing the edge of a large piece of wood calls for the wood plane. Also, the tool can be used to make cut edges square with the top and bottom. The longer the plane, the smoother the edge. The amount of wood removed with each stroke of the plane is adjusted by setting the height of the cutting blade.

The grain orientation has a great impact on the planing direction. Don't plane into the slope of the grain. For example, if the board's grain slopes downward to the right, plane from left to right. Otherwise, the blade will catch the cement lines and dig in.

A plane is nothing but a chisel on a carrier. Hence, use the chisel in situations where lines of wood must be removed.

But chisels won't leave the smooth surface that a plane will. And some cases just require moderate shaping and smoothing. Sandpaper and files are the obvious choices.

Files work best where inside corners must be sharp, or where hard glue deposits must be eliminated. Their design lets them reach into tight crannies for those last minute adjustments.

Sandpaper's various grit sizes allow better control of the amount of material removed. Unless sanding a curved surface, use a sanding block. Wrap a sheet of sandpaper around a piece of 2 × 4 or 1 × 2. This evens out the hand pressure; otherwise the paper directly under each finger will dig in more.

Several power tools are used for final shaping. Belt sanders can square off ends before pieces are glued together, and can quickly smooth down rough exterior shapes. Pad or rotary sanders are used for similar purposes, but work more slowly. Dremel has a tiny sanding drum accessory for their Moto-Tool, which is dandy for tight, precision shaping.

These power sanders share the main problem of all power tools: if not watched carefully, they can strip too much material too fast. In addition, watch the edge effects of using power sanders. If you don't work on spreading out the action, the tool can leave indentations at the edge of the working area.

One type of final shaping is the rounding of all edges that won't be joined to other pieces. Nice sharp edges might look good, but they are a weak point for several reasons. They aren't supported very well, and a bit of impact will break off pieces. This isn't dangerous from the point of view of structural strength, but it is unsightly. The part might stay just as strong, but broken-off edges make your workmanship look embarrassingly crude.

Also, the splinters that get started can propagate and disrupt gluing surfaces. And if the corner breaks away after varnishing, the exposed wood surfaces can be a starting point for deterioration.

The tools used for rounding vary with the size of the piece involved. Small pieces need only a bit of sanding. Too much tool, and you stand the chance of breaking the piece.

Larger pieces leave a number of options. The power sanders work great. If you can, use a plane to cut the corner off, then use the sanders to round the remainder. This leaves a beautiful, durable edge.

BENDING

One of the reasons wood has always been popular as a homebuilt aircraft material is its ease in forming smooth curves. One only has to look at say, a Falco and a T-18 to see the difference. Metal *can* form complex shapes. But it's usually beyond the capability of the average homebuilder.

It's easy to see that wood is a bit flexible. We'd expect a 10-foot piece of 2-inch-square spruce to be able to take the moderate bend required of a fuselage longeron. But some pieces must be bent more than the wood might naturally take.

The basic method is soaking and forming.

Soaking and Forming

Did you ever see a piece of unprotected plywood left at the mercy of the elements? After awhile, it ends up severely warped. When wet, wood loses some of its stiffness. A little bit of pressure will bend it. When dried out, the wood regains full strength and tends to hold its new shape.

And that's exactly what you'll do with some of the wood pieces of your kitplane. The process calls for soaking the wood, forming it into the desired shape, then letting it dry.

The first requirement is to get the wood wet. Hot water works best; the hotter the better. Boiling water works best, even steam if you can rig up a way to produce and deliver it. Small, long pieces can be placed in a metal tube fed with the exhaust from a teakettle.

This is pretty extreme. You might have some smaller pieces to bend, but the likely target will be plywood. The easiest soaking place is the bathtub. It makes sense because it's long, generally wide, and already set up to deliver hot water.

Household water heaters are generally set to around 130 degrees or so. More heat would be nice, but make sure the system can take it. Make sure, also, that nobody is planning on washing dishes or taking a shower in another bathroom. Not only will they be robbing you of hot water, but the extra high temperature of the tap supply might be an unpleasant surprise. Family relations in homebuilder households are generally bad enough without broiling one's spouse or children.

Pieces bigger than the tub present another problem. You can try to swab down the sheet with hot water, but it isn't as good as soaking. Although you might need a large area, the depth isn't important. Try to rig up a temporary tub with 2 × 4s and plastic sheeting, or perhaps an old waterbed frame. Deliver hot water via a hose to an inside faucet, or heat pans/buckets on the stove, or glass jars in the microwave, whatever's necessary.

Make sure the sheet is in contact with the hot water on both sides. If not, it might try to take a contrary bend. It'll probably try to float, push it back down with sticks. If you weight it down, have some scraps underneath it to keep from sitting flat on the bottom.

Soak time will depend upon the temperature used and the thickness of the piece. Thirty minutes should be plenty. Then it's time to force the piece into the desired shape and hold it until dry.

For simple bends, a former can be built from scrap 2 × 4s as shown in Fig. 9-11. The plywood is bent by hand and slid between the two boards. The boards should be positioned so that the radius held by the plywood is slightly smaller than the final radius, as the sheet will have a slight amount of springback when removed. Leave it in place until dried, a process that might take a day or so.

On some airplanes, the wet pieces are immediately glued in place. This elim-

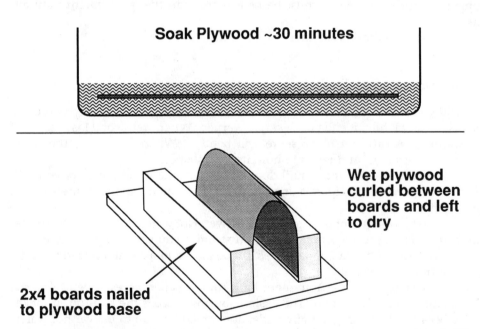

Fig. 9-11. After soaking thin plywood in hot water for a half hour, curl it between two boards. When dry, it will retain the angle, less a little springback.

inates additional formers. The glues used by wooden kitplanes are still effective while the piece is wet. Details on gluing and clamping are provided later.

Laminating

The soak and flex method works for thin pieces and plywood. But what if you need a large piece of wood with a bend in it?

The solution is to take a lot of small pieces, bend them to the final shape, and glue them together. It's not what I consider a process that the builder of a kitplane should have to do. Parts should come prelaminated. But let's take a brief look at the procedure just for the sake of education.

It begins with a form in the shape of the desired final product. This might consist of a continuous surface, or just a few blocks set up at key locations. The wooden strips that will be laminated together are coated on both sides with glue, placed together, and clamped into the form. When the glue has cured, the laminate is removed.

Doesn't sound too tough, does it? But getting the form right is the hard part. When removed from the form, the laminated assembly will spring back slightly. Thus the form must take this into account, relying on the builder's eye more than anything.

If your kit calls for laminating an assembly, the plans should give exact dimensions for the form. Clamping must be evenly spread out over the whole surface. Don't rely on just the shoe of the clamp, use blocks of wood between the clamps and the laminate.

GLUING

After all the work cutting and trimming the pieces, actually gluing them together is simple.

Joint Preparation

The amount of allowable joint gap depends upon the type of glue being used. *Casein* and *resorcinal* glues require tight joints, while epoxies like T-88 can fill a gap up to $1/16$ inch. When cut, the wood surfaces tend to dry out. If possible, glue the parts within a few hours of cutting.

If something gets between the two surfaces being glued, the glue will bond to the contaminant instead. Hence, the surfaces of the joints must be clean. Sawdust is a major culprit. Sanding makes fine tight-fitting joints, but the dust tends to fill the wood's pores and interfere with bonding. Blow out the dust with an air compressor.

Get rid of oil and grease with acetone or lacquer thinner. Let the surfaces dry before gluing.

Glue Application

If you've been through the chapter on composite construction, you read about the need to use glues sparingly to keep the weight down.

This doesn't apply to wood construction. Wood tends to soak up glue. While this is bad from the weight standpoint, it's worse from the bond strength aspect. If the wood absorbs the glue, less is left to actually form the bond. Hence, the builder must apply enough glue to ensure enough remains on the surface.

The situation is relieved in one way: wood is always glued under pressure, the excess is squeezed out of the joint. It can be wiped away and removed.

To avoid glue starvation, spread the glue thickly on both surfaces using a brush. If a piece of lumber has been cut directly across the end, the exposed "end grain" is extremely porous and thirsty for glue. Dab some glue onto the end, wait 20 minutes, then add more glue and clamp the pieces together.

In some cases the pieces aren't joined immediately. Sometimes the plans or the glue manufacturer requires that the glue air-dry for awhile before clamping.

Many glues can be applied while the pieces are damp from the soak and bend cycle. Check the instructions. Work the glue well into the surfaces to drive out the water.

Clamping

The parts must be clamped during the curing process. It's not quite as critical for epoxy glues, but casein and resorcinal glues require at least 125 pounds per square inch of pressure.

Clamping should make glue ooze from the joint. If it doesn't, separate the pieces and add more glue (Fig. 9-12). Wipe up the excess with acetone or thinner. Smooth the joint with a fingertip or popsicle stick, leaving a small fillet.

Clamping methods vary. Ordinary C-clamps and spring clamps are the most common. C-clamps apply very localized pressure, so add scrap pieces of wood under the shoes to spread the force over a larger area. Spring clamps can't apply enough pressure for traditional glues, but are just dandy for epoxy-glued structures (Fig. 9-13). There are other types of clamps as well; the only requirement is the ability to supply the required pressure without damaging the wood. Even clothespins can have their uses; they're cheap, too.

Any hardware store should carry a stock of various types of woodworking clamps. If the budget allows, pick up a sample or two in advance. Otherwise, keep their stock in mind as you face various jobs. It's always better to use the right tool than to struggle along with the wrong one. But sometimes exactly the right clamp just isn't made, and you have to come up with a solution (Fig. 9-14).

Large plywood pieces require pressure over a large area, which is tough to supply with conventional clamps. Enter the aircraft nail. The nail is used as a clamp, in either a temporary or permanent installation. They can be used instead of clamps in other applications, as well. The nails should be four times as long as the sheet of plywood being glued.

The nails don't add to the strength of the joint, so it's nice (but not necessary) to remove them once the glue has cured. Individually, they're devils to extract. But installing them through a thin strip of wood makes it easier. Either lift up on the strip, or break it away to expose the heads to a pair of pliers.

Fig. 9-12. The center joint shows about the right amount of epoxy squeezed out, but this excess should be wiped off. The lower joint is starved; there's no extruded glue showing.

Staples are an alternative to nails. Don't use a standard commercial gun, because the staples are thicker than necessary. Buy one intended for aircraft use. The staples can be removed using the sharpened blade of a screwdriver (Fig. 9-16).

Scarf Joints

One of the necessary evils of building a wooden airplane is the need for *scarf joints*. Often, two pieces of wood must be joined end-to-end. For instance, two four-foot pieces of lumber are used to make a single eight-foot unit, or two sheets of plywood are laid side-by-side.

An ordinary butt joint isn't strong enough. Only the mediocre shear strength of the glue holds the pieces together, and the contact area is small to begin with.

Instead, aircraft use scarf joints. The edges of the two pieces are beveled, then glued. Because the bevel slope must be at least 10-to-1, the contact area is dramatically increased. The overlap adds dramatically to the shear strength.

Some kits supply components prebeveled, others require you make your own. For boards, the slope can be marked along the edge and cut with a bandsaw. Plywood is a little more difficult. The basic requirement is to feed the sheet

Fig. 9-13. Spring clamps supply adequate pressure for epoxy adhesives. Here, the geodesic structure of a Loehle 5151 Mustang vertical stablizer is clamped.

Fig. 9-14. Sometimes you have to clamp the clamps. Here, one clamp is used to help the other clamp maintain pressure. Note the thin strips of scrap wood under the pads of the clamps to protect the components.

Fig. 9-15. Loehle recommends holding plywood in place with small staples until the glue dries. Here, a sharpened screwdriver demonstrates how to remove the staples from the rudder of a 5151 Mustang.

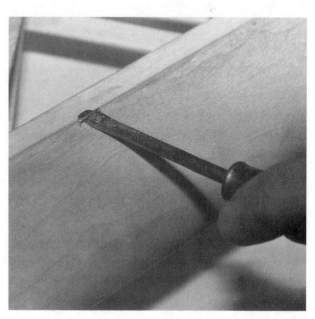

Fig. 9-16. Remove excess glue with files and/or sandpaper. The glued area should be faired in with the rest of the component.

to a cutting tool that is set to the prescribed angle. Your plans should show how, or check with your EAA technical counselor. There are dozens of ways to scarf plywood.

The biggest problem is to ensure the scarf is properly oriented in relation to the grain. The scarf should approximately parallel the grain; when completed, only one or two grain lines should show on the cut area. The second piece should be cut in the same fashion. When joined, the two grains should appear continuous, except for a slight joggle at the scarf line (Fig. 9-17).

PROTECTION

As mentioned earlier in the chapter, wood's moisture content must remain at about 12 percent for maximum life. Too much wetter, and rot can begin. Too dry, and it becomes brittle and weak.

That's the reason you must varnish wood components. It's nothing but a water barrier to maintain the optimal moisture content. It sheds dampness; it seals existing water into the wood.

If the varnish protects the wood, why do older wooden aircraft have rotting problems?

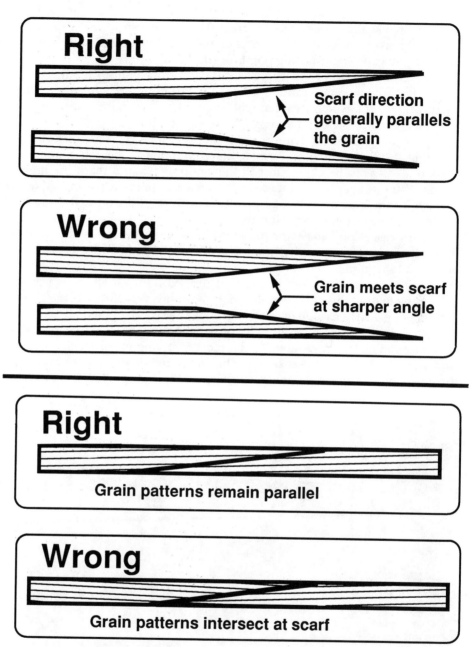

Fig. 9-17. Scarf joints are used to join wood end-to-end.

The main problem is old-fashioned varnishes. Wood flexes under load; that's one reason it's strong. The older varnishes are not flexible, they crack when the wood bends. Eventually, the cracks go though to the wood and problems begin.

Modern urethane and epoxy varnishes are flexible, which results in less cracking. Most are two-part systems that must be mixed before use.

They're usually applied with a brush. The skeletal structure of most wooden airplanes makes spraying it on a waste of material. The goal is complete coverage because any gaps are a starting point for rot.

The best opportunity to apply varnish is before gluing. It's hard to get to some of the nooks and crannies before the structure goes together. However, keep varnish off any surfaces that will be glued. As previously mentioned, contaminants disrupt the bond and result in a weak joint.

The result is often a compromise. If access will be good after assembly, glue first (Fig. 9-18). Otherwise, varnish the components, leaving a good margin around all areas to be glued. Test-fit the components and mark the limits with pencil. You can keep a careful eye on the marked areas while brushing on the varnish, or the glue areas can be protected with masking tape. Be sure to clean off any residue of the tape's adhesive.

In any case, don't shave the margins too closely. The varnish tends to ooze a little bit. Let the coating dry, then glue. When cured, add a coat of varnish around all the joints.

Smaller components like ribs can be dipped into a shallow pan of varnish then hung to dry. This gives excellent coverage with very little additional weight. Again, place a strip of tape over the portions that will be glued later.

Wood is the traditional aircraft construction material. While probably the slowest-building, it's undoubtedly the most forgiving of mistakes. Personally, I find woodwork the most satisfying. The components are smooth and warm to the touch, and rarely have dangerous sharp edges. Wood is the easiest to cut and

Fig. 9-18. The cockpit area of this Fisher Classic is easy to get to after assembly; hence varnishing can wait.

shape, and the shop's smell rivals that of a kitchen on bread-baking day. There's nothing prettier than a wooden airplane just before the fabric is applied.

FABRIC COVERING

Fabric covering methods and materials have grown with aviation. Early pioneers used linen coated with ordinary varnish. Later on, various grades of cotton fabric became standard. But all these materials had one big problem: poor longevity. Even when kept indoors, a cotton-covered aircraft had to be recovered approximately every 5 years; recovered every year or so, if the airplane were left outside.

It didn't bother people too much because it was just another cost of owning an airplane. Everybody had to do it. Fabric longevity didn't become an issue until the strides in aircraft design during the '30s and '40s, when aviation's explosive growth brought about lightweight aluminum alloys and the acceptance of monocoque construction.

After the war, the technology trickled down to the lightplane world. In 1946, the Cessna 120/140 aircraft featured metal monocoque fuselage and fabric-covered wings. Only the wings had to be recovered every few years. Later-model Cessna 140s metalized the wings as well.

Suddenly, one of the basic costs of ownership evaporated. In 10 years, a Piper PA-12 owner might have to recover and repaint three or four times. A Cessna might lose a bit of shine, that's all.

All-metal airplanes didn't replace the tube-and-fabric models overnight. Outfits like Piper and a big investment in welding and fabric-working equipment. But tube-and-fabric is labor intensive; it's not very compatible with lower-cost assembly-line methods. In the long run, all-metal airplanes were cheaper to build.

Piper's steel-tube Colts and Tri-Pacers couldn't hold back the Cessna tide, so the Cherokee was introduced in the early 60s. Champion and Bellanca survived (for awhile) by filling specialized market niches.

While all this was happening, a fabric revolution occurred. Synthetic cloth had been around for a hundred years, but DuPont's polyester (trade name Dacron) was the first to exhibit the peculiar qualities needed for aircraft use.

But the all-metal tide was unstoppable. While durable and easy to apply, Dacron couldn't change the fact that metal airplanes were perceived as the wave of the future. Under a variety of names, it quickly took over the fabric market. Cotton is used only by those antiquers interested in exact duplication of the original processes.

Why do some kitplanes use fabric, anyway? Why fiddle with obsolete technology? Why not make them composite or all-metal and be done with it?

In the first place, just because metal has advantages for assembly-line production doesn't necessarily make it best for a person building a single airplane. A factory worker gets lots of experience at riveting aluminum skins, but the homebuilder generally gets only one crack at it. An Avid Flyer builder applies glue to the structure, lays down the fabric, and then twitches it back and forth to eliminate any wrinkles.

Strength? The P-51 Mustang is a metal monocoque design, but the rudder is fabric covered. If it's strong enough for a 400+ mph fighter, it's sufficient for a Kitfox.

Weight? Fabric covering a typical kitplane might weigh 10 pounds. The .040-inch aluminum skins on an all-metal airplane can weigh 20 times as much, although they also provides structural strength, for which the fabric airplane needs a beefier frame.

Drop a wrench on a metal airplane, and you'll get a dent. It'll probably just bounce right off a fabric surface. And if the surface gets damaged or must be removed to access internal components, fabric can be easily stripped away. Patched areas are almost invisible.

Durability? During the 1960s, one local homebuilt had its tail surfaces recovered with nurses' uniform cloth from the local fabric store. The material was Dacron. The plane is still flying today, with the same fabric on the tail.

Some might prefer the antiseptic sheen of a composite speedster, or the mirrorlike shine of an unpainted metal airplane. Personally, I love the satiny glow of doped fabric; and the delicate curves and dips as it wraps tightly around the underlying structure.

Pretty as it could become, it won't get that way without some work on your part. The following sections give a little background on the techniques and problems of applying a covering to your aircraft:

- Fabric selection.
- Surface preparation.
- Application.
- Shrinking and doping.

Fabric is applied to both wood and metal surfaces. For instance, the Fisher Classic (Fig. 9-19) and the 5151 Mustang's have all-wood structures. The RANS Sakota is a typical all-metal tube structure. The Avid Flier is somewhat of a crossbreed, where the wing fabric must be attached to wood ribs and metal-tube

Fig. 9-19. Fabric is perfect for light wood structures.

spars. Some wooden airplanes use fabric as an exterior cover, with no aerodynamic purpose, because it reduces the need to fill nail holes, and the like.

Whatever the underlying surface, fabric is installed nearly the same way. Let's take a look at some of the basics.

THE BASICS

Aircraft coverings, be they fabric, metal, or whatever, are used to define the exterior shape of the aircraft. They lessen drag by enclosing the fuselage structure, and produce lift by assuming an airfoil shape.

Therefore, aircraft coverings must be airtight (to be able to redirect the wind), and should tightly conform to whatever structure they are applied to.

It's easy enough with aluminum. But aircraft fabric is another matter. It's little different than the material in the shirt you're wearing because the breeze moves through it, and it wrinkles and sags at a whim. Obviously, builder effort is necessary to turn it into a proper covering.

The airtight process is the easiest. The weave is filled with a flexible coating called *dope*. Flexibility is the key because the airstream will cause the fabric to vibrate. A stiff coating will crack.

Conforming to the underlying structure is a bit more difficult. Attachment is by a variety of methods, from lacing to dope to glues. But to conform to the desired shape, the excess fabric must be removed between the attachment points. Once the fabric is attached to the skeletal structure of the aircraft, then the bagginess is removed by causing the fabric to shrink between the attachment points.

The shrinking method depends on the fabric material. Cotton is shrunk by the same dope used to fill the weave, while heat is used for polyester. It might seem that cotton is a more efficient system, but the polyester allows better control because it is attached, shrunk, and doped in separate steps.

Processes

The major polyester-fabric systems (Stits Poly-Fiber, Ceconite, HIPEC) generally follow the steps described in the previous section, but get there in slightly different ways. Marketers offer low-cost videos on installing their product, and often distribute installation instructions at no charge.

The major difference between the processes is the method of gluing down the fabric. Stits and HIPEC use a single-part glue, while Ceconite has a cement activator that might allow better control. Any of them will produce an acceptable result. The processes allow some optimization based on aircraft type and mission.

Whichever way you decide, *don't mix processes*. They are not necessarily compatible. You might end up with a finish that develops cracks and flakes away. Materials might not be compatible between processes; you might apply dope and find it dissolves the other company's fabric glue. If you decided on one of the major players, use only their components, except where substitutions are specifically permitted.

If your kit doesn't include the covering materials, count on spending at least $1,000 for the fabric, dopes, and accessories. You can save about 25–35 percent by buying generic Dacron and dope. But it's easier for a first-timer to work within the comforting framework of a well-designed commercial process.

ADDITIONAL TOOLS

Primary tools will be brushes, mixing cups, *pinking shears*, and a 1,100-watt (or higher) iron.

Pinking shears are scissors that cut a zig-zag line. When cut, fabric edges start to unravel. When cut with the pinking shears, the material can only unravel to the next zig. These tools are available through fabric stores or the home-builder's catalogs.

Irons are different. An ordinary clothes iron will do, but don't swipe the wife's, unless you buy her a new one. The homebuilder's iron will get all grungy with glue and dope. If you're really being cheap, check the local Goodwill or Salvation Army stores for a used iron.

Units designed for aircraft use are sold for $20 or so. Hobby shops sell small irons (for applying plastic coverings to RC models) that work great for tight corners.

The primary requirement is the ability to hold a particular temperature. Clothing irons are labeled for material types, not actual heat setting. Buy a thermometer at a hardware store and a little tube of silicon heat sink compound from Radio Shack. Place the thermometer in a pan of boiling water and make sure it reads 212°F (or whatever the boiling point is at the altitude of your location). If the scale is off, adjust it to read correctly and glue it in place.

Place a dab of the compound on the bottom of the iron and insert the bulb of the thermometer. Turn on the iron, and adjust its dial till the thermometer reads 200 degrees. Make a mark on the iron's dial, and repeat for the other setpoints required by your covering process.

Some processes use a heat gun. These cost about $75.

MATERIAL, FASTENERS, AND SAFETY

There's very little high-tech involved in fabric operations. Most of it is traditional, but there are a few modern quirks.

Material

The primary material is the polyester fabric. You can still use cotton if you desire, but as mentioned, it will need replacement regularly.

In addition to being strong and long-lasting, polyester has another trait that makes it well suited for covering airplanes: heat-induced shrinkage. The fabric is applied to the surface, then the wrinkles are taken out by applying heat.

Most fabric-covered kitplanes include the covering materials. In a few cases like the RANS Coyote, the covering is a sailcloth slipcover similar to that used by ultralights. Otherwise, your kit includes at least a few bolts of cloth.

Fabric's strength and durability is determined by its weight, in ounces per square yard. The heavier the cloth, the stronger and more durable it is. Typical weights run from about one and a half to four ounces.

The need for strength is obvious enough because you don't want the fabric ripping way at high speeds. Durability is another aspect. When first introduced, polyester was marketed as a lifetime fabric. Experience over the years has shown that lifetime, in this case, is about 20 years.

But there are other aspects to durability. Wear resistance is an important one. The typical fabric-covered kitplane is a taildragger, likely to be flown from rough grass strips. While a lighter fabric might be strong enough, the designer may pick heavier grade in order to make the finished surface more resistant to puncture damage.

If you run out of fabric, two factors are important when resupplying:

1. Order the proper weight. One size heavier is OK, except from the weight standpoint, but don't install lighter than the designer specifies.
2. Order from the same fabric manufacturer. A number of companies produce similar polyester fabric under their own trade names: DuPont's Dacron, Celanese's Fortrel. Government regulations only specify the chemical content. Each company develops its own process. Do not substitute another company's product for the one supplied with your kit.

One timesaver is a covering envelope. Some aftermarket suppliers sew fabric into a sock that slips over airframe components (Fig. 9-20). The builder merely glues them down and shrinks them; all the fitting and cutting is eliminated. Envelopes can save a considerable amount of time. They add about a third more to the *fabric* cost, which works out to about a 10 percent increase in total covering cost.

Fabric is also used in tape form, for smoothing sharp corners and reinforcing high-stress areas. You could spend hours cutting strips from your cloth, but why bother? Order tape rolls instead. The two types are reinforcement tape, which is narrow but thick, and finishing tape, which is thin, has a very smooth weave, and is two or three inches wide.

Dopes are the other major material. Dope has two major qualities. First, it seals the weave of the cloth to make it airtight. Second, it's flexible, so the normal drumming of the fabric won't make it break away. Some processes have their own trademarked name instead of the term dope. While the formulation might be different, the function is the same.

At least one coat of dope must contain blockers to stop ultraviolet ray deterioration due to the sun. This usually consists of aluminum powder preadded to the dope.

Temperature is important to the doping process. Too cool, and the dope gets thick and stiff to apply and takes forever to dry. Every 10-degree drop in temperature *doubles* the drying time. The longer it stays wet, the more dust, dirt, and wayward flies end up imbedded in the surface.

Hot weather isn't good, either. If the dope dries too fast, it gets a rough, dusty appearance. Retarders are available that slow the drying process.

Fig. 9-20. A fabric envelope would slip right over this wing panel, eliminating the need for measuring and cutting. Again, it's the usual question: Save money by doing it the hard way, or save time by spending money on envelopes?

Again, when using a particular process like Stits or HIPEC, *always* use *only* the materials approved by the process designer. In addition, make sure the surface coating—varnish, primer—of the aircraft structure is compatible with the covering process. There normally isn't any problem, but it's better to be safe.

Fasteners

Fabric must be attached to the underlying structure. The primary concern is making sure the fabric doesn't separate from the top surface of the wing. The traditional method is rib stitching, or making loops with a lacing cord around each rib through the top and bottom fabric.

But rib-stitching is time-consuming. The modern alternative is gluing the fabric to the ribs. This isn't just some off-the-wall alternative, the Stits process is STC'd, and it replaces most of the traditional lacing procedures with gluing.

The Stits process requires the underlying surface be at least 1 inch wide. Some FAA inspectors won't approve glue in cases where the ribs are round tubing, such as the RANS series and the CIRCA Nieuports. In these cases, alternatives to rib-stitching include small pop rivets or stainless-steel sheet metal screws.

Safety

The health hazards of fabric work are the usual: use with adequate ventilation and avoid contact with bare skin. (I swear I've seen the same warning on a bar of soap.)

Anyway, any fabric glue or dope will make you giddy, so get some fresh air in the shop. Too much wind will blow dust around; you might have postpone doping in these conditions. Avoid skin contact by using brushes and rubber gloves. As usual, a respirator should be worn for any type of spraying. The solvents used in the glues and dopes are toxic as well, so follow the manufacturer's recommendations.

A big hazard is fire. Dopes are flammable; dope fumes are even worse. Eliminate all sources of ignition when applying dope, or provide extraordinary ventilation. Friction or even static electricity can ignite freshly-doped surfaces. Provide air circulation and leave the parts alone until the dope is dry.

COVERING PROCEDURE

The specific steps and sequences of covering your kitplane should be included in the plans, and specific information on using the covering systems should be available at low cost, or even free, from the process manufacturer. Let's look at the type of tasks involved.

Protection

Traditional varnishes, spar varnish, for example, are not compatible with the solvents used in the covering process. Therefore, don't use spar varnish where fabric will be attached. Check with the process manufacturer, but any two-part coating (urethane, epoxy) will probably be acceptable.

Once the airframe is protected from the fabric, protect the fabric from the airframe. Any little bump, any sharp edge might cut the surface. A classic example is the wire trailing edge used on the entire Avid Aviation line. Without protection, the cable could cut through the fabric. Watch bold heads, and pop rivet heads as well. Round off wooden corners prior to varnishing. The edge of most flat aluminum sheets is another problem area. These points and sharp edges must be blunted.

There are several ways. The traditional one is to glue antichafe fabric tape over the offending points (Fig. 9-21). Masking and duct tapes have the advantage of being self-adhesive, but deteriorate over time. Electrical tape is more hardy, but it's expensive and awkward to use due to its stretchiness; it's admirably suited for narrow areas where more thickness is necessary, like along the thin aluminum edges (Fig. 9-22).

Whichever method is used, apply as many layers as necessary and taper successive ones to smooth the transition. Don't bury bolts that might eventually need removal.

Another factor in protection is the reinforcement of areas where large holes must be opened in the covering. An example is wing strut fittings. The usual

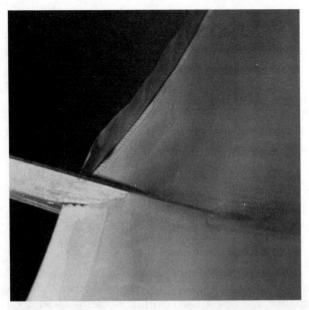

Fig. 9-21. Most Avid aircraft have a taut cable for a trailing edge. This would tend to cut its way through fabric, so tapes are applied before and after the main covering to reinforce the area.

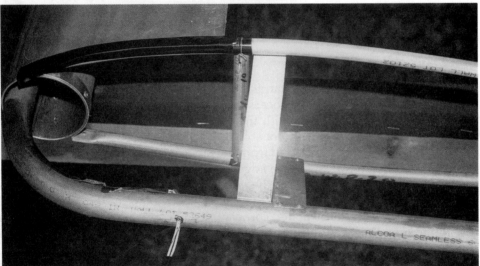

Fig. 9-22. Electrical tape on the edges of the aluminum sheet protects the fabric. This is the leading edge of a RANS Sakota wing.

practice is to pop rivet an aluminum frame (of thin sheet) around the position of the opening. When the covering is applied, it is glued to the sheet. The fabric is cut out of the center of the frame when the glue has dried (Fig. 9-23).

On many kitplanes, rudder cables must pierce the fuselage covering. These smaller holes don't need frame-type reinforcement attached to the structure; rather, the fabric itself is reinforced locally. However, determining the exact exit point is difficult once the covering is in place.

Fig. 9-23. Typical hole through fabric. Note the additional fabric added for reinforcement.

Temporarily install the cables prior to applying the fabric. Make a cardboard template to match the structure, and cut the proper exit location on it. Note that if you make a mistake on the cardboard, just tape up the hole and cut a new one. Once the fabric is installed, place the template in the proper position and cut the hole.

Or a straightedge across the bare truss can establish the plane of the fabric and the intersection point of the cable. Measure the distances to several reference points on the truss. When the covering is installed, tie a pencil to a string and draw short arcs of the proper radius from the reference points. The intersection of the arcs will be the exit point.

Fabric Application

Like most glues, the fabric attaching stuff doesn't work well on smooth, shiny surfaces like some epoxy metal primers. Buff such areas with 500 grit sandpaper. Don't cut through the protection; just dull the surface a little. Wipe the dust off with a clean rag or paper towel.

The different portions of the aircraft require variations in the basic technique. Fabric pieces should be as large as practical. Most kitplane wings, for instance, require only two pieces: one top, one bottom.

Fuselages are another case. The complex shapes involved, such as the vertical stabilizer intersection, might need a smaller piece or two. Again, follow your plans.

The first step is to cut the fabric to size, with considerable extra to account for errors. Clamp it in place, and cut any slots required for fittings to pass through. These areas will subsequently be reinforced with tape. Then glue the material down in the sequence recommended by the builder's and fabric processor's manuals.

Don't sling the glue wildly around. It can drip to the other side of the structure and distort its fabric. Don't add glue to the exterior of the fabric unless called for. Otherwise, it might actually slow the curing process and require extensive cleanup before doping.

Apply the fabric as tightly as possible. The shinking performed after installation is not intended to make up for sloppy work. The more the fabric is shrunk, the weaker it gets. Don't become fixated on a tight surface before shinking; folds are all right but try to eliminate large sags.

When gluing fabric to tubing, it must wrap between 180 and 270 degrees around the tube. In other words, at least halfway around the tube. Not farther than three quarters of the way, though, because it becomes tough to get the point of the iron far enough inside to seal the edge.

When fabric must overlap on a tube (such as covering a control surface or the horizontal stabilizer), there must be at least 1 inch of overlap between the two sheets. Add a 2-inch fabric tape as well.

When the joints between fabric sheets cannot wrap around the structure, the two sheets must overlap by about 2 inches if gluing is the only means used to join them. Such a seam can only be made if 2 inches of structure are underneath. The seam can't be made over a wing rib, for instance, unless that rib is two inches wide over the entire length of the seam. Sewing is the only approved method if there isn't enough structure underneath.

However, wings are covered lengthwise. Unless you're scrimping with short stretches of fabric remnants, there's no need to join fabric at anywhere but the leading and trailing edge.

The top and bottom surfaces should overlap at least 2 inches at the leading edge. This shouldn't be any problem, as every kitplane covers the leading edge with thin aluminum sheet that adds strength and impact resistance as well as leaving a dandy surface for gluing fabric.

In all cases, seams should be covered with fabric tape wider than the overlapped area.

Shrinking

Polyester fabric is shrunk under heat, and an ordinary iron is the typical shrinking tool. As mentioned earlier, calibrate its dial and make sure it can hold a constant temperature. While the household iron is perfect for large expanses like wings and fuselage sides, smaller RC model irons are best suited for precision work. Some processes use heat guns, but they're not really precision devices. Don't use one unless recommended by the process manufacturer.

One thing is sure: You will be surprised and pleased at the degree to which the fabric shrinks. As mentioned in the last section, overshrinking isn't good. But careful attention to eliminating the biggest sags during fabric installation will reduce the danger. One type of wrinkle the iron might not take out are small ones that are set in glue. Watch these, and straighten them out before the glue dries.

Another glue-oriented problem is fabric sticking too far around a tube. For instance, the surface-side of the covering might become stuck around much of the tube's circumference, so there's a long indentation where the fabric meets the tube. Slip your fingers behind the dented area and push outwards while using your other hand to keep from pushing too far. Tighten the freed fabric with the iron.

Use care when shrinking the fabric, as excessive tension can warp the underlying structure on some of the lighter ARV-class vehicles. However, most of the processes require heat application at a certain high level, typically 300 to 500 degrees, to "set" the fabric. Otherwise the fabric might become baggy in cold weather. You haven't much choice in the matter.

When possible, equalize the shrinking process by shrinking both sides sequentially. Shrink the top side of an aileron a bit, then the other side. Then shrink the top surface a bit more, and so forth. Contact the kit and/or covering process manufacturer for more details.

Fortunately, any distortion is not generally permanent. The fabric can be stripped away for another try.

When the fabric is installed and shrunk, finishing tapes are glued over all seams and areas that require reinforcement, like wingtips and sharp edges. The finishing tapes cover irregularities and make the entire job look smoother.

Rib Attachment

Certified aircraft must have positive attachment of the upper wing fabric to each rib. The traditional method is *rib stitching*, which is running a loop of cord through the fabric around each rib.

Typically, kitplane manufacturers substitute glue for rib stitching. According to the Stits manual: "Over the years, the question of direct substituting rib lacing with cement bonding to the ribs has come up regularly, and the answer is no."

Basically, Stits feels the peel strength of the glue is overly affected by the surface preparation, solvent penetration from the finishing coats, and aging of the glue. The top of the wing is a severe low pressure area (it must be to supply lift) and fabric separation would be disastrous.

Fig. 9-24. Rib stitching demonstration at an Oshkosh forum. A good place to learn the Seine knot.

This isn't a condemnation of the Stits process. It's one of the standards of the industry; and Leroy Stits is highly regarded by all of aviation.

Why, then, do the kitplane manufacturers specify cement instead of stitching?

Because instances of fabric separation are rare. In other words, gluing works. Rib stitching is a tedious, thankless job. No matter what you do, the lacing shows through the fabric and detracts from the desired smooth appearance.

It's your decision. My club's Fly Baby has glued fabric and has flown for eight years without a problem. But tube-type ribs, like the CIRCA Nieuport or RANS series, don't leave much surface for fabric attachment. In these cases, some sort of additional security is called for.

As mentioned, the traditional method is lacing with cord. Rib stitching is one of those ye olde aviation arts; The lacing cord runs along the lower surface of the wing, loops over the ribs at intervals, and is tied with a *seine* knot (Fig. 9-24). I've seen many a diagram on the seine knot, none of which made it clear. It's actually fairly simple, but is best learned by seeing it demonstrated. Talk to your EAA counselor.

One problem with rib stitching is the manpower required because a partner is usually needed. Other attachment methods can be easily done solo. The most popular are screws (either wood or sheet-metal depending upon the rib material) and pop rivets (for tube-type ribs). Use a 1/2-inch diameter aluminum washer to apply the force over a wide area.

The minimum attachment spacing for a given method depends upon whether the area is within the propeller slipstream. For aircraft with redline speeds less than 150 knots, spacing is 2.5 inches in the slipstream and 3.5 inches everywhere else.

Penetrating the fabric weakens it, hence a strip of reinforcement tape should be glued to the target areas before beginning. When completed, the laces or screw/rivet heads are covered by a strip of finishing tape.

Access Panels and Drain Grommets

One nice thing about fabric covering is the ability to install large access panels without affecting the structural integrity. Oversized access panels are awkward, but it's quite easy to put in nice hand-sized panels.

Of course, you can't cut big holes willy-nilly. They must be reinforced to prevent the fabric from tearing. There's a standard circular inspection plate that works quite nicely. It's convex in shape (like a contact lens) with an aluminum strap riveted across the back. When installed, the rim pinches the fabric to both ends of the strap, holding the plate in place. But apply pressure to the center, and the center snaps inward, releasing the pressure and allowing the plate to be removed.

They require a 3.5-inch hole in the fabric, which is reinforced by a plastic ring. Glue the ring to the inside of the fabric. Cut away the material in the center when the glue is dry, then use tapes around the ring to further anchor it.

Fig. 9-25. Wide access panels in the tail of this Sonerai allow good access to the workings of the tail feathers.

While it's easiest to use these available commercial items, access panels of any size can be made (Fig. 9-25). Make them and similarly-shaped reinforcement rings from aluminum. Drill a few mounting holes between the plate and the ring. Glue the ring to the fabric and cut out the center area. Sheet metal screws then attach the plate to the ring.

Drain grommets are easier. Although the fabric covering should be nearly watertight, moisture can leak in or simply form due to humidity and temperature changes. The water will accelerate wood rot and/or aluminum corrosion unless grommets permit drainage.

The grommets are nothing more than plastic washers. Glue them to the inside of the fabric at points that will be low when the plane is at rest. One should be installed underneath the uphill side of the trailing edge of each rib, for example. Then punch out the center with the tip of a soldering iron or a hot nail.

The next step is dope application, which is discussed in the next chapter in the painting section.

Otherwise, the fabric covering process is complete. It's not especially difficult, but quality is important. Not from the strength point of view, but because the fabric is the exterior surface of your aircraft. No matter how meticulous you might be on the interior, wrinkles and sags in the fabric will turn your diamond back into a lump of coal. Take your time; be careful.

10

Completion

BUILDING A KITPLANE OF ANY TYPE TAKES A LONG TIME. The structure itself is the most daunting, but the subsystems (engine, electronics, brakes, controls) require just as much attention and construction time. Even a simple Rotax engine installation will take 50 hours or more.

But the subsystems work has several advantages. For one thing, much of it is simple bolt-together work. Unlike structural integrity itself, subsystem operation is easily tested after completion. Changes and adaptations can then be made as problems are found.

Probably the best reference for subsystems installation are three books by Tony Bingelis: *The Sportplane Builder, Sportplane Construction Techniques*, and *Firewall Forward*. These books are filled with details on brake systems, electronics, interiors, occupant protection, and a myriad other subjects for which you'll eventually have questions. The books are available through EAA and homebuilder's supply companies.

This chapter concentrates on the finishing details: painting, testing, certification, and first flight.

PAINTING

Painting is a major construction landmark because in most builder's minds, it's when the project turns into an airplane. Don't let the gleam in your eye result in a less-than-gleamy aircraft.

When?

Most builders are eager to get the bird looking really sharp for the first flight. We want glittering wings for the post-flight photos. So once everything's together, we paint the aircraft and haul it to the airport.

There's one problem many builders don't face: the airplane, and their flying of it, probably won't be perfect. The test flight period isn't just a formality. Few airplanes make it through the 25 or 40 hours without the need for some work.

It might be minor. Wheel shimmy takes its toll on wheel pants. Engine cooling often is insufficient, and cowling modifications are necessary.

Problems occasionally are more serious. Got much taildragger time? Even 10 hours or so in a Citabria might not be enough to prepare you for the lighter, lower-powered ARVs. Minor cosmetic damage is a common occurrence during the test period.

I once saw a video of a builder taxi-testing a Kitfox. The plane spent more time dragging wingtips than running in a straight line. By the time the builder found his gear bungee cord tension was incorrect, he'd ground several inches off his drooped-down wingtips. Didn't do his nice paint job much good, either.

The only thing more difficult than a classy paint job is matching the paint after repairs. It's often a good idea to wait until the test period is completed.

This causes some problems. First, getting the plane to the airport is usually a major operation. Folding-wing airplanes aren't too bad, but the rest need the wings removed. So if you fly the plane before painting it, you end up repeating the process two more times; one round trip from the airport to home (for painting) and back. Painting the airplane in your hangar might seem like an ideal solution, but many airports prohibit it.

Second, some materials must be protected before the first flight. Ragwings should have UV-block silver dope applied. Aluminum airplanes should be primed. The engine and plexiglass must be protected anytime paint is being applied. Waiting means you'll have to mask them off for priming, then mask again after flight testing. Why do it twice?

Lastly, waiting to paint involves more thorough preparation when the time comes. Exhaust deposits must be cleaned off. A minor oil leak means an extensive scrubbing before paint can be applied.

It's kind of a tossup. Decide which way suits you.

Who?

Should you paint the plane yourself, or hand it over to a pro? It depends on your goals, confidence, and finances.

Are you just looking for a decent-looking paint job? Then you probably can do an acceptable job in your shop or yard. Thousands of homebuilders have.

But what if you want more than just an OK finish? Do you feel up to it? Do you have access to a good paint booth and professional equipment? Award-winners have been painted in the owner's garage, but if you're interested in only the best and can afford it, consider having the aircraft professionally painted.

After all, no one can be an expert at everything. And the external finish is how 99 percent of the people will evaluate your plane. You might be a magician in fiberglass, or your Falco might be a cabinetmaker's dream. But all your good work comes to naught without a smooth hand on the paint gun.

Professional paint jobs cost from $2,000 to $4,000, depending upon complexity of the scheme and the size of the shop. Some independents will do it for less. Typically, the painter charges $1,000 or so plus materials. Because the materials usually run $500 or so, you'll save about 25% over the minimum professional charge.

Should you go with an independent? Ask for some references; look at other planes he's painted. Call the owners and see if they are satisified.

Cost of professional painting can be reduced in another way. Professional paint shops spend most of their time preparing the aircraft by sanding the surface and masking off the things that shouldn't be painted. Most places are willing to reduce the charge if the owner does most of the busy work.

There's no real craft to covering the plexiglass with aluminum foil, nor any skill required to buff with fine sandpaper. Deliver the airplane ready for painting in the truest sense—all the painter has to do is load his gun and spray—and most places will cut the fee considerably. Or apply the primer coats yourself and turn the plane over to the pro for the fancy finish (Fig. 10-1).

Part of the busy work you'll leave to them is the masking off for the various trim colors, the N-number, and the like. But there are ways around it. Some graphics companies make hi-tech decals that can replace trim paint. The painter can apply the base color, and you'll add the trim later, perhaps after flight testing. The cost is less than paint, especially for complex schemes.

These graphics systems go beyond simple colored tape. Complex, fairly large patterns are designed and precision-cut by computer. The backing paper is actually on the front side of the pattern; prepare the surface, press the pattern in place, and peel off the paper.

Fig. 10-1. The hard part of painting the aircraft is applying the perfect finish coat. By masking and priming yourself, then hauling it to the local paint shop, you can get a first-class paint job at a discount price.

Painting Facilities and Equipment

The most obvious need is spray painting equipment, such as compressor, paint gun, air hose, moisture trap, and respirators. Enamel coats must be heavy to build a gloss, so find a professional model gun rather than a hobby one. Such guns require about 8 cubic feet per minute of compressed air, which in turn requires at least a two to three horsepower compressor.

The airplane doesn't have to be painted as one piece, but you'll need enough space to store pieces while other parts are being sprayed. Not having enough space to set up the entire plane means you'll have to wait until pieces are dry before you can move them out of the way. It'll take a bit of room; where will you keep the fuselage while the wings are being painted?

Keep in mind you need more than just room for the airplane. There must be space for the painter to safely walk around the components, with the usual foot or two clearance between the gun and the surface. If the booth is too small, there's a danger of dragging the air hose across newly painted surfaces.

A common practice is to wall off a section of the garage or shed with plastic sheet. Be careful about ventilation. The respirator cleans the air going to your lungs, but doesn't ensure it contains enough oxygen. But paint fumes are flammable, and ordinary fan motors can ignite them. If you must use an ordinary fan, use it to *inflate* your painting area, instead of pulling fumes out. Explosion-proof exhaust fans are available; see if anyone in your EAA chapter has a portable fan.

One common paint area is the great outdoors. Wait for windless days (or paint at dawn and dusk), and be prepared to get bugs and dust in your finish. Overspray drifting across the fence won't endear you to the neighbors, either.

Good lighting is a must. Years ago, I helped paint some Civil Air Patrol jeeps in an old hangar. The old color was olive drab, the new color was Air Force blue. All we had for lighting was a couple of old fluorescents on the ceiling. In the darkness and spray, we couldn't tell where the old paint ended and the new began. When the jeeps were rolled outside, the effect was startling.

The same thing could happen in your garage. You could finish the trim colors, roll it outside, and discover the paint is too thin over the lower fuselage. Then you have to push it back in, add another couple of coats, tape the trim strips again, and so on.

Make some portable light stands from PVC tubing and fluorescent light fixtures. Stick some clear plastic wrap across the front of the lights to keep the overspray from the tubes. Use glass or heat-resistant plastic if using incandescent lighting.

The best light source available is the sun. Roll the airplane briefly outside between coats. Get low and squint at the surface.

What Kind of Paint?

There are four basic kinds of paint: *Enamels, lacquers, polyurethane,* and dope.

Enamels are inexpensive and hardy, but gradually oxidize and need waxing more often than other types. They take a long time to dry, so tend to pick up a bit of dust.

Lacquers cost a bit more and can be shined to a deep gloss, but a bit more susceptible to chipping. They're quite thin and require multiple coats (on the order of 5 to 20) and require plenty of rubbing to build a shine. They dry fast, though, so there's little problem with dust.

Polyurethane wet look paints are very shiny, very durable, and very expensive. Like enamels, they take a long time to dry.

Dopes are the exclusive province of fabric aircraft: a lacquer paint with softeners added.

Let's look at specific examples.

While composite airplanes always must be painted, metal-airplane builders have an option: They can leave the airplane bare metal (Fig. 10-2).

Fig. 10-2. While it's certainly feasible, few metal-aircraft builders leave their plane in bare metal. The owner of this RV-4 painted it after flight testing.

I don't really recommend it. For one thing, few of us are able to do a 100 percent perfect job on the exterior skin. We might end up with a few scratches, or a couple of dings that need a little bondo filler. They're hard to hide without paint.

Second, unpainted metal airplanes are a lot of work. Polyurethane-painted jobs can get by with a coat of wax or so every year. But all-aluminum planes have to be polished continuously.

Dopes have always been the traditional choice for fabric airplanes for a very good reason: They are flexible enough to withstand fabric vibration without cracking. The other types dry to a hard shell, which is not very compatible with a flexible fabric covering.

However, automotive technology comes to the rescue. Now that so many cars incorporate flexible plastic bumpers and fenders, flex agents are available for these other paints. Some even say they aren't needed for polyurethane paints.

Dope shares one problem with lacquer paint: To get a really glossy finish, 20 coats or more must be applied. Each coat must be buffed before the next is added. The multiple coating isn't a requirement because sufficient protection and coverage are achieved with just a few coats. Most ragwing owners are satisfied with the satin finish that a couple of coats brings.

Traditional dopes cannot be applied to metal surfaces, making the painter switch to enamel for areas like cowlings. More modern dopes don't have this problem. Check with the manufacturer.

One big advantage of using dope is repairs are simple and almost invisible. New dope blends almost invisibly with old. Overspray of the other coating types will get dull around the repaired area, while new dope overlaps with nary a trace.

Overspray causes problems with gloss paints, especially polyurethane. When a dried, painted surface is dusted with airborne paint, it dulls the shine. Buffing enamels and lacquers reduce or eliminate the problems, but it doesn't work very well with polyurethane.

The typical problem area is atop the fuselage. The painter sprays the left side and left wing, then walks around to the right. If the paint on the left side is still wet, everything's all right. But if it's dried, the new paint won't blend in with the cured stuff. It curls around the top of the fuselage and dusts the left side. Later the painter notices the dull area, and repaints it. You guessed it, now the *right* side looks bad. Professional shops get around this problem by using two painters at the same time.

Your kitplane's paint job is important. It's how most people will judge your plane. A well-executed simple scheme will gain more credit than a sloppy pseudo-tiger with technicolor fangs.

Not only is it important for appearances' sake, but there can be safety issues as well. Control surfaces of fast airplanes must be balanced; too much weight aft of the hinge line could cause flutter. It's not unknown for an overly-thick coat of paint to change marginal flutter stability to an outright hazard.

It's quite possible to get acceptable results in the garage, even if you've never painted anything large before. But there are different techniques for different paints; even for painting different areas of the airplane. This is one area where good advice and training are vital. Find an experienced hand to get you started.

Don't expect to be done in a couple of days. Painting can take 100 hours or more.

RIGGING AND CENTER OF GRAVITY

A modern kitplane is a pretty foolproof item. You have to really foul up before the airplane becomes structurally unsafe.

However, a moderate inattention to detail can result in poor handling characteristics and poor performance. The two major causes of these problems are mis-rigging and incorrect *center of gravity* (CG).

Rigging

Ever wonder why seemingly identical airplanes can fly so differently? If you have hundreds of hours in your own 172, every other 172 feels . . . wrong. Alien.

Like wearing somebody else's shoes. Something's not right but there's nothing you can put a finger on.

Or one plane handles crisp and stalls clean, and another of the same model flies cockeyed and stalls viciously. What's the difference, anyway?

"It's out of *rig*," most pilots would guess. But what do we mean by rig? Like many aviation terms, it has a nautical origin. In the days of sail, a ship's speed and handling depended upon the set of the standing and running rigging. The rake of the masts and the bracing of the yards made the difference between a clipper and a hooker.

Rig on aircraft has to do with the relationship of the airfoils, stabilizers, and control surfaces to the relative wind.

Production aircraft, and some kitplanes, have the ability to individually adjust the lift of each wing to make up for slight differences in lift. These differences usually come from manufacturing variations. And, as you might expect, kitplane building is rife with manufacturing variations.

A degree's worth of difference in incidence between the left and right wing can cause noticeable wing heaviness in flight, getting worse as speed increases. In contrast, a weight difference between the wings becomes less noticeable at cruise. Hence the need for precision during construction.

But it's impossible to be perfect. The ability to compensate for these problems varies. Some kitplanes offer independent incidence adjustment, like production aircraft. Wire-braced and strut-braced homebuilts are prime examples of adjustable airplanes. Others don't provide for it, and handle problems with fixed external trim tabs.

Before the first flight, the aircraft should be rigged to the neutral position, or wherever else the plans specify. When this is done depends on the aircraft. Ragwing airplanes are usually rigged before the covering is applied (Fig. 10-3).

The first step is to set the fuselage to an in-flight attitude. The plans should specify a reference point from which level is determined. On tube-type airplanes, it's generally a certain point in an upper-fuselage longeron. It might be the cockpit sill or the top of the cowling, depending upon the airplane.

Once a bubble level is in place, the airplane's pitch attitude is adjusted to center the bubble. Most kitplanes (even trigear) have a nose-up attitude on the ground, so the tail must be raised. It's easy enough with conventional gear. Place the tailwheel on a stool and add shims until level. For trigears, roll the mains onto an elevated platform, and shim the nosewheel.

The instructions should indicate the correct adjustment, and the method of measurement. The usual practice uses a carpenter's level and shims. For instance, the horizontal stabilizer setting might be indicated as -2 degrees. The bubble on a 24-inch level should be level when the front end of the tool rests on shims equaling about $13/16$ of an inch.

The positions of the surfaces are set by various means. Some merely add washers under bolts. (Make sure the bolt is long enough to accommodate the number of washers which might be required.) Turnbuckles set the incidence and dihedral on wire-braced wings. Wing struts incorporate threaded ends to change their length.

Fig. 10-3. Ragwing builders usually rig the aircraft prior to covering to minimize wrinkles. This Fisher Classic is having its lower wing incidence checked.

Turnbuckles are also used to set the tension for control cables. A *tensiometer* (tension-measuring gauge) is nice; unless otherwise specified in the plans, set the tension to 20 – 30 pounds or so. Check the tensions at the full excursions of the controls to ensure the cable doesn't go slack in either direction.

As a final note on rigging, take a look at the ailerons. Make sure they are both in the neutral position when the stick is centered. It's quite possible for both to be up or down slightly; adjust them if necessary.

Center of Gravity

As pilots, we all learned the importance of center of gravity location. Every airplane's paperwork includes an actual empty weight and CG, to which we add the weight and moment of the passengers, fuel, and baggage. The CG must fall within a particular range of the airfoil's *center of pressure* (CP).

Because we are the manufacturer of our particular homebuilt, it's up to us to determine the basic figures. The kitplane's plans might include a typical empty weight and moment. These are just samples, you *must* calculate your own.

The procedure is simple: The aircraft is leveled and the weight at each wheel is measured. This, along with the position of the wheels relative to some reference point (usually the firewall), determines the CG's location.

If the result falls within the allowable range, all is well. If it doesn't, you've got your work cut out for you. Unfortunately, most solutions require adding weight.

A forward CG location is easiest to solve. The tail cone has a long moment, so a little weight in back goes a long way. One common fix is to shift the battery box from the cowling to the baggage compartment. This adds very little weight, but rerouting the heavy cables through the cabin is a hassle.

Aft CG locations are more common, and harder to solve. There isn't much moment arm forward of the CP. It's possible. I've seen a couple of homebuilts with steel or lead blocks bolted to the engine mount. Aft-mounted equipment is rare, but see if any can be shifted forward.

One traditional solution is to lengthen the engine mount. If the aircraft is close to completion, this is an act of desperation. Imagine, everything connecting the engine to the airframe has to be lengthened. Wiring. Fuel lines. Throttle, mixture, and carb heat cables. Scrap the cowling, and scratch-build a new one. The exhaust system must be rerouted. Not to mention fabrication of the new engine mount.

Often, the only practical solution is to limit the useful load so the aft CG limit isn't breached. This typically amounts to placarding a baggage area for a lower weight limit.

A Few Words About Weight

So you rolled it up on the scales and it came out a few pounds heavy, didn't it?

It isn't rare. You can chalk up the advertised "empty weight" alongside those other great myths of aviation.

But how does it *really* affect you?

As far as the kit manufacturer is concerned, not at all. Because you are naturally going to limit the gross weight to the book value, right?

Let's face it. Few kitplane owners do. Most load the seats; load the tanks; load the baggage. One homebuilt was designed for a 450-pound gross weight. One builder won grand-champion honors with his version, which he flew at *650* pounds, almost 50 percent heavier than the prototype.

Let's look at how overloads affect the operation of the aircraft.

In the first place, throw all performance specifications out the window. It'll take more room to take off and land. Cruise speed and climb rate will be lower. It'll stall at a faster speed, and probably more sharply as well.

The CG is a little trickier, as the load point isn't even on the graph.

These are all factors a pilot would expect. However, consider the effect on load factors. For instance, assume the plane mentioned earlier is stressed to withstand 6 Gs. That means the structure can take a load of 6 × 450, or 2,700 pounds. But with a 650-pound gross weight, this translates to a limit of a little over 4 Gs.

During normal flying, this won't make much difference. But if the builder likes aerobatics. . . . Well, let's just hope he wears a parachute.

Let's make one thing clear: I do not condone flying any airplane weighing more than the manufacturer intended.

Buy flying beyond the limit is nothing new. The FARs even allow certain older production aircraft operated in Alaska to operate at 115 percent of their certified gross weight.

Just remember that homebuilt designs don't incorporate the generous margins of factory airplanes. Watch the CG especially, project out the CG envelope to your personal gross weight. The sides of the envelope are usually linear. And fly gently, gently.

THE LEGALITIES

At some point before completion, you should initiate the paper chase that will result in an airworthiness certificate. The forms mentioned below are available through either the FAA or EAA.

N-number and Registration

You'll need a registration, or N-number, before painting the aircraft, so you can paint the registration number at the same time as the trim. As far as the FAA is concerned, the number must be on the aircraft at the time of inspection, so you can wait if necessary.

Then again, there's nothing illegal about requesting a number before starting construction. But some states cross-check the FAA's aircraft list to find owners who haven't paid taxes or registration fees. Request a number early, and you might get a bill from the state. They don't care if the aircraft is just a pile of parts in the garage, it's registered, so you have to pay.

An N-number is obtained by submitting a notarized affidavit of ownership to the FAA's Aircraft Registry department along with a short cover letter requesting assignment of an identification number for an amateur-built aircraft. The affidavit is available as *AC Form 8050-88*. The form itself isn't necessary, as long as you provide the following:

- Builder's name.
- Model (Glasair III, Fisher FP-404, etc.)
- Serial number. (Usually assigned by the kit manufacturer, otherwise make up your own.)
- Class. (Airplane, rotorcraft, glider, etc.)
- Type of engine. (Reciprocating, unless you have a jet.)
- Number of engines.
- Manufacturer, model, and serial number of engine.
- Land or sea operation, or both, if an amphibian.
- Include the following statement: "This aircraft was built from parts by the undersigned and I am the owner," and sign the form.

The form must be notarized, and hence should include the usual legalese: county and state, notary's commission expiration date, and date of notarization.

If you want a custom N-number, your cover letter should include the request for a special identification number. List your primary choice and a number of alternatives. You can request up to five characters after the "N" prefix. The last two characters can be letters.

Traditionally, homebuilders tend to request the year the airplane is completed, followed by their initials: N1986TD, N82EU, and the like. The FAA is loath to hand out short numbers, such as N4R, but will do so if your aircraft has limited space for full-sized registration numbers.

Custom N-numbers cost $10. The reservation is only good for one year, so if your airplane hasn't been licensed yet, you'll have to send another check to hold the number.

To summarize the N-number process, place the following in an envelope.

- A cover letter requesting an identification number for an amateur-built aircraft. You may either take a randomly assigned number, or request a specific one (include alternates).
- A check for $10, *if requesting a specific N-number*. Randomly assigned numbers are free.
- A notarized affidavit of ownership, as described earlier.

Send the envelope to:
FAA Aircraft Registry
Department of Transportation
P.O. Box 25504
Oklahoma City, OK 73125

Aircraft Registry will return a form with your N-number, and an aircraft registry application, *AC Form 8050-1* (Fig. 10-4). Fill it out, and mail it back along with a check for $5. Keep the pink copy of the multipart form because it is your temporary registration.

The N-number must be applied to both sides of the fuselage between the trailing edge of the wing and the leading edge of the horizontal stabilizer. It can also be applied to the vertical stabilizer. The minimum size is 3 inches; however, the situations under which 3-inch N-numbers are legal are rapidly diminishing.

If any of the following statements are true, the ID must be at least 12 inches high:

- You plan on flying within an Air Defense Identification Zone.
- You plan on flying to another country (including Canada and Mexico).
- Your kitplane cruises at 180 knots or faster.

In my opinion, 3-inch letters are an endangered species. You might as well plan on the 12-inch size. The style or font of the lettering is restricted to simple block styles; a little slant is OK, but no Old English or italic styles, please. Color choice is left open, but it must contrast with the surrounding paint in order to be easily readable.

Airworthiness Certificate

Three things are required to gain an Airworthiness Certificate: completion of *FAA Form 8130-6*, completion of *FAA Form 8130-12*, and a *signoff inspection*.

UNITED STATES OF AMERICA DEPARTMENT OF TRANSPORTATION
FEDERAL AVIATION ADMINISTRATION-MIKE MONRONEY AERONAUTICAL CENTER
AIRCRAFT REGISTRATION APPLICATION

CERT. ISSUE DATE

UNITED STATES
REGISTRATION NUMBER **N**

AIRCRAFT MANUFACTURER & MODEL

AIRCRAFT SERIAL No.

FOR FAA USE ONLY

TYPE OF REGISTRATION (Check one box)

☐ 1. Individual ☐ 2. Partnership ☐ 3. Corporation ☐ 4. Co-owner ☐ 5. Gov't. ☐ 8. Foreign-owned Corporation

NAME OF APPLICANT (Person(s) shown on evidence of ownership. If individual, give last name, first name, and middle initial.)

TELEPHONE NUMBER: (   )   –

ADDRESS (Permanent mailing address for first applicant listed.)

Number and street: _____

Rural Route: _____ P.O. Box: _____

CITY	STATE	ZIP CODE

☐ **CHECK HERE IF YOU ARE ONLY REPORTING A CHANGE OF ADDRESS**

ATTENTION! Read the following statement before signing this application.

A false or dishonest answer to any question in this application may be grounds for punishment by fine and / or imprisonment (U.S. Code, Title 18, Sec. 1001).

CERTIFICATION

I/WE CERTIFY:

(1) That the above aircraft is owned by the undersigned applicant, who is a citizen (including corporations) of the United States.

(For voting trust, give name of trustee: _____), or:

CHECK ONE AS APPROPRIATE:

a. ☐ A resident alien, with alien registration (Form 1-151 or Form 1-551) No. _____

b. ☐ A foreign-owned corporation organized and doing business under the laws of (state or possession) _____, and said aircraft is based and primarily used in the United States. Records of flight hours are available for inspection at _____

(2) That the aircraft is not registered under the laws of any foreign country; and
(3) That legal evidence of ownership is attached or has been filed with the Federal Aviation Administration.

NOTE: If executed for co-ownership all applicants must sign. Use reverse side if necessary.

TYPE OR PRINT NAME BELOW SIGNATURE

EACH PART OF THIS APPLICATION MUST BE SIGNED IN INK.

SIGNATURE	TITLE	DATE
SIGNATURE	TITLE	DATE
SIGNATURE	TITLE	DATE

NOTE: Pending receipt of the Certificate of Aircraft Registration, the aircraft may be operated for a period not in excess of 90 days, during which time the PINK copy of this application must be carried in the aircraft.

AC FORM 8050-1 (1-83) (0052-00-628-9005)

Fig. 10-4. AC Form 8050-1, Aircraft Registration Application. Fill this out and send it to the FAA after receiving an N-number.

FORM APPROVED
O.M.B. NO. 2120-0018

US Department of Transportation
Federal Aviation Administration

ELIGIBILITY STATEMENT
AMATEUR-BUILT AIRCRAFT

Instructions: Print or type all information except signature. Submit original to an authorized FAA representative. Applicant completes Section I thru III. Notary Public completes Section IV.

I. APPLICANT INFORMATION

Name ____________________

Address ____________________
No. & Street City State Zip

Telephone No. ____________________
Residence Business

II. AIRCRAFT INFORMATION

Model ____________________ Engine(s) Make ____________________

Assigned Serial No. ____________________ Engine(s) Serial No.(s) ____________________

Registration No. ____________________ Prop./Rotor(s) Make ____________________

Aircraft Fabricated: Plan ☐ Kit ☐ Prop./Rotor(s) Serial No.(s) ____________________

III. MAJOR PORTION ELIGIBILITY STATEMENT OF APPLICANT

The major portion of the aforementioned aircraft was fabricated and assembled for education and/or recreation. I have evidence to support this statement and will make it available to the FAA upon request.

Date of Statement	Applicant's Signature

IV. NOTARIZATION STATEMENT

FAA Form 8130-12 (11-83)

☆ U.S. Government Printing Office: 1985—461-823/21541

Fig. 10-5. With FAA Form 8130-12, the builder officially states that he or she constructed the majority of the aircraft.

U.S. DEPARTMENT OF TRANSPORTATION
FEDERAL AVIATION ADMINISTRATION

APPLICATION FOR AIRWORTHINESS CERTIFICATE

INSTRUCTIONS — Print or type. Do not write in shaded areas; these are for FAA use only. Submit original only to an authorized FAA Representative. If additional space is required, use an attachment. For special flight permits complete Sections II and VI or VII as applicable.

I. AIRCRAFT DESCRIPTION

1. REGISTRATION MARK	2. AIRCRAFT BUILDER'S NAME (Make)	3. AIRCRAFT MODEL DESIGNATION	4. YR. MFG.	FAA CODING
5. AIRCRAFT SERIAL NO.	6. ENGINE BUILDER'S NAME (Make)	7. ENGINE MODEL DESIGNATION		
8. NUMBER OF ENGINES	9. PROPELLER BUILDER'S NAME (Make)	10. PROPELLER MODEL DESIGNATION	11. AIRCRAFT IS: NEW USED IMPORT	

II. CERTIFICATION REQUESTED

APPLICATION IS HEREBY MADE FOR: (Check applicable items)

A — STANDARD AIRWORTHINESS CERTIFICATE (Indicate category)   NORMAL   UTILITY   ACROBATIC   TRANSPORT   GLIDER   BALLOON

B — SPECIAL AIRWORTHINESS CERTIFICATE (Check appropriate items)

LIMITED

PROVISIONAL (Indicate class)   CLASS I   CLASS II

RESTRICTED (Indicate operation to be conducted)
- AGRICULTURE AND PEST CONTROL
- AERIAL SURVEYING
- AERIAL ADVERTISING
- FOREST (Wildlife conservation)
- PATROLLING
- WEATHER CONTROL
- OTHER (Specify)

EXPERIMENTAL (Indicate operation(s) to be conducted)
- RESEARCH AND DEVELOPMENT
- AMATEUR BUILT
- EXHIBITION
- RACING
- CREW TRAINING
- MKT. SURVEY
- TO SHOW COMPLIANCE WITH FAR

SPECIAL FLIGHT PERMIT (Indicate operation to be conducted, then complete Section VI or VII as applicable on reverse side)
- FERRY FLIGHT FOR REPAIRS, ALTERATIONS, MAINTENANCE OR STORAGE
- EVACUATE FROM AREA OF IMPENDING DANGER
- OPERATION IN EXCESS OF MAXIMUM CERTIFICATED TAKE-OFF WEIGHT
- DELIVERING OR EXPORT
- PRODUCTION FLIGHT TESTING

C — MULTIPLE AIRWORTHINESS CERTIFICATE (Check ABOVE "Restricted Operation" and "Standard" or "Limited," as applicable.)

III. OWNER'S CERTIFICATION

A. REGISTERED OWNER (As shown on certificate of aircraft registration)   IF DEALER, CHECK HERE ⟶

NAME

ADDRESS

B. AIRCRAFT CERTIFICATION BASIS (Check applicable blocks and complete items as indicated)

AIRCRAFT SPECIFICATION OR TYPE CERTIFICATION DATA SHEET (Give No. and Revision No.)

AIRWORTHINESS DIRECTIVES (Check if all applicable AD's complied with and give latest AD No.)

AIRCRAFT LISTING (Give page number(s))

SUPPLEMENTAL TYPE CERTIFICATE (List number of each STC incorporated)

C. AIRCRAFT OPERATION AND MAINTENANCE RECORDS

CHECK IF RECORDS IN COMPLIANCE WITH FAR 91.173

TOTAL AIRFRAME HOURS (Enter for used aircraft only)

EXPERIMENTAL ONLY (Enter hours flown since last certificate issued or renewed)

D. CERTIFICATION — I hereby certify that I am the owner (or his agent) of the aircraft described above; that the aircraft is registered with the Federal Aviation Administration in acccordance with Section 501 of the Federal Aviation Act of 1958, and applicable Federal Aviation Regulations; and that the aircraft has been inspected and is airworthy and eligible for the airworthiness certificate requested.

DATE OF APPLICATION

NAME AND TITLE (Print or type)

SIGNATURE

IV. INSPECTION AGENCY VERIFICATION

A. THE AIRCRAFT DESCRIBED ABOVE HAS BEEN INSPECTED AND FOUND AIRWORTHY BY: (Complete this section only if FAR 21.183 (d) applies)

FAR PART 121 OR 127 CERTIFICATE HOLDER (Give Certificate No.)

CERTIFICATED MECHANIC (Give Certificate No.)

CERTIFICATED REPAIR STATION (Give Certificate No.)

AIRCRAFT MANUFACTURER (Give name of firm)

DATE

TITLE

SIGNATURE

V. FAA REPRESENTATIVE CERTIFICATION

(Check ALL applicable blocks in items A and B)

THE CERTIFICATE REQUESTED

A. I find that the aircraft described in Section I or VII meets requirements for: AMENDMENT OR MODIFICATION OF CURRENT AIRWORTHINESS CERTIFICATE

B. Inspection for a special flight permit under section VII was conducted by:

FAA INSPECTOR

CERTIFICATE HOLDER UNDER: FAA 65   FAR 121 OR 127   FAR 145

DATE   DISTRICT OFFICE   DESIGNEE'S SIGNATURE AND NO.   FAA INSPECTOR'S SIGNATURE

FAA Form 8130—6 (4-80) SUPERSEDES PREVIOUS EDITION

☆ U.S. GOVERNMENT PRINTING OFFICE: 675-964/147

Fig. 10-6. The front side of FAA Form 8130-6, the application for an airworthiness certificate. The FAA inspector or DAR signs this form upon completion of his inspection.

VI. PRODUCTION FLIGHT TESTING

A. MANUFACTURER

NAME	ADDRESS

B. PRODUCTION BASIS *(Check applicable item)*

	PRODUCTION CERTIFICATE *(Give production certificate number)*
	TYPE CERTIFICATE ONLY
	APPROVED PRODUCTION INSPECTION SYSTEM

C. GIVE QUANTITY OF CERTIFICATES REQUIRED FOR OPERATING NEEDS ⟶

DATE OF APPLICATION	NAME AND TITLE *(Print or type)*	SIGNATURE

VII. SPECIAL FLIGHT PERMIT PURPOSES OTHER THAN PRODUCTION FLIGHT TEST

A. DESCRIPTION OF AIRCRAFT

REGISTERED OWNER	ADDRESS
BUILDER *(Make)*	MODEL
SERIAL NUMBER	REGISTRATION MARK

B. DESCRIPTION OF FLIGHT

FROM	TO	
VIA	DEPARTURE DATE	DURATION

C. CREW REQUIRED TO OPERATE THE AIRCRAFT AND ITS EQUIPMENT

PILOT	CO-PILOT	NAVIGATOR	OTHER *(Specify)*

D. THE AIRCRAFT DOES NOT MEET THE APPLICABLE AIRWORTHINESS REQUIREMENTS AS FOLLOWS:

E. THE FOLLOWING RESTRICTIONS ARE CONSIDERED NECESSARY FOR SAFE OPERATION *(Use attachment if necessary)*

F. CERTIFICATION — I hereby certify that I am the registered owner (or his agent) of the aircraft described above; that the aircraft is registered with the Federal Aviation Administration in accordance with Section 501 of the Federal Aviation Act of 1958, and applicable Federal Aviation Regulations; and that the aircraft has been inspected and is airworthy for the flight described.

DATE	NAME AND TITLE *(Print or type)*	SIGNATURE

VIII. AIRWORTHINESS DOCUMENTATION (FAA use only)

A. Operating Limitations and Markings in Compliance with FAR 91.31 as Applicable	G. Statement of Conformity, FAA Form 8130-9 *(Attach when required)*
B. Current Operating Limitations Attached	H. Foreign Airworthiness Certification for Import Aircraft *(Attach when required)*
C. Data, Drawings, Photographs, etc. *(Attach when required)*	I. Previous Airworthiness Certificate Issued in Accordance with
D. Current Weight and Balance Information Available in Aircraft	FAR _____ CAR _____ *(Original attached)*
E. Major Repair and Alteration, FAA 337 *(Attach when required)*	J. Current Airworthiness Certificate Issued in Accordance with
F. This Inspection Recorded in Aircraft Records	FAR _____ *(Copy attached)*

Fig. 10-7. The back side of FAA Form 8130-6.

Form 8130-12 (Fig. 10-5) is the eligibility statement. The aircraft is generally described, and the builder must certify that he or she built the major portion of the aircraft. This form must be notarized.

FAA Form 8130-6 is the biggie: Application for Airworthiness Certificate (Figs. 10-6 and 10-7). This form describes the aircraft in detail (including a three-view diagram), what it will be used for, and the restrictions placed upon its operation. The kitplane manufacturer should supply the three-view; in fact, it's probably included in the instructions.

Upon successful completion of the signoff inspection, the inspector will sign this form and give you a temporary airworthiness certificate. More information on this process is given later in the chapter.

Radio Station License

If you're going to use a transmitter from your aircraft, you need to apply for a station license from the Federal Communications Commission (FCC). Note the operative word is *use* because even if the only radio is a hand-held transmitter, a license is necessary.

As an interesting aside, it's illegal to operate a hand-held aviation transceiver outside the aircraft or other facility. You can't just get a license for your person; it must be registered to some sort of station.

Anyway, the application is *FCC Form 404-2*, which is only available from FCC offices. Check the phone book under U.S. Government, or write to The Federal Communications Commission, Gettysburg, PA 17325.

Repairman's Certificate

Part of the reason you built the aircraft was to save maintenance costs, so don't forget to apply for a Repairman's Certificate. Submit an *FAA Form 8610-2* (Fig. 10-8), which is the same form A&P candidates use to apply for their certificates.

AIRCRAFT PREPARATIONS

The first flight requires exhaustive preparations to ensure the aircraft is ready. Critical elements must be checked and verified.

Fuel System

The best workmanship in the world won't help if the engine can't get any fuel. Building an airplane results in a lot of grit and garbage; need it be said that you don't want said garbage in your fuel system? Seventy percent of engine failures in homebuilts are caused by crud in the tanks and lines.

The elimination process begins during construction. Clean the tanks thoroughly after completion, and cover the inlets and ports to prevent entry of foreign material. The same applies to the lines, valves, and *gascolators*. Before adding gas to the tanks the first time, vacuum them out. Rig up a reducer on the end of your shop vac so you can slip a smaller hose into the filler opening.

U.S. Department of Transportation
Federal Aviation Administration

AIRMAN CERTIFICATE AND/OR RATING APPLICATION

☐ MECHANIC
   ☐ AIRFRAME
   ☐ POWERPLANT

☐ REPAIRMAN

_____
(Specify Rating)

☐ PARACHUTE RIGGER
  ☐ SENIOR   ☐ MASTER
  ☐ SEAT   ☐ CHEST
  ☐ BACK   ☐ LAP

APPLICATION FOR: ☐ ORIGINAL ISSUANCE ☐ ADDED RATING

I. APPLICANT INFORMATION

A. NAME (First, Middle, Last)

K. PERMANENT MAILING ADDRESS

B. SOCIAL SECURITY NO. | **C. DOB** (Mo.,Day Yr.) | **D. HEIGHT** | **E. WEIGHT**

NUMBER AND STREET, P.O. BOX, ETC.

F. HAIR | **G. EYES** | **H. SEX** | **I. NATIONALITY** (Citizenship)

CITY

J. PLACE OF BIRTH

STATE   ZIP CODE

L. HAVE YOU EVER HAD AN AIRMAN CERTIFICATE SUSPENDED OR REVOKED?
☐ NO
☐ YES (If "Yes," explain on an attached sheet keying to appropriate item number)

M. DO YOU NOW OR HAVE YOU EVER HELD AN FAA AIRMAN CERTIFICATE?
  ☐ NO ☐ YES
SPECIFY TYPE:

N. HAVE YOU EVER BEEN CONVICTED FOR VIOLATION OF ANY FEDERAL OR STATE STATUTES PERTAINING TO NARCOTIC DRUGS, MARIJUANA, AND DEPRESSANT OR STIMULANT DRUGS OR SUBSTANCES? . . . . ☐ NO ☐ YES ⟶
DATE OF FINAL CONVICTION

II. CERTIFICATE OR RATING APPLIED FOR ON BASIS OF —

☐ **A. CIVIL EXPERIENCE**   ☐ **B. MILITARY EXPERIENCE**   ☐ **C. LETTER OF RECOMMENDATION FOR REPAIRMAN** ((Attach copy)

☐ **D. GRADUATE OF APPROVED COURSE**
(1) NAME AND LOCATION OF SCHOOL
(2) SCHOOL NO. | (3) CURRICULUM FROM WHICH GRADUATED | (4) DATE

☐ **E. STUDENT HAS MADE SATISFACTORY PROGRESS AND IS RECOMMENDED TO TAKE THE ORAL/PRACTICAL TEST (FAR 65.80)**
(1) SCHOOL NAME | NO | (2) SCHOOL OFFICIAL'S SIGNATURE

☐ **F. SPECIAL AUTHORIZATION TO TAKE MECHANIC'S ORAL/PRACTICAL TEST (FAR 65.80)**
(1) DATE AUTH. | (2) DATE AUTH. EXPIRES | (3) FAA INSPECTOR SIGNATURE | (4) FAA DIST. OFC.

III. RECORD OF EXPERIENCE

A. MILITARY COMPETENCE OBTAINED IN ⟶
(1) SERVICE | (2) RANK OR PAY LEVEL | (3) MILITARY SPECIALTY CODE

B. APPLICANTS OTHER THAN FAA CERTIFICATED SCHOOL GRADUATES. LIST EXPERIENCE RELATING TO CERTIFICATE AND RATING APPLIED FOR. (Continue on separate sheet, if more space is needed)

DATES—MONTH AND YEAR		EMPLOYER AND LOCATION	TYPE WORK PERFORMED
FROM	TO		

C. PARACHUTE RIGGER APPLICANTS: INDICATE BY TYPE HOW MANY PARACHUTES PACKED ⟶

SEAT	CHEST	BACK	LAP	FOR MASTER RATING ONLY	PACKED AS A —
					☐ SENIOR RIGGER ☐ MILITARY RIGGER

IV. APPLICANT'S CERTIFICATION

I CERTIFY THAT THE STATEMENTS BY ME ON THIS APPLICATION ARE TRUE

A. SIGNATURE   **B. DATE**

V. I FIND THIS APPLICANT MEETS THE EXPERIENCE REQUIREMENTS OF FAR 65 AND IS ELIGIBLE TO TAKE THE REQUIRED TESTS.

DATE | INSPECTOR'S SIGNATURE | FAA DISTRICT OFFICE

FOR FAA USE ONLY

Emp.	reg.	D.O.	seal	con	iss	Act	lev	TR	s.h.	Srch	#rte	RATING (1)	RATING (2)	RATING (3)	RATING (4)

LIMITATIONS

FAA Form 8610-2 (2-85) SUPERSEDES PREVIOUS EDITION

Fig. 10-8. Apply for your Repairman Certificate with FAA Form 8610-2.

There'll still be some debris in the system. Your main fear should be the blockage of a fuel line or valve, as it can occur unseen and takes much effort to trace down and find it. Filters and screens make contaminants easy to find and eliminate, and ensure any particles making it through can't block the downstream lines or valves.

To start with, each tank should incorporate a finger screen inside the tank at the outlet port. These thread into a fitting on the base of the tank, and the fuel line then threads into them. The fine mesh catches the debris, and the long shape delays the point where the screen gets clogged.

Every aircraft must have a gascolator. It's the last redoubt against contamination reaching the carburetor or fuel pump. The gascolator also incorporates a quick fuel-drain to draw off any water or particles in the fuel. It's always installed on the engine side of the firewall, as low as possible. Any water will then tend to collect in it.

The finger screens and the gascolator are your main lines of defense against fuel starvation due to blockage. They should be removed and cleaned at least once before the first flight, and at gradually increasing periods thereafter. The finger strainers can only be checked when the tank is empty, so removal and cleaning must be carefully planned.

If air can't get into the tank, fuel can't flow out of it. Tank vents should be clear, too. Some can be cleaned with a short piece of wire.

Gravity-flow fuel systems require verification of flow rates. These systems rely upon *fuel head* (pressure resulting from the fuel tank being higher than the carburetor) for flow, and if there's insufficient fuel head the engine will be starved. Verification isn't necessary with engines using fuel pumps; it's only required for gravity-flow systems.

Fuel flow testing is done in the worst-case situation: Simulating a climb at full power with near-empty tanks. The fuel system must be capable of supplying 150 percent of the fuel required by the engine at full power. This information should be available from the engine manufacturer. Typically, the full-power rate runs between twice and three times the cruise power consumption.

The first step is to set up the aircraft in a climb attitude. This angle varies between aircraft, but it's typically more than the typical taildragger ground attitude. Even pulling a trigear's tail down to the ground might not produce the right climb attitude.

Roll the main gear up a ramp to sit atop platforms. Taildraggers are then ready; the tails of trigear airplanes must be pulled down and tied, or a block placed under the nosewheel. The kitplane manufacturer should have the required attitude, but the platform height will probably be around 12 inches or so.

Ensure the tanks are empty and disconnect the fuel line at the carburetor. The disconnected end must stay at the same general level as the carb; don't let it hang down. Tie in it place with wire or tape. Slip a piece of neoprene fuel hose on the end and drape it down to an empty fuel can.

Slowly add fuel to the tank under test until it comes out of the fuel line in a steady stream. Keep track of how much you've added; this is the value for unusable fuel in that tank.

Once you have a steady-state flow established, turn off the fuel valve, empty the drain can, and dump exactly one gallon into the tank. Position the newly-emptied can under the engine, stick the neoprene drain line into it, and turn on the fuel valve and start the timer.

When the can is filled to the start value, stop the timer. Compute the actual fuel flow, convert it into gallons per hour, and compare to the required value. Here's some sample values for one gallon:

1 minute	60 gph
2 minutes	30 gph
3 minutes	20 gph
4 minutes	15 gph
5 minutes	12 gph
6 minutes	10 gph

During this process, watch for leaks. The rubber used in some of the lines tends to age when exposed to air for a long period (like during construction) and can become brittle and crack. If so, replacement will be necessary.

Instruments

It's tough enough to make the first flight of a new aircraft without being faced with inoperative instruments. Altimeter failure wouldn't be too tasking. But what if the airspeed needle remains solidly on zero? How would that affect your landing? What if the oil temperature gauge is inoperative and the engine overheats?

You'd really like to check out the operation of your instruments before the first takeoff.

Operation of most of them can be verified in common-sense ways. The oil temperature gauge, for instance: Remove the sensing bulb from the engine and dunk it into boiling water. Or use water heated to a lesser degree with a candy thermometer alongside to provide a reference. Ditto for the coolant-temperature gauge for Rotaxes.

Ordinary hand vacuum pumps can test fuel and oil pressure gauges. But don't use them on pitot-static instruments because the pump is too strong. An FAA-approved repair station can check the entire system for leaks and accuracy. It's not a bad precaution, if you have the bucks. Don't spend $80,000 or more on your Cirrus and begrudge the $100 or so a pitot-static system check might cost.

But you can do some checks yourself. The airspeed indicator is easy, and a little fun, to test. Stick a couple of hoses on the pitot and static ports and go for a drive. Have a passenger hold the extended pitot tube out the window and check the reading of the instrument. Remember, the airspeed might not match the speedometer reading.

Similarly, an altimeter or VSI can be checked by going flying with someone else. Just hold the instruments on your lap and compare them to the aircraft's gauges. There'll be some differences because your units won't be connected to the aircraft's static system.

However, most homebuilt-aircraft problems arise in the pitot-static system itself. That's the advantage of the repair station's equipment because they check the whole system from end to end. It's difficult to get that degree of accuracy in the home workshop, but you can at least ensure the system reacts to pressure changes as it should.

The pitot system can be tested by slipping a piece of rubber hose over the end of the pitot tube. Roll up the end of the tube while a helper watches the gauge. This compresses the air in the tube and simulates the gentle pressures involved.

When the reading reaches the anticipated cruise speed, pinch the hose shut to hold the reading. The airspeed needle should hold constant for at least a minute. If it falls, there's a leak in the system. It could be due to the attachment of your rubber test hose, so check it first.

Checking the static system requires negative pressure. Block off all the static ports. Tightly roll up a couple of feet of neoprene tubing and connect the end to a static line. Unroll the tubing and watch the VSI. It should indicate a climb. The altimeter should rise as well. When it reads 2,000 feet or so, pinch off the tube. Again, the indication should hold constant for at least a minute.

Any leaks in the pitot-static system can be chased down with a small pressure source, but *disconnect all instruments first*. Plug the ends that go into the gauges, apply pressure to the other end of the line, and wipe soapy water over all connections. The bubbles indicate the leaks.

Testing vacuum-operated instruments can wait until the flight phase because they include none of your required VFR instruments. However, if you can arrange a source of sufficient vacuum, test these instruments during construction before the panel and cockpit are closed up. Problems will then be a lot easier to fix.

The aircraft electrical system and various masses of metal affect the magnetic field of the airplane. Aircraft compasses include small magnets that are adjusted to compensate for each particular installation. Perfect correction isn't possible, so errors are tabulated and listed on a correction card mounted near the compass.

The procedure for determining these errors is called *swinging the compass*. It must be performed after the engine has been broken in. The aircraft is aligned with the magnetic cardinal directions (north, east, south, and west) with engine and radios operating. Many fields have a compass rose painted onto the asphalt for exactly this purpose. If you've got a good orienteering compass you can tape lines on the asphalt for your own compass rose.

The differences between compass heading and actual heading are noted, and the N-S and E-W compensating magnets are adjusted as listed in the compass instructions. Once all possible error has been eliminated, the correction card is filled out.

You'd like to swing the compass with the aircraft in flight attitude. Easy to work with trigear planes, it's almost insurmountable for taildraggers. It's not that tough to prop the tail on a sawhorse or whatever, but running the engine in this position can be a bit dicey. If you can come up with a safe way to do it, great.

Controls

Controls must operate freely over their entire ranges. Any binding or scraping must be eliminated. Try to simulate flight conditions. For instance, have someone apply pressure to the surfaces while you move them with the stick.

Have the helper lift up a wingtip while the ailerons are moved. Twenty years ago, a small problem with a certain homebuilt design was discovered: When the wing started developing lift during the takeoff run, the structure pinched-down on the aileron torque tubes. The controls were free and easy on the ground, but hard to move in the air. Make every effort to ensure your controls are free and clear in any combination of attitude or loading.

Ensure all other controls work smoothly as well. Throttle, mixture, carb heat, flaps, brakes, and the like, should move freely throughout their ranges.

Make sure there's no interference between controls. Hold full left rudder, and move the elevator throughout its range. Repeat with the right rudder, then check the flaps and ailerons the same way.

Propeller Tracking

Each propeller blade should track along the same path; otherwise vibration and failure might result. Mistracking is usually caused by improper propeller installation or a bad prop. Tracking should be checked at some point prior to the engine's first start.

For safety's sake, remove one plug from each cylinder. Set the plane at flight attitude. Move the prop until the blades are vertical. Place a block of wood on the floor so that one edge is just barely touching the front of the bottom blade. Then rotate the prop until the other blade meets the block.

If it just barely brushes it, tracking is dead on. Otherwise, some gap exists between the second blade and the block or the blade brushes the block away. Measure the amount of difference. If blade tracking is off by more than $1/16$ inch, check the installation and contact the propeller maker.

ENGINE START

Your next actions can affect the aircraft's operation for years. Abuse the engine in the first few hours of operation, and it might use oil, run roughly, even fail prematurely.

First Start

There comes a time when the engine must be started for the first time. The FAA requires at least one hour of ground running before they'll allow the aircraft to fly. Don't pencil whip this requirement.

The first startup is not to be done alone. You want at least one other person around to help; preferably your EAA tech counselor or another experienced builder.

Your primary desire is safety. If you're starting it up at home, make sure the propeller area is blocked off to prevent dogs and kids from wandering into it. Keep the propellor plane clear as well, in case the prop shatters.

The area around the engine should be spotless and afford nonslip footing. It'll be necessary to examine the engine while it's running, so you don't want a slippery floor. Keep a bag or two of kitty litter handy to soak up any spills.

Don't fill up all the tanks, and keep stored gasoline away. Have fire extinguishers handy; make sure everyone knows where they are. You're most likely to need a Class B (burning liquids) certified extinguisher, but get the garden hose out as well for nongasoline or oil fires.

Ensure everyone has earplugs or hearing protectors because they are going to be working in close proximity to a very loud engine. Eye protection of some sort should be worn as well. At the very least, the propeller will be stirring up a lot of dust.

Make sure the oil tank is full, and that all engine controls are free. The engine manual should specify which type of oil should be used. Two-cycle engines might require a different oil/gas mix for break-in; check the manual.

Fill the coolant tank for liquid-cooled engines (Rotax 532/582 and auto conversions). Rock the plane fore and aft and side to side, to purge air from the coolant lines, tank, and radiator. Top off as necessary.

Go over the engine compartment, tightening anything you can fit a wrench to and making sure everything that should be safety-wired is. Check the mag timing one more time and make sure you've properly set the valves. Nothing should be loose in the engine compartment, even if it has nothing to do with the test run. Have the tech duplicate the whole examination as a double-check.

Three instruments are vital: an oil pressure gauge, an oil temperature gauge, and a cylinder head temperature gauge. The oil gauges should already be installed on the panel and should have been checked as mentioned earlier in the chapter. An add-on calibrated set should be used to compare their readings to. Because so many break-in operations are controlled by engine temperature, accurate gauges are vital. Liquid-cooled engines need a reliable coolant-temperature gauge as well.

You might or might not have a cylinder head temperature gauge installed. In any case, you'll want one for the testing phase, even if it's temporarily mounted. The temperature senders (thermocouples) must be solidly attached to the cylinder heads. Aircraft engines include a threaded hole for the thermocouple to screw into; these are called *bayonet mount*. Another kind is built into a spark-plug gasket. The gasket type is adaptable to practically any engine.

One issue that arises is how many to install. The best way is to add a thermocouple to every head and wire them into a selection switch. Many just install one sender, in this case, it should be mounted to the aftmost cylinder on the right side. Two-cylinder Rotax engines should mount the thermocouple to the rear cylinder. In any case, the sender should be removed and installed on other cylinders on a regular basis to make sure none are running hot.

There should be one person to monitor the gauges, and at least one other watching the engine. Tie down the tail and wings, and chock the wheels. Coordinate a system by which any person can signal for a quick shut down, even if it's just frantic waving and yelling. Considering the noise and the earplugs everyone should be wearing, perhaps whistles or an aerosol-can boat horn might not be a bad idea.

Some criteria for emergency shutdown include:

- Visible flames of any sort.
- Spraying liquids.
- Continuous smoke from other than exhaust (grease spots and the like might burn away, sending up momentary puffs).
- Moderate, continuous smoke from the exhaust (some black smoke might appear due to improper mixture, which is normal).
- Loose components, especially those threatening to come loose and foul the prop.
- Excessive vibration (some must be expected at first, but the engine should smooth out).
- Any other perceived safety problem (kids wandering too closely).
- Any instrument condition indicative of engine faults.

Ready to start.

Crank it by hand a few times to loosen it and distribute the oil. Make sure the mags are off, of course. Follow the engine manual's instructions for first start. Watch the oil pressure gauge like a hawk and if you do not have pressure within 30 seconds, shut the engine down, pronto.

The other problem to watch for is overheating. There are many causes, but one is easy to fix: Don't run an air-cooled engine very long with the cowling off. On any design where the cylinders *don't* stick out of the cowling (a la J-3 Cub) the cowling is a necessary part of the cooling system. Running without it isn't a bad idea for the first start or so, but most aircraft require them in place for break-in.

Break-In

The engine manufacturer should have supplied a break-in plan. The actual procedure varies. Usually the engine is revved up to a particular rpm, held for a specified time, then power is either increased to the next level or dropped to idle. The entire break-in might take a couple of hours total running time. Usually they can be separated into several sessions, but Rotax engines must be run to a particular rpm profile. If the sequence is interrupted for any reason, it must be rerun from the start.

Monitor the engine gauges throughout the process. Change the oil after two hours of accumulated running, and watch for the appearance of metal in the filter or strainer. Don't forget to log all times and maintenance operations in the engine log.

When the basic break-in procedure is complete, it's time to operate it under actual conditions.

GROUND TESTING AND FAA INSPECTION

With the engine apparently operating normally, the aircraft's ground handling can be investigated.

Taxi Testing

Taxi testing has two objectives. The first is the pilot's familiarization with the ground handling of the aircraft, and identification of any problems. The second is to predict the aircraft's trim during takeoff and landings.

During familiarization, you'll get a feel for how the runway looks with the aircraft on the ground, to be able to judge height during landings. Learn, too, how to control the airplane on the ground. Check the steering and brakes.

Don't forget to watch the engine gauges, to forestall damage from overheating or oil exhaustion. Operate the engine with full cowling; now is the time to find out if the cooling system needs modification.

Predicting the aircraft's trim requires some high-speed passes down the runway. Work up to a maximum of 80 percent of the predicted stall speed in 5-mph increments. Test the ailerons by rocking the wings. See how effective the rudder is. At 80 percent of stall speed, the elevators should be powerful enough to allow the aircraft to assume takeoff attitude: tail down in trigears, tail up in taildraggers. If it isn't, you might have a CG problem.

From the distance required to reach 80 percent of stall, predict where the actual takeoff point will be. Run a simulated takeoff abort, measure the distance required, and mark it off from the opposite end of the runway.

One of the slight drawbacks of high-speed taxi testing is the, well, let's call it the "unfortunate" tendency to leave the throttle forward until the plane "accidentally" leaves the ground. And because, by golly, you're unsure how much runway it'll take to stop, well, then, "naturally" it's better just to go around.

FAA inspectors have heard it all before. Believe me. *The aircraft isn't authorized for flight until the application for the airworthiness certificate is approved.*

The Inspector

You have two options for persons to perform the airworthiness inspection: a genuine *FAA inspector* from a local Engineering Manufacturing District Office, or an independent *Designated Airworthiness Representative* (DAR).

The FAA inspector is free, but is usually booked months in advance, and a DAR is available, but charges perhaps $200.

DARs are essentially trusted agents of the FAA; they have passed rigid requirements and have years of experience with small aircraft. They are often A&P mechanics. Having your airplane inspected by a DAR isn't getting the second team. They wouldn't be trusted if they couldn't do the job.

But they are kind of expensive. Not only must you pay for their inspection, but for travel and insurance as well. There are ways to reduce this cost, for instance, move your kitplane closer to the DAR's work location to reduce his travel time. Builders with folding-wing kitplanes have been known to show up at the inspector's door and set up their airplanes in the parking lot.

The EAA recommends you arrange for an inspector at the time the project is started. This way you can discuss changes and options with the inspector prior to implementing them.

The Flight Test Area

The inspector sets the flight test area. The main concern is the protection of the public; hence, he might require that the initial flight testing can be performed at some out-of-the-way airport.

In my home area, a lot of local builders are based out of a small airpark on the outskirts of town. The runway is nothing more than a 3,000-foot gash in the trees with a power substation to the south, and housing developments strewn in all directions. Instead, the FAA usually specifies the first 10 hours of flight testing be performed from an ex-Air Force field 60 miles to the north. The runway is 5,000 feet long and it's in the middle of farmland.

In my case, it takes 80 minutes to drive to the airport instead of five. It's awkward. I don't like it. But if the engine fails after takeoff. . . . In any case, the FAA has been very reasonable about switching to the local airport after the aircraft has proven itself.

As far as the flight test area is concerned, the aircraft is generally restricted to a 25 mile radius of the test airport. This area is modified by cultural and terrain features and you might only get half the circle due to surrounding cities. Or the circle might bulge a bit to allow testing over a lake or other suitable area.

If you have legitimate reasons to change the shape or enlarge it, speak up. A fast airplane like the Glasair III can cross a 25-mile test area in five minutes; hardly enough time to get into stable cruise flight. You might negotiate a corridor instead of a circle.

The flight test period is either 25 or 40 hours, depending upon the engine-propeller combination. If you have an approved combination—an exact model engine as used in a particular certified aircraft, combined with the exact model propeller used in that aircraft—the test period can be set to 25 hours. Some DARs will assign fewer than 25 hours for a licensed homebuilt with ultralight-like characteristics. A Sorrell Hiperlight, for example, was assigned a 10-hour test period.

The flight test period is referred to as the *Phase I operating limitations*. The inspector will also assign *Phase II limitations*, those which you'll have to abide by after the test flight period. In most cases, the limits are those set by FAR 91.42: No operation for hire, no operation over congested areas or in congested airways (except as specifically authorized and/or for takeoffs and landings).

Neither the FAA nor the inspector keeps track of how many hours you've put on the aircraft. When you reach the required number, make a log entry to that effect and the aircraft is cleared to Phase II limitations.

Keep in mind that the numbers quoted above are guidelines, not set by regulation. The inspector certainly won't assign *less*. But he or she might assign more.

For instance, if the aircraft features some questionable design elements, the inspector might restrict the aircraft to the immediate proximity of the airport. Or if you are using a *real* off-the-wall powerplant, he might decide that 40 hours aren't sufficient, and double the period of your Phase I operating limitations.

It's the inspector's responsibility. Once the safety and reliability of your kitplane are proven, there shouldn't be any problem getting your limits extended.

The Inspection

How can someone examine a completed aircraft and detect all problems or errors?

It's not possible. It used to be a little easier back when an inspection was required prior to closing up the interior (covering with fabric, riveting the skins, etc.). But nowadays, the FAA merely recommends periodic inspections by knowledgeable individuals, such as EAA technical counselors and A&P mechanics.

It's important to understand the purpose of the inspection. It's not to determine if the aircraft is safe. It's to ensure that it meets the regulations governing the flight testing of experimental aircraft. The safety of the aircraft will be proven in the air; it's the inspector's job to ensure adequate verification before turning the aircraft loose on the airways.

One major area is to check that the plane meets all regulations applicable to experimental aircraft, and has the following:

- Basic VFR instruments, all marked with appropriate operating ranges (redlines, max/min pressure or temperature, etc.).
- All controls labeled with function and action (ON/OFF signs for the switches, OPEN/CLOSED for the throttle, COLD/HOT for the carb heat, etc.).
- Properly registered, with the N-number applied according to regulations.
- A metal identification plate listing the builder's name and address, model designation, builder's serial number, and date of manufacture.
- A plate visible to both occupants, explaining the experimental nature of the aircraft. (The PASSENGER WARNING placard available from EAA.)
- The word EXPERIMENTAL in letters at least 2 inches high, visible near each entrance to the cockpit.
- A plate near the tail, giving the manufacturer's name, the model, and serial number of the aircraft. (DEA plate: The builder's plate is adequate, if mounted near the tail.)
- Seat belts and shoulder harnesses installed for all occupants.
- Current and correct weight and balance paperwork.

If you lack any of the above, the inspector will not sign off your aircraft.

The inspector will thoroughly examine the aircraft. He's not looking for basic flaws, although there will be no hesitation to point any out. Rather, the inspection is aimed at the immediate airworthiness: all pins installed and safetied, bolts head-up and forward, cables properly tensioned, and the like.

He'll also want to examine the plans/instructions, your builder's log, and the photos you took during construction, to satisfy the issue as to who built the aircraft. Have your receipts standing by, in case the inspector asks to see them. Be prepared to explain any deviation from the plans.

The inspector will probably find a few things wrong. Most of them will be minor and can be corrected on the spot. Have your tools handy.

Once the inspector is satisfied, he or she will sign your log and airworthiness certificate. The aircraft is cleared to fly.

How to Keep Your FAA Inspector Happy

The airworthiness inspection isn't an IRS audit. You and the inspector are on the same side: You want to fly a safe airplane, and he wants the airplane you fly to be safe.

- Have your EAA technical counselor inspect the plane before setting up the FAA appointment. Correct all the deficiencies he finds prior to the inspector's arrival.
- Have the aircraft ready when the inspector arrives: in a hangar, with all inspection panels removed.
- Document all deviations from the plans in your builder's log.
- Use aircraft-quality materials throughout. Technically, the aircraft is experimental and doesn't have to use approved materials. But overreliance on such materials might make the inspector question the airworthiness of the aircraft as a whole. Aircraft-quality parts and hardware tells the inspector you're more interested in safety than saving money, and that's the attitude he wants to see.
- Be prepared to compare any section of the aircraft with the plans. If the inspector doesn't like the design of a particular area, you'd like to prove that it was constructed per the instructions. If the change was your idea, be prepared to justify the redesign.
- Make sure all your taxi tests and engine runups are documented in the appropriate log. The inspector will want to see log entires equaling at least one hour of engine operation.
- Have a flashlight or trouble light handy.

PILOT PREPARATIONS

The airplane's ready. Is the pilot?

All too often, homebuilders stop flying during the construction process. And far, far, too often, the aircraft's first flight is the builder's first flight in months. If not years.

If the aircraft is an easy flier, and if everything works, both the aircraft and the pilot might come out of it in one piece. The decision is the pilot's. I suppose he or she has a perfect right to take those kind of chances. But I hate seeing perfectly good aircraft dinged up.

A pilot probably doesn't lose the basic skills over a long layoff. I quit flying for seven years at one point, then bought a 150. An hour and a half of dual later, the instructor turned me loose. Flying an airplane is like riding a bicycle, except it's harder to put playing cards in the spokes.

But any lapse in skill results in more than skinned knees. And while the primary stick-and-rudder talent might not fade very much, the reflexes for handling emergencies aren't there any more. Plus you've probably lost the ability to do multiple things simultaneously.

Example 1: You take off carefully, mindful of every motion of the aircraft. You're totally concentrated on control during which, unnoticed, the oil pressure drops and the temperature rockets. The engine seizes on downwind. Forced landings are hard enough for those who fly regularly.

Example 2: The pitot line vibrates free of a connection soon after takeoff. The airspeed indicator drops to zero. A current pilot is less likely to dive into the ground chasing an inoperative gauge, and has a better chance of executing a safe landing.

Example 3: You're intent, flying smoothly, and monitoring the engine instruments. You neglect your navigation and wander into a restricted area or TCA, and the FAA pulls your license for 60 days.

Even if you manage to survive the test flying period, your less-than-skillful handling of the aircraft can result in extra wear and tear.

There are two solutions: Finding someone to fly the initial tests, or getting current yourself.

Who Ya Gonna Call?

Who do you trust to make the first flight of your pride and joy? Or perhaps more to the point, who trusts *you* enough to make the first flight in your homemade airplane?

Your preferred choice would be someone who has a lot of hours in the same type of aircraft. Not only would they be familiar with any handling quirks the type possesses, but they'd be able to quickly tell you where your plane is deficient. If you've had a mentor through the construction of your plane, he or she might consent to making the first flight.

Experienced flight instructors are another good source. Flying skills are second nature, and they practice emergency procedures on a daily basis. You'd prefer one with experience in a wide variety of aircraft. Don't be insulted if they send someone to inspect the aircraft before they agree to take on the job.

Or does your EAA chapter have its equivalent of Chuck Yeager? My chapter's tech counselor specializes in first flights. He's not an ex-military pilot, nor does he have a commercial certificate or even an instrument rating. What he has is 30 years experience in building, inspecting, and flying homebuilt aircraft. He's made the first flight on Dragonflies, Fly Babies, T-18s, Jodels, and a host of other aircraft.

Pilots from any of these three categories can do an adequate job of test piloting. Or they might not be up to the challenge. You're taking a chance, as well. The best bet would be someone who fits all categories: An experienced instructor who has built the same type of aircraft and has flown the test programs for variety of homebuilts. Lotsa luck.

Doing It Yourself

In most cases, the builder decides to make the first flight himself. This isn't a bad decision, as long as the pilot is legal and ready.

Legal means two things: current medical and current biannual/annual flight check. The inspector might request proof when he or she signs off the aircraft.

As a private pilot, you can't carry passengers unless you've made three landings in the last 90 days. But you can't carry passengers during your test period, anyway. So you can legally make the first flight as long as you have a current biannual flight review (or annual, if you have fewer than 400 hours). Legal maybe. Smart, no.

There are three phases to first-flight preparation. The first is knocking the rust off your basic skills, which are probably adequate, but a refresher on basic airmanship will lay a good groundwork for the next phases. Stall recognition and recovery should be extensively covered.

The second is preparation for the special challenges presented by your kitplane. A few bump-and-goes won't prepare you for flying a Venture or White Lightning. Try to get some stick time on the same type, ride right seat in someone elses' Glasair, for example. If one isn't available, take some dual in the closest available production equivalent. Some recommend flying a variety of aircraft, just to become accustomed to differences. The nimble handling of the Grumman Trainer (Fig. 10-9) is similar to many light homebuilts. These aircraft are still rented by a variety of clubs and FBOs.

The classic example of this type of training is the conventional-gear checkout. Most of the smaller ARV-class aircraft are taildraggers. Taildraggers aren't really harder to fly, they do not tolerate inattention during takeoff and landing.

If you've never flown a taildragger before, count on needing 5 to 10 hours of dual. It can be combined with the dual needed to brush up your basic skills, so

Fig. 10-9. A few hours of dual in one of the Grumman TR-2/Trainer/Lynx series is good training for flying a kitplane. After all, it was originally designed as a homebuilt.

it's not an *additional* 10 hours. If you've got some Cub or Champ time in your history but nothing recent, an hour or two of refresher will probably suffice.

The last phase of preparation is training to handle the various emergencies that might occur. Even if current, any pilot can benefit from emergency drill.

Most of this should be done dual; again, it can be made part of the general refresher course. Simulated engine failures in all modes of flight are one aspect, of course. Practice aborted takeoffs on long runways with a marker or runway light simulating the real end.

Rigging problems can be duplicated by running elevator and rudder trim to the limit. Have the instructor block use of rudder or ailerons to simulate control jamming, especially during maneuvers. Control problems like these should only be practiced at altitude; don't risk an aircraft down low.

Take some aerobatic dual, if available. Concentrate on spins and unusual attitude recovery. Make some flights from the test airport, if it's one you aren't familiar with. Make note of the landmarks and possible emergency landing fields.

Finally, practice emergencies while sitting in the kitplane. Can you find each control and switch without looking at it? Can you get out quickly if you need to? Try it with a parachute on because they can make a difference. If it's a retractable, jack it up and practice emergency extension.

Develop emergency checklists and commit them to memory. Repeat them aloud while sitting in the cockpit performing the actions.

You might not do everything I've outlined. And you'll probably be OK. But all this isn't just to get your reflexes tuned. It's mental preparation as well. You are soon going to be in a very scary situation: A first flight.

You've been through it once already, on the day you soloed. Preflight jitters are alleviated (but not eliminated) by *confidence*. Confidence in the airplane, and confidence in yourself. Your first solo flight occurred in an airplane you had a number of hours on. The instructor probably rode through a couple of touch and goes before turning you loose. No question that the aircraft was in tiptop shape, and that you were fully capable of flying it.

It's a similar experience for the first flight of a homebuilt. You've been working on the aircraft for months or years. The plans were followed religiously. Your EAA tech counselor visited several times and checked your work. Weight and balance are spot-on. The FAA inspector actually *smiled* as he signed off the paperwork. The engine purrs like a lion cub. The plane should be all right.

Which leaves *you*. You've flown for 10 hours or more in the past month. The instructor failed the engine more times than you can count. You know how the plane is supposed to fly, because you got a few hours practice in someone else's. You can reach all the controls blindfolded and half-asleep. You know how to handle jammed controls, engine fires, stalls, spins, the whole panoply of aviation disasters.

With practice comes skill, and with skill comes confidence.

You'll still worry because you're a good pilot, and good pilots *know* they can't anticipate everything. But you'll be eager, too. Ready to take on whatever the plane has to give.

I'll be darned. You're ready to fly.

THE DAY

Don't tell anyone!

You need your tech counselor or another experienced builder to help check the plane. Call in someone else to record the great event in pictures or on video-tape, if you like. But don't invite the world.

The problem is the space shuttle Challenger Syndrome: The feeling you *must* fly, because everyone is expecting you to fly. So you ignore the crosswind: shrug off the rough-idling engine: decide to fly a high-speed pass to show off the plane.

You must be willing to scrub the flight at any deficiency. A bunch of kitbitzers complicates matters. There's the human tendency to please the crowd. And the 10-year-old inside you doesn't want to be called chicken.

So do yourself a favor and keep the flight time secret. This can get a bit uncomfortable. You might have to round up a bunch of volunteers to unload your kitplane at the airport and assemble it (Fig. 10-10). They'll all want to know when the first flight will be. But for own protection, decline to name the specific date. If they are EAAers, they'll understand.

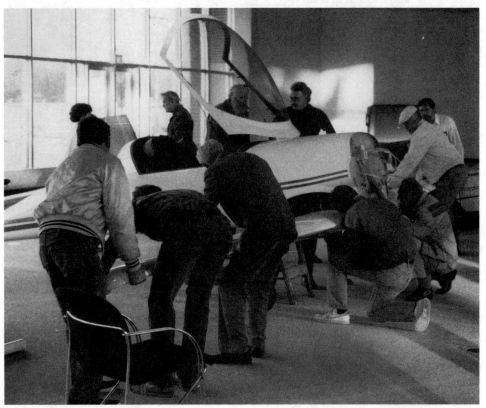

Fig. 10-10. It's hard to keep an upcoming first flight a secret, especially when it takes half an EAA chapter just to assemble your aircraft.

The flight is often made immediately after the FAA inspection. The inspector might hang around, but it isn't required that he witness the event.

Everything should be perfect: weather, equipment, pilot readiness and condition. Weather won't be perfect, of course. But you'd like a reasonably high ceiling and light winds straight down the runway.

Inspect the plane carefully. Enough gas should be carried for two hours' worth of flight. Oil and coolant should be topped off. Work the controls; make sure they are not binding. Set the flaps and trim to the takeoff position. Have the counselor duplicate your inspection, to be on the safe side. Try to have the CG towards the forward limit.

Run the engine for 10 minutes or so and ensure that full power is available (Fig. 10-11). Make sure all latches work properly. Phone the tower, and let them know you'll be making a first flight. No, you *don't* want the equipment standing by; just let them know you might need expedited handling if any problems occur.

Fig. 10-11. Everything should be perfect before the first flight, especially the engine. Jim Smith listens carefully to the Revmaster VW conversion on his Sonerai.

(By the way, Figs. 10-11 through 10-13 are from the actual first flight of a Sonerai II owned and flown by Jim Smith, Auburn, Washington.)

You should be well rested and alert. Eat a snack if hungry, but no large meals. Have the flight well planned from takeoff to landing. Anticipate the worst. Mentally review your reactions to various emergencies. Wear leather shoes (preferably boots), not sneakers. Borrow a Nomex flight suit if possible; if not, wear natural fibers, not synthetics. Check with any motorcycle-riding buddies for a helmet loan.

Deciding whether to wear a parachute is often the hardest part. Is it possible to bail out of your aircraft? Is it a pusher? Can the doors be jettisoned? Can seat cushions be removed to allow you to sit normally? Can you egress quickly with a chute on?

You might not have to buy a chute because your tech should know where you can borrow or rent one. Often, all you'll have to pay is just the rigger's fee for routine repacking.

Eventually, everything's ready. Fire up the engine and taxi to the runway. Run up the engine one more time, and check the controls. Roll into position. Treat the exercise like a high-speed taxi test. Feed the power slowly. Keep to the centerline. Listen to the engine. Feel the controls. Abort at the first sign of trouble.

Rotate. Watch the attitude as the plane leaves the ground because overcontrolling is common (Fig. 10-12). Move the controls deliberately to prevent *pilot induced oscillation* (PIO). PIO is the fancy name for overcontrolling; if the nose is too high, you shove it down hard, then the nose is too low, and you pull it back up even harder. Eventually, you're a half-cycle behind the airplane. Make all inputs gentle and gradual.

Fig. 10-12. First takeoff, Sonerai N227S. The aircraft is slightly nose high because the sensitive controls took Smith slightly by surprise. Rather than overreacting, he carefully stopped the pitch-up and eased the nose back to a more normal climb attitude.

Climb out at the manufacturer's best rate of climb speed or 1.3 times the predicted stall speed. Don't touch anything until the aircraft has climbed 1,000 feet. Keep the gear down. Don't throttle back, unless an engine problem arises.

Once you have some altitude, fly to the test area and execute the test plan. Basically, the first flight consists of controllability, trim, and engine checks. You'll check control pressures, rigging, and engine operation. You won't actually stall the airplane, but you'll practice slow flight and approaches to landings.

Fig. 10-13. All smiles after a 30-minute test flight within the confines of the airfield. Contrast Mr. Smith's expression in this picture with Fig. 10-11.

Don't plan on staying up all day. A 30-minute first flight is plenty. Your andrenalin will be a'pumping, and fatigue sets in quickly. At any sign of aircraft problems, start heading back. Keep your eyes open for emergency landing fields.

Return to the airport. All you have to do is one good landing. Fly your approach at about 1.3 to 1.4 times the stall speed. Go around if necessary, more than once, if required.

The plane might handle a bit different in ground effect, so be prepared. It's not necessary to do a perfect landing; don't get into PIO trying to grease it on.

Taxi back, shut down, climb out, and have your picture taken (Fig. 10-13).

Congratulations!

You no longer have a kit, you have an airplane.

The above is merely an abbreviated description of your first flight activities. FAA Advisory Circular AC-90-89, "Amateur-Built Aircraft Flight Testing Handbook," developed in cooperation with the EAA, covers homebuilt testing from first flight and beyond. It can be ordered from the U.S. Government Printing Office, Washington, D.C. 20402. Cost is about $2.50.

11

Get 'em Flying

I MET AN INTERESTING REACTION while working on this book. Many kit-builders were reluctant to have pictures of their projects taken for publication. My photo sessions were punctuated with "Don't shoot that!" and "Let me move that piece out of the picture." One builder was aghast when I wanted to take a picture of his portable workstand: "That's just something I threw together. That's not a normal piece of hardware!"

Well, Merrill, it's in chapter 4.

We homebuilders are no different than anyone else. Articles about grand champions give us an inferiority complex: all those pictures of homebuilts with acres of avionics, laser-sharp paint, and gadgets galore.

So progress slows. We're just a little embarrassed about that crooked seam, or little bit of fiberglass weave showing, or the fact that we're building the simpler fixed-gear version instead.

Build the airplane that's right for you. Don't pick a plane because it's the latest thing. Know your mission, and buy a kit that will support it.

And it's not called the amateur-built category for nothing. Expect to make mistakes, and plenty of them. You're getting an education. Expect a few B minuses among the As. Airworthiness should be your top priority—don't go overboard on the accessories and the cosmetics. Airplanes are meant to fly. I'd rather see one average-quality homebuilt in the air than look at five future grand champions in the workshop.

Let's get'em flying!

Building a kitplane is not an easy task; nor is it a quick one. There's a lot more to it than slapping some glue on some parts and wrapping them with rubber bands.

Arduous. Yes.

Difficult? At times.

Impossible? Not hardly.

Thousands of men and women built their own aircraft even before the advent of the kit.

But there will be times when it overwhelms you. The enormity of it. The apparent slow progress. The stabilizer you glued on upside down.

One cure for the homebuilder's blues is your local chapter of the Experimental Aircraft Association.

This is not another plug for EAA. I have no official connection, other than being a member of two very fine chapters.

But within the chapters you'll find support: Somebody else building the same airplane to help you through unclear portions in the plans: A technical counselor with advice on problems: Leads toward finding a low-cost engine: Willing hands and strong backs to help move projects from the garage to the airport.

Building an aircraft, even a modern kitplane, has never been a one-person job. Even if only one pair of hands touches the tools, a successful homebuilt depends on cooperation and coordination between the designer, kit manufacturer, FAA, EAA, and the builder. Your local chapter can help you tap into a mother lode of assistance and support. Contact national EAA headquarters:

Experimental Aircraft Association
EAA Aviation Center
Oshkosh, WI 54903-3086
414-426-4876 (for chapter information)

Appendix

Selected Kit Aircraft

THE FOLLOWING TABLES PRESENT basic data for a variety of aircraft kits. This is not, by any means, a comprehensive list—there are many more kits on the market than are listed. The intent is merely to compare characteristics between a number of aircraft.

Every December, *Kitplanes* magazine includes a complete directory of kit and plans-built aircraft. Issues may be ordered by writing:

Kitplanes Back Issues
P.O. Box 6050
Mission Viejo, CA 92690

All dimensions are approximate. Obtained performance might vary. Unless otherwise stated, aircraft has two seats, fixed gear, and is available in a complete kit. Most Rotax-powered aircraft include the engine with the kit. The Rotax 582 engine was just released at press time; it is expected that many kits using either the Rotax 503 or 532 will be converting to this engine.

COMPOSITE

Name	Manufacturer	Engine	Length (Feet)	Span (Ft)	Cruise (MPH)	Range (SM)	Useful Load (Lbs)	Comments
Cirrus VK30	Cirrus Design Corp.	Lyc TSIO-540	26	39	250	1300	1150	Four seat pusher.
Dragonfly MK1	Mosler	HAPI VW	20	22	175	550	540	
Glasair IIFT	Stoddard-Hamilton	Lyc O-360	19	23	231	1400	650	Retractable
Glasair III	"	Lyc IO-540	21	23	282	1550	900	Retractable, also version using Lyc O-235
Lancair 320	Neico Aviation	Lyc O-320	21	24	230	1100	685	
Pulsar	Aero Designs	Rotax 532	20	25	120	400	440	
Sea Hawker	Aero Composites	Lyc O-320	22	24	150	1200	750	Biplane amphibian
Seawind 2000	Seawind International	Lyc IO-360	27	35	150	1300	1150	Four seat amphibian

METAL MONOCOQUE

Name	Manufacturer	Engine	Length (Feet)	Span (Ft)	Cruise (MPH)	Range (SM)	Useful Load (Lbs)	Comments
Capella XS	Flightworks Corp.	Rotax 532	18	27	92	654	300	Also single-seat version
Moni-ARV	Mosler	KFM 107	15	28	110	320	340	Single Seat
Prowler	Prowler Aviation	Buick V-8	21	25	200	1200	750	Retractable
Sonerai II	Mosler	VW	19	19	155	350	650	Tube-type fuselage
RV-4	Van's Aircraft	Lyc O-320	20	23	186	650	590	Also RV-3 & RV-6
Venture	Questair	Cont. IO-550	17	28	276	1000	740	Retractable
Zodiac	Zenair	Rotax 912	19	27	105	420	500	Has several designs for various engines

TUBE-AND-FABRIC

Name	Manufacturer	Engine	Length (Feet)	Span (Ft)	Cruise (MPH)	Range (SM)	Useful Load (Lbs)	Comments
Avid Flyer	Avid Aircraft	Rotax 532	17	30	85	350	431	Wings fold, multiple variants
Avid Amphibian	"	Rotax 532	19	36	75	430	600	Three-seat ampib., wings fold
Carrera 180	Advanced Aviation	Rotax 532	22	31	60	650	540	Several designs available
Christen Eagle	Christen Industries	Lyc AEIO-360	19	20	165	380	450	Competition biplane
CIRCA Nieu. 11	Leading Edge Air Foils	Rotax 503	16	22	80	200	325	Replica biplane, 1 seat, Materials Kit
CIRCA Nieu. 12	"	Rotax 532	22	27	70	180	450	Replica biplane, Materials Kit
Drifter MU 532	Maxair Aircraft	Rotax 532	19	30	80	150	540	Single seat versions available
Hiperbipe SNS-7	Sorrell Aviation	Lyc 1O-360	21	23	160	500	650	Negative-stagger biplane
LM-2X-SP	Light Miniature A/C	Rotax 503	18	32	75	250	425	Several designs available
Kitfox	Denny Aerocraft	Rotax 582	18	31	85	260	530	Wings fold
Mark II	Kold Company	Rotax 503	22	30	65	700	430	Wings fold
N-3 Super Pup	Mosler	Mosler MM-CB	17	26	80	533	300	
PT-2	ProTech	VW	18	32	100	275	550	
Renegade Spirit	Murphy Aviation	Rotax 532	19	21	80	250	460	Biplane
S-10 Sakota	RANS	Rotax 532	18	23	100	250	475	

WOOD

Name	Manufacturer	Engine	Length (Feet)	Span (Ft)	Cruise (MPH)	Range (SM)	Useful Load (Lbs)	Comments
Classic	Fisher Flying Prod.	Rotax 532	17	22	85	225	450	Biplane
Falco	Sequoia Aircraft	Lyc IO-320	21	26	190	1000	680	Retractable
5151 Mustang	Leohle Aviation	Rotax 503	23	27	75	325	325	Retract. gear avail.
KR-2	Rand-Robinson	VW 2100CC	13	21	180	700	420	Wood structure, composite exterior
MiniMAX	TEAM, Inc.	Rotax 277	16	25	55	120	235	Single seat ultralight
Super Kitten	Hipp's Superbirds	Rotax 447	16	28	70	120	280	Single seat

Glossary

AC 43.13 FAA Advisory Circular "Acceptable Methods, Techniques, and Practices—Aircraft Inspection and Repair."

aircraft manufacturer According to regulation, the builder of the kitplane (you) is considered the manufacturer. *See* KITPLANE MANUFACTURER.

aircraft-quality materials Those components that meet federal quality and characteristics standards.

ALCLAD A coating of pure aluminum applied to the surface of an aluminum alloy as an anticorrosion method.

alloy Description of the makeup of a metal. Usually a numerical code, 4130 Steel, 2024 Aluminum.

AN Army-Navy (Standard).

ARV Air Recreational Vehicle. Generally means an aircraft built using ultralight design methods but certified in the experimental category.

band saw A floor or table-top power tool that cuts using a toothed band.

bastard Calm down. It's really a descriptive term for a type of file.

bias Term used in describing the direction of the weave in cloth, including fiberglass.

bid Slang term for bidirectional fiberglass cloth. It has nearly equal strength in all directions.

bonding agent *See* RESIN MIXTURE.

brake, bending brake Tool for making long, even, bends in metal sheet.

bucking bar Tool used to make the shop end of driven rivets.

bulkhead Vertical sheet (metal, composite, or wood) that provides structural rigidity to the fuselage.

bungee cord Long strips of rubber wrapped in cloth. Essentially, a big rubber band. They're used for shock absorption in small aircraft.

burr A rough protrusion of metal. Equivalent to a splinter in wood.

cad, or cadmium Anticorrosion treatment for metal hardware.

capstrip Flat piece of wood around the periphery of a wing rib.

chamfer A sharp 90-degree angle converted into two 45-degree angles with a narrow flat section between.

chip chaser Tool for removing burrs between metal sheets.

chro-moly Slang term for 4130 steel (Chromium-Molybdenum steel).

chisel Tool with a wedge-shaped edge, used in shaping wood. A cold chisel is specially hardened for cutting metal.

cleco A temporary fastener, inserted from one side using special pliers.

clevis pin Thick metal pin that closes a fork or shackle.

composite Building material consisting of resin-soaked fiberglass layups with foam inserts to add strength.

compression load A force that tries to collapse an object, like a brick atop a soda straw.

compression strut Structural element designed to withstand compression loads. Usually found between wing spars.

cotter pin Small piece of doubled wire that acts as a safety device to prevent a bolt, clevis pin, nut, or other object from accidental disengagement.

countersink Tool for cutting shallow cone-shaped depressions. Used to prepare metal for flush riveting and to eliminate burrs from drilled holes.

DAR Designated Airworthiness Representative. Person authorized to perform FAA airworthiness inspections.

deburring tool Device to smooth away burrs. *See* BURR.

diagonals Structural element on truss-type fuselages, running diagonally between longerons.

dimple Shallow depression. Can be applied by a dimpling tool for dimpling die to prepare sheet for flush riveting.

draw filing Very long file strokes made by pulling the file along the edge of the metal. Described in chapter 5.

drift punch A thin rod of metal of a given diameter, used to remove jammed rivets, bolts, and the like.

driven rivet A rivet where the shop head is made by ramming the rivet shank repeatedly into a bucking bar.

dry micro Very thick mixture of glass bubbles and epoxy/vinylester, used for rough surfaces. A ratio of about 5-to-1 bubbles to resin.

edge margin Minimum spacing for holes. No closer than three times the diameter between holes, and two times the diameter between a hole and the edge of the metal.

emery cloth Special kind of sandpaper. Available at any hardware store.

epoxy Modern two-component (resin and hardener) bonding material.

etchant Mild acid solution used to prepare aluminum for painting.

fuel head Amount of fuel pressure developed at the carburetor due to the distance of the tank above the engine.

exotherm Heat-producing chain reaction caused by excessively large mixtures of epoxy or vinylester.

eye, cable eye Loop in the end of an aircraft cable.

fabric Cloth used to cover the skeleton of an aircraft.

fairlead Device used to redirect or support aircraft cable.

fill yarns Threads that hold the weave together on unidirectional fiberglass cloth.

fill Orientation of the weakest axis of a woven material. Most commonly used in composite construction.

fluting pliers Tool for applying special dent on the edge of metal.

former Non-structural component that defines the exterior cross-section of the fuselage. STRINGERS run between formers to define the three-dimensional shape.

4130 A steel alloy commonly used in aircraft.

grain Direction of primary strength.

green cure A state where resin-soaked fiberglass can easily be trimmed with a razor blade or other sharp knife.

grip length Unthreaded area on shank of bolt.

hardware-store parts Components that aren't certified as having met federal standards (AN, NAS, MS, and the like).

head Portion of rivet or bolt that acts as a stop to prevent the item from passing completely into a hole. *See* SHOP HEAD and MANUFACTURED HEAD.

hot wire cutter Device that slices styrofoam through the use of a heated wire.

jig A structure that holds aircraft components while they are being attached.

jig saw A hand-held tool with a reciprocating saw blade. Also called a saber saw.

joggle Slight discontinuity in the edge of mating parts which produce a smooth, sturdy seam when joined.

kitplane manufacturer The company which produces the kit for the aircraft. *See* AIRCRAFT MANUFACTURER.

laterals Structural element on truss-type fuselages, running sideways between longerons.

layup A single layer of fiberglass and associated resin. Single layups are rare; most require multiple layers of cloth applied sequentially.

line oil Preservative mixture used in corrosion-proofing steel-tube fuselages.

longeron Primary fuselage structural elements for truss-type designs. Form the corners of the fuselage box.

manufactured head The head the rivet comes with. See SHOP HEAD.

micro Shorthand for microspheres, an obsolete material. The term is also used to refer to its modern replacement, glass bubbles.

MS Military Standard.

NAS National Aerospace Standard.

Nicopress System for attaching fittings to aircraft cable, developed by the National Telephone Supply Company.

ovaling When friction or other forces change a round hole into an oblong shape.

peel ply Dacron cloth used to smooth the surface of a fiberglass layup.

pop rivet *See* PULLED RIVET.

pressure cowling A cowling that completely encloses the engine, and hence must include baffling to route cooling air past the cylinders.

pulled rivet A rivet that can be emplaced from one side, using a tool to pull a mandrel to distort the shop end.

Q-cells Tiny glass bubbles.

ragwing Slang term for a fabric-covered airplane.

radius The distance from the center of a circle to the edge. Also refers to the rounding off of a sharp corner.

resin mixture Shorthand used within this book for composite bonding materials in the ready-to-use condition, also called bonding agent. For epoxy operations, it refers to the appropriate mixture of resin and hardener. In the vinylester world, it means resin with catalyst added.

rib Wing structural element that defines the airfoil shape.

rib stitch Wrapping cord around a wing rib to hold the fabric in place.

rivet set Tool that transmits hammering of rivet gun to the rivet itself.

rivet A piece of metal stuck through a hole between two objects and distorted to hold the parts together. *See* PULLED RIVET and DRIVEN RIVET.

saber saw *See* JIG SAW.

scarf joint Method of joining two pieces of wood end-to-end.

set *See* RIVET SET.

shackle U-shaped piece of metal for attaching a cable eye to structure. The shackle uses a CLEVIS PIN to close off the open end.

shank Portion of bolt or rivet below the head.

shear load A sideways force, like that imposed on a nail used to hang a picture.

sheet stock Thin metal, aluminum or steel.

shop head In riveting, the mushroom-shape applied to the end of the rivet to hold it in place.

6061 Aluminum alloy commonly used in aircraft.

slurry A one-to-one mixture of glass bubbles and epoxy/vinylester.

spar Main structural element of the wing, running from root to tip.

spar cap On an I or C shaped spar, the top and bottom short components. The cap withstands the stresses and shouldn't be drilled unless the plans specify to do so.

spar web On an I or C shaped spar, it is the vertical component. Serves to maintain the caps at the desired distance. Web can be drilled to some extent without weakening it.

springback Tendency of bent metal to return to some intermediate position.

stop nut A nut that incorporates a particular method to prevent self-rotation and accidental disengagement.

stringer Thin strip of wood or metal used to define an exterior surface on a fabric-covered airplane: nonstructural.

swage To crush down under extreme pressure.

table saw A metal table with a circular saw blade sticking vertically out of the top.

tack weld A spot weld intended to temporarily hold together two pieces of steel. It has no strength to speak of.

tang A protrusion of metal intended for attachment of a cable fitting.

tension load A pulling force, like a claw hammer pulling a nail out of a piece of wood.

thick micro, thick mix *See* WET MICRO.

thimble A teardrop-shaped device for making an eye from aircraft cable.

tube-and-fabric   Construction method where the aircraft structure is composed of steel or aluminum tubing covered by fabric.

turnbuckle   Device for tightening aircraft cable.

turtledeck   Top of the fuselage between the cockpit and the vertical stabilizer.

2024   Aluminum alloy commonly used in aircraft.

uni   Slang term for unidirectional fiberglass cloth. Most of its strength is in one direction.

uprights   Structural elements on truss-type fuselages, running vertically between longerons.

vinylester   A type of two-part bonding material.

Wanttaja   An aviation writer. Pronounced Wahn-TIE-ah.

warp   Orientation of strongest axis of a woven material. Most commonly used in composite construction.

wet micro   Approximately three parts glass bubbles to one part epoxy/vinylester. Also called thick micro or thick mix.

wet mix   *See* SLURRY.

zinc chromate   The traditional anticorrosive finish on aluminum structures.

Index

I

IA, 11, 40, 84
IFR flight, 105
Inspection Authorization (IA), 11, 40, 84
inspection, 372-373
instrument costs, 105
instruments, 104-105, 365-366
insurance costs, 72
insurance, 73-75
intercoms, 112

J

Javelin Ford V-6, 97
jig saw, 144
joggles, 210-212

K

kit components
  storing, 133-138
kitbuilding
  disadvantages, 29-30
  glamour vs. drudgery, 25-28
  inspection by builder, 28-29
  modifications, 29
  personal costs, 24-25
  process, 16-30
  skills required, 30
  time, 24-26
kitplane
  advantages, 15-16, 37
  affordability, 71-79
  alternatives, 35-42
  aluminum monocoque construction
    techniques, 238-272
  aluminum-tube construction tech-
    niques, 273-285
  building considerations, 24
  building process, 16-30
  buying partially-completed, 39-42
  buying used, 39-42
  composite construction techniques,
    200-237
  construction types, 16-20
  costs, 15-16, 23, 31
  disadvantages, 13-15, 37
  features, 44-47
  manufacturers, 23
  metal monocoque construction tech-
    niques, 238-272

operational expenses, 38-39
painting, 347-352
performance requirements, 42-43
plans, 59-61
registration, 356
selection, 32-70
sheet-metal construction techniques,
  238-272
steel-tube construction techniques,
  273-285
tube-and-fabric construction techniques, 273-285
wooden construction techniques, 310-
  347
kits
  complete, 47, 49, 50
  composite, 50
  construction time, 54-59
  delivery problems, 117-119
  factory pickup, 115-117
  inspection, 119
  materials, 47, 48, 49
  reducing costs, 49-50
  shipping, 115-119
  subkits, 47, 48, 49, 50
  types of, 47-49
KR-1, 8, 17
KR-2, 8, 17
Kubelwagen, 93

L

Lancair, 55
layup
  basic, 228
  multiple, 230
  smoothing, 234-237
layup operations, 223-233
legalities, 356-362
liability insurance, 74
license, 362
lighting, 123
lights, 107
Loehle 5151 Mustang replica, 19
Long-EZ, 17
Lycoming engines, 81-82, 84

M

magnetor failure, 14
manufactured head, 243

R

rachet wrench, 142
racing aircraft, 11
radio installation, 113-114
radio kit, 113
radio station license, 362
radius bends, 304
Rand, Ken, 8
RANS S-9, 39
razor saw, 313
red-tagged propeller, 103
registration, 356
repairman's certificate, 12, 39, 362
replica aircraft
  construction time, 56
replica aircraft, 56-57
Request for Proposals (RFP), 32
research and development aircraft, 11
resin mixture, 215, 220-222
resorcinal glue, 315
RFP, 32
rib stitching, 343
rigging, 352-354
rivet cutter, 241
rivet gun, 260
rivet spacer, 256
riveting, 255-267
riveting errors, 267
rivets, 242 (*see also* specific names of)
  countersinking, 257-259
  determining placement, 255-256
  dimpling, 257-259
  drilling, 256
  removing, 264-267
  specifications, 262
rotary-type power sander, 312
Rotax 503, 99
Rotax 532, 99
Rotax 582, 99
Rotax two-strokes engine, 98-100
rulers, 148
Rutan, Burt, 7

S

SA-6B Flut-R-Bug, 5
saber saw, 144
safety, 16
safety goggles, 149
safety procedures, 149-150

sandblasting, 285
saws, 144
screwdrivers, 143
Sea Hawker, 34
self-insurance, 74
self-locking nuts, 159
Sequoia Falco, 19
shape bends, 304
sheet metal aircraft
  construction techniques, 238-272
shipping, 115-119
shop head, 243
  making, 259-264
shop supplies
  miscellaneous, 151-152
shop tools, 138 (*see also* specific names
  of)
side cutters, 143
Since Major Overhaul (SMOH), 85
slurry, 220
SMOH, 85
snips, 143
socket sets, 142
space, 120-121
specialty tools, 138 (*see also* specific
  names of)
SPO, 32
Sport Trainer, 22
spring clamp, 144
stability, 12-13
STC, 11
steel
  bending, 173-178
  cutting, 169-170
  drilling, 178-181
  purchasing, 167-168
  specifications, 166-167
steel-tube aircraft
  construction techniques, 273-285
  fasteners, 277-278
  materials, 277
  safety procedures, 279
  tools, 277
steel-tube fuselage
  corrosion-proofing, 284-286
stereos, 112
Stits, Ray, 5
storage costs, 72
styrofoam, 206
  cutting, 217-219

Y

Other Bestsellers of Related Interest

THE PILOT'S HANDBOOK OF AERONAUTICAL KNOWLEDGE:
Revised and Expanded Edition—Paul E. Illman
All the technical and operational information you'll need to prepare for private and commercial pilot certification is contained in this easy-to-read book. Based on official government documents—and updated for the 1990s with the latest technical data, flying tips, and general aviation guidance—it reviews everything from navigation technologies and flight planning through radio communications and air traffic control.
ISBN 0-07-155191-3 $19.95 Paper

CUTTING THE COST OF FLYING—Geza Szurovy
More than 200 practical, easy-to-follow suggestions on how to get more flying time for less money. Savings ideas are in alphabetical order for easy reference.
ISBN 0-07-062993-5 $14.95 Paper

BE A BETTER PILOT: Making the Right Decisions—Paul A. Craig
Why do good pilots sometimes make bad decisions? This book takes an in-depth look at the ways pilots make important preflight and in-flight decisions. It dispels the myths surrounding the pilot personality, provides straight-forward solutions to poor decision-making, and determines traits that pilots appear to share—traits that affect they way they approach situations.
ISBN 0-07-157664-9 $15.95 Paper
ISBN 0-07-157665-7 $24.95 Hard

THE ILLUSTRATED BUYER'S GUIDE TO USED AIRPLANES—3rd Edition
—Bill Clarke
In this popular guide, Clarke provides the practical guidance and realistic advice used aircraft buyers need to make smart purchases. He covers all aspects of buying and owning a used airplane, describing what's available, how to choose the right aircraft, and where to find it at an affordable price. Readers will also find useful tips on aircraft insurance, storage, maintenance, and protection.
ISBN 0-07-011266-5 $19.95 Paper

101 THINGS TO DO WITH YOUR PRIVATE LICENSE—2nd Edition
—LeRoy Cook

Beat the post-training blues with this entertaining collection of tips and ideas for sharpening your skills and making every flying hour a rewarding experience. Have you thought about planning your first cross-country flying vacation, fine-tuning your weather forecasting skills, joining a flying club, or even turning your license into profit? If so, this book outlines opportunities like these and much more.

ISBN 0-07-155914-0 $16.95 Paper

THE PILOT'S GUIDE TO AFFORDABLE CLASSICS—2nd Edition
—Bill Clarke

A model-by-model analysis of aircraft by Aeronca, Cessna, Ercoupe, Luscombe, Piper, Stinson, and Taylorcraft is offered to the flyer on a budget. Includes plane reconditioning, maintenance suggestions, avionics descriptions, and explanations of insurance and paperwork.

ISBN 0-07-011268-1 $16.95 Paper
ISBN 0-07-011267-3 $26.95 Hard